Shortcut to Superconductivity

Armen Gulian

Shortcut to Superconductivity

Superconducting Electronics via COMSOL
Modeling

 Springer

Armen Gulian
Advanced Physics Laboratory
Institute for Quantum Studies
Chapman University
Burtonsville, MD, USA

Capitol Technology University
Laurel, MD, USA

ISBN 978-3-030-23488-1   ISBN 978-3-030-23486-7   (eBook)
https://doi.org/10.1007/978-3-030-23486-7

This Springer imprint is published by the registered company Springer Nature Switzerland AG
The registered company address is: Gewerbestrasse 11, 6330 Cham, Switzerland

To my Teachers

Preface

Why a 'Shortcut'?

The idea for writing this book originated from my experience over the past few years of teaching a one-semester course to graduate students in the Department of Electric Engineering of Capitol Technology University (Laurel, Maryland). I noticed that the students were always interested in learning (typically, from scratch) about both superconductivity and COMSOL Multiphysics. The problem, however, was not only that they had very little knowledge of solid-state physics and quantum physics (since they had been prepared for careers in engineering). Most importantly, they had very little time for reading textbooks on these subjects; almost all of them were supporting their studies (and families) with part-time or even full-time jobs. My task was to deliver the knowledge to them so clearly and concisely that whatever they understood in the classroom would stay with them for the rest of their lives. I glossed over trivial content, such as coefficients in formulas; my goal was never to cover everything, but rather to introduce the unusual concepts of superconductivity in the most 'user friendly' and transparent manner, always taking shortcuts to essentials.

The Google dictionary defines 'shortcut' as *a shorter alternative route* or *an accelerated way of doing or achieving something*. A similar meaning is provided by the Oxford English dictionary: *an accelerated way of doing or achieving something*. The Cambridge English dictionary agrees with that: *a quicker way of doing something in order to save time or effort*. These definitions pinpoint our goal and should resonate with the reader's intention: to achieve professionalism in superconductivity faster than expected. Paradoxically, the more developed a society is, the less time is left for fundamental studies: too many temptations are around. The accelerated progress of science aggravates the situation even more: the volume of knowledge becomes increasingly large, and researchers should digest all of it for their own new findings to occur, which of course increases this volume further… It is hard to imagine what could have happened if computers and the Internet would not give us a hand. Each time I am finding unbelievable to find answers at the

Internet, I recall my version of a common Western adage: "God created mean equal; the Internet made them equal!". In view of Internet, there is no list of references in Part I of this book: the readers are advised to use Google.

Computer power does miracles for those who tame it. Many young people have computer addiction. Regretfully, many waste their time by playing computer games. Access to a finite element modeling software, the COMSOL basic package, will let them switching to a big fan with a real science, the science of superconductivity with this book. Their time will be used much more productively when new scientific results will start coming out under their fingers on keyboards!

There is another, older group of very serious people involved in science and technology. They already have understood the value of superconductivity, but have no time to devote for learning the sophisticated mathematical background of this rather unusual area of condensed matter physics. Nor have they time to learn COMSOL. Even if they do, COMSOL has no superconductivity module to help in their effort. It has many other modules, for example, the thermoelectricity module, the optics module, but not yet a superconductivity one. Thus we provide a detailed description of how to use the mathematical module of the basic COMSOL package to model certain, sometimes rather sophisticated superconducting tasks, and elevate the results to the level of animations. These animations are very exciting: they look similar to visualized laboratory experimental results obtained without spending many months of time and tons of money. Moreover, you can watch the picosecond-scale dynamics on timescales convenient to human eyes, and understand the details of evolution…

Our treatment of superconductivity in Part I culminates with the application of time-dependent Ginzburg-Landau (TDGL) equations to various dynamic effects important for superconducting electronics. They also allow us to trace the evolution of superconducting state under the influence of various fields. COMSOL knowledge is not assumed. The codes are described step by step in the book. Also, the readers can download these codes, except the very last one, from the Springer website, create shortcuts and use them in direct accordance with the shortcut's 'computer science' meaning as *a quick way to start or use a computer program*. Pretty soon they will be able to explore yet unexplored problems in the areas of superconducting electronics, publish articles and share the results with the third parties. If they do not have the COMSOL package, they can make use of COMSOL exe-files, which also come with this book, and run them on any PC. Those who prefer to deal directly with the animations without changing physical parameters of the solutions, can just download and play the avi-files.

At this point I feel obligated to warn the readers that the word 'shortcut' has one more facet: *a method or means of doing something more directly and quickly than and often not so thoroughly as by ordinary procedure* (the Merriam-Webster dictionary). This warning is for those who would like to deepen studying of superconductivity for reaching advanced level of knowledge. They are invited into Part II of this book, which contains the full palette of the traditional theory of superconductivity, culminating with the derivation of the full set of time-dependent Ginzburg-Landau equations for finite-gap superconductors. Its simplified form is used in Part I as the instrument of superconductivity exploration. One last remark: in a book with about 1000 equations, it is difficult to avoid typos and occasional mistakes. Rely on the book, but also on your intuition and judgment.

To all those who would like to start enjoying the beauties of quantum nature of the world: *Welcome to the 'Shortcut to Superconductivity'*!

Burtonsville/Laurel, MD, USA   Armen Gulian
February 2020

Acknowledgements

I would like to express deep gratitude to the US Office of Naval Research for continuing support of my research since 1994 which, in particular, made it possible to obtain the results included in Part I of this book. Part II of the book is mainly based on the materials I generated during my Ph.D. studies and subsequent postdoctoral research at V. L. Ginzburg's Department of Theoretical Physics in Moscow. It was originally part of the book *Superconductors in External Fields* published in Russian by Nauka Publishing House (1989) in coauthorship with G. F. Zharkov under partial support of the Scientific Council on HTSC Problem of the USSR. Later, that book's extended version was published by Kluwer Academic/Plenum Publishers (currently Springer) under the title *Nonequilibrium Electrons and Phonons in Superconductors* (1999). I would like to use this opportunity and express my deep gratitude to all my Teachers, Colleagues, Students, Friends, and Family Members for their help, comments, and creative remarks.

Contents

**Part I   Mastering Superconductivity with Computers
via Time-Dependent Ginzburg-Landau Equations**

1   Basics 3
   1.1   What is a Superconductor? 4
   1.2   London Brothers Approach 4
   1.3   Superconducting Disk in a Magnetic Field: COMSOL
        Example 8
   1.4   Seminar 1. Rotating Superconductor: London's Moment 17
   1.5   Ginzburg and Landau Approach 23
   1.6   Josephson Effects 28
   1.7   SQUIDs 31
   1.8   Time-Dependent Ginzburg–Landau (TDGL) Theory 34
   1.9   Seminar 2. When will We Have Superconductors
        at Ambient Conditions? 38
        1.9.1   Dielectric Function 39
        1.9.2   BCS Attraction 42
        1.9.3   Phonon Resonance 42
        1.9.4   Coulomb Potential and Tolmachov Logarithm 43
        1.9.5   Superconductivity at $\mu > \lambda$, but $\lambda > \mu^*$. 46
        1.9.6   Little's Model of High-T_c Superconductors
                and Interplay Between ω, μ and λ 47

**2   Exploring Superconductivity with COMSOL via TDGL
Equations** 51
   2.1   General Notions of TDGL Equations 52
   2.2   Disk in a Magnetic Field: Ginzburg–Landau Approach 55
   2.3   Disk in a Magnetic Field: Simpler Approach to Boundary
        Conditions 62
   2.4   Penetration of Vortices into 3D Washer 65

2.5   Dynamics in Current-Carrying Superconducting Wires   73
2.6   Current Flow in Thin Superconducting Strips: Annihilation
        of Abrikosov Vortices.   86
2.7   Generation of SFQ Pulses in SNS Junctions.   95
2.8   Cloning of SFQ Pulses   97
2.9   Discovering New Effects with COMSOL-TDGL   100
2.10   Final Remarks on COMSOL Modeling   107

Part II   Derivation of Time-Dependent Ginzburg Landau Equations

3   Stationary Ginzburg–Landau Equations.   113
    3.1   Introductory Concepts.   113
            3.1.1   Infinite Conductivity.   114
            3.1.2   Ideal Diamagnetism   114
            3.1.3   Energy Gap   115
            3.1.4   Bogolyubov–De Gennes Equations: Analogy
                      with Relativistic Quantum Theory.   116
            3.1.5   Andreev Reflection.   118
            3.1.6   Electron Density of States.   119
            3.1.7   Coherence Factors   120
    3.2   Phenomenological GL Theory: Triumph and Limits
            of Human Imagination   121
            3.2.1   Free Energy Functional   122
            3.2.2   London Penetration Depth   125
            3.2.3   Coherence Length   126
            3.2.4   Sign of Surface Energy   126
    3.3   BCS-Gor'kov Theory   129
            3.3.1   Equations for Ψ-Operators   129
            3.3.2   Off-Diagonal Long-Range Order   131
            3.3.3   Spin-Singlet Pairing   132
            3.3.4   Solutions in Momentum Representation.   132
            3.3.5   Self-Consistency Equation   133
            3.3.6   Isotope Effect.   134
            3.3.7   Gauge Invariance   134
            3.3.8   Description at Finite Temperatures   135
            3.3.9   Weak-Coupling Ratio $2\Delta(T = 0)/T_c$.   136
    3.4   Self-Consistent Pair-Field: Microscopic Justification
            of G–L Equations.   138
            3.4.1   Iterated Equations   138
            3.4.2   Magnetic Field Inclusion   140
            3.4.3   Slow Variation Hypothesis   140
            3.4.4   Computation of Phenomenological Parameters   142

	3.4.5	Flux Quantization	143
	3.4.6	Failure of "Quantum-Mechanical Generalization" for Time-Dependent Problems	145
References			146
4	**Superconductors with Impurities**		**149**
4.1	Scattering on Ordinary Impurities		149
	4.1.1	Magnetic and Nonmagnetic Impurities	149
	4.1.2	Diagram Expansion and Spatial Averaging for Normal Metals	150
	4.1.3	Born's Approximation	152
	4.1.4	Equations for a Superconducting State	154
	4.1.5	Anderson's Theorem	156
	4.1.6	"Londonization" by Elastic Scattering	156
4.2	Magnetic Impurities		157
	4.2.1	Averaging over Spin Directions	158
	4.2.2	Spin-Flip Time τ_S	158
	4.2.3	Reduction of Transition Temperature	159
	4.2.4	Energy-Gap Suppression	162
	4.2.5	Gapless Superconductivity	163
4.3	Nonstationary Ginzburg–Landau Equations		164
	4.3.1	Causality Principle and Nonlinear Problems	164
	4.3.2	Equations on Imaginary Axis	165
	4.3.3	Analytical Continuation Procedure	166
	4.3.4	Anomalous Propagators and Dyson Equations	168
	4.3.5	Regular Terms	170
	4.3.6	TDGL Equations for Gapless Superconductors	172
References			173
5	**General Equations for Nonequilibrium States**		**175**
5.1	Migdal–Eliashberg Phonon Model		175
	5.1.1	Fröhlich's Hamiltonian	175
	5.1.2	Migdal Diagram Expansion	177
	5.1.3	Eliashberg Equations in Weak-Coupling Limit	178
	5.1.4	Comparison with BCS-Gor'kov Model	179
5.2	Equations for Nonequilibrium Propagators		180
	5.2.1	Phonon Heat-Bath: Applicability	180
	5.2.2	Expansion over External Field Power	180
	5.2.3	Analytical Continuation: Causal Propagators	181
	5.2.4	Phonon Heat-Bath: Consequences	182
	5.2.5	Analytical Continuation: Anomalous Functions	184
	5.2.6	Complete Set of Equations	186
	5.2.7	Keldysh Technique Approach	187

    5.3   Quasiclassical Approximation 187
        5.3.1   Eilenberger Propagators 188
        5.3.2   Eliashberg Kinetic Equations 189
        5.3.3   Normalization Condition........................ 190
        5.3.4   Gauge Transformation Rules..................... 191
        5.3.5   Electron and Hole Distribution Functions 192
        5.3.6   Kinetic Equations: Keldysh Option 193
        5.3.7   Expressions for Charge and Current 194
    References 195

6 Electron and Phonon Collision Integrals...................... 197
    6.1   Collision Integral Derivation 197
        6.1.1   Spatially Homogeneous States.................... 197
        6.1.2   Separation of Real and Virtual Processes 198
        6.1.3   Nondiagonal Channel 199
        6.1.4   Impurities 199
        6.1.5   Effective Collision Integral 200
    6.2   Inelastic Electron-Electron Collisions 201
        6.2.1   Diagram Evaluation of Electron-Electron
               Self-Energy 201
        6.2.2   Analytical Continuation 202
        6.2.3   Transition to Energy-Integrated Propagators.......... 203
        6.2.4   Derivation of the Canonical Form 204
        6.2.5   Essence of Elementary Acts 207
    6.3   Kinetic Equation for Phonons 207
        6.3.1   Application of Keldysh Technique 207
        6.3.2   Quasiclassical Approximation 209
        6.3.3   Phonon Distribution Function 210
        6.3.4   Polarization Operators in Keldysh's Technique 211
        6.3.5   Polarization Operators: Analytical Continuation
               Technique 213
        6.3.6   Equivalence of Keldysh and Eliashberg Approaches.... 215
        6.3.7   Transition to Energy-Integrated Propagators.......... 216
    6.4   Inelastic Electron-Phonon Collisions 217
        6.4.1   Electron-Phonon Self-Energy Parts 218
        6.4.2   Canonical Form for Electron-Phonon Collisions 219
        6.4.3   Canonical Form for Phonon-Electron Collisions 221
    6.5   Seminar 3. Cooling by Heating 221
        6.5.1   Gap Enhancement 221
        6.5.2   Negative Phonon Fluxes........................ 226
    References 229

7 Time-Dependent Ginzburg–Landau (TDGL) Equations 231
  7.1 Order Parameter, Electron Excitations, and Phonons 231
    7.1.1 Basic Kinetic Equations 232
    7.1.2 Normalization Condition......................... 233
    7.1.3 Definition of Order Parameter 234
    7.1.4 Nondiagonal Collision Channel................... 234
    7.1.5 Spectral Functions R_1, R_2, N_1, and N_2 235
    7.1.6 Gap-Control Term 236
    7.1.7 Local-Equilibrium Approximation 237
    7.1.8 Determination of f_1-Function...................... 237
    7.1.9 Determination of f_2-Function...................... 239
    7.1.10 Order Parameter Equation....................... 239
    7.1.11 Contribution of Nonequilibrium Phonons............ 240
    7.1.12 Galayko's μ^2-Term............................ 240
    7.1.13 Phonons and Order Parameter Dynamics 241
  7.2 Interference Current 243
    7.2.1 Usadel Approximation 244
    7.2.2 Normal Flow Contribution to Interference Current 249
    7.2.3 Condensate Contribution to Interference Current 250
    7.2.4 Boundary Conditions 252
  7.3 Fluctuations 252
    7.3.1 Ginzburg's Number 253
    7.3.2 Paraconductivity.............................. 256
    7.3.3 Aslamazov-Larkin Mechanism 258
    7.3.4 Maki-Thompson Mechanism..................... 259
  7.4 Longitudinal Electric Field in Superconductors.............. 261
    7.4.1 Tinkham-Clark Gauge-Invariant Potential 262
    7.4.2 Normal Metal–Superconductor Interface 263
    7.4.3 New Characteristic Length in Superconductors........ 265
    7.4.4 Carlson-Goldman Modes 266
    7.4.5 Dispersion of Charge-Imbalance Mode 267
  References 268

Index 273

Part I
Mastering Superconductivity with Computers via Time-Dependent Ginzburg-Landau Equations

Chapter 1
Basics

This Chapter will introduce, in a simple way, the basic concepts of superconductivity required to start modeling of superconducting electronics devices. Also, the first very basic COMSOL example will be given in full details—no preliminary knowledge of COMSOL is required. From a conceptual point of view, the most important goal is to understand what makes superconductors the quantum objects. From the practical point of view we will learn (i) what is replacing Ohm's law in superconductors; (ii) how to understand Meissner effect in Londons' approach; (iii) what is the magnetic field screening length; (iv) how to describe superconductors in a gauge-invariant way; (v) Ginzburg–Landau Ψ-function approach to superconductivity and the role of the Ψ-function phase; (vi) flux quantization and related hint on electron pairing (Cooper condensation) in superconductors; (vii) difference between Cooper and Bose condensates; (viii) Josephson effects, and (ix) SQUIDSs. This Chapter will introduce the simplified version of the time-dependent Ginzburg–Landau (TDGL) equations valid for so-called gapless superconductors. In analogy with the Schrödinger equation, these equations will be postulated without derivation for solving practical tasks in the next chapter. The rigorous quantum mechanical derivation of TDGL equations will require all of Part II of this book.

Two seminar-type discussions are added to this Chapter. One of them discusses magnetic moment of rotating superconducting ball—the so-called London's moment. This example, though very useful educationally, is absent in usual textbooks. The second seminar discusses how high the superconducting transition temperature can be—this is one of the topics which I have found to always be interesting for curious students.

Electronic supplementary material The online version of this chapter (https://doi.org/10.1007/978-3-030-23486-7_1) contains supplementary material, which is available to authorized users.

1.1   What is a Superconductor?

Do not be afraid of the most straightforward answer: a conductor with no resistance. This, a bit naive, definition will lead us to many logical puzzles, and we will decipher them one by one.

What does it mean 'a conductor with no resistance'? The resistance R of a conductor with a length L and a cross section S is characterized by the well-known expression $R = \rho L/S$, where ρ is the resistivity, an intrinsic property of the material. In general, ρ depends on temperature T. We will recall that in metals, resistivity $\rho(T)$ typically increases when the temperature goes up; in semiconductors, it increases when the temperature goes down. In dielectrics, ρ is very large, and $\rho(T)$ makes no sense; the conductivity $\sigma \equiv 1/\rho$ is taken to be zero: $\sigma = 0$. For superconductors, as was just declared, $R = 0$, which means $\rho = 0$. That creates problems when working with the Ohm's law. We recall that on the microscopic scale this law can be written as

$$\mathbf{j} = \sigma \mathbf{E}, \tag{1.1}$$

where $\mathbf{j}$ is the current density, which is related to the electric field $\mathbf{E}$ by the conductivity σ. Then $\sigma = \infty$ at $\rho \equiv 1/\sigma = 0$. How should one handle the infinitely large current density (1.1)? It appears that for superconductors, the relation (1.1) should be replaced by a more appropriate one. Let us determine this expression.

1.2   London Brothers Approach

When a constant electric field $\mathbf{E}$ is applied to a metal, electrons begin drifting in this metal, and a stationary flow is possible because of the viscous nature of this drift. Indeed, the equation of viscous motion for an electron is:

$$\ddot{\mathbf{x}} = e\mathbf{E}/m - \gamma\dot{\mathbf{x}} \tag{1.2}$$

(here m is the electron mass, e is its charge, γ is the damping coefficient, and $\mathbf{x}$ is the coordinate of the electron – for simplicity, we consider motion along this coordinate only). This is Newton's Second law with damping taken into account. From (1.2), it follows that the *stationary drift velocity* ($\ddot{\mathbf{x}} = 0$, $\dot{\mathbf{x}} = const$) *is proportional to the acting force:*

$$\dot{\mathbf{x}} = e\mathbf{E}/(\gamma m). \tag{1.3}$$

This result corresponds to the Aristotelian notion that a motion with constant velocity requires a constant force—that statement is true in viscous media. The electric current density can now be written as:

$$\mathbf{j} = en\dot{\mathbf{x}} = \frac{e^2 n}{\gamma m}\mathbf{E} \tag{1.4}$$

(where n is the number of electrons per unit volume). One can notice by comparing (1.1) and (1.4) that we have derived the expression for conductivity:

$$\sigma = ne^2 \tau / m, \tag{1.5}$$

where $\tau = 1/\gamma$ is the characteristic damping time of electrons motion. Equation (1.5) is valid for normal metals. In superconductors, the electronic motion has no damping: $\gamma = 0$, $\tau \to \infty$, so that

$$\sigma\,|_{\tau \to \infty} = \infty. \tag{1.6}$$

At $\gamma = 0$, (1.2) converts into $\ddot{x} = eE/m$, i.e., into a purely Newtonian motion: *the acceleration is proportional to the acting force*. To deal with it further, we need to recall the definition of the electric field via the potentials $\mathbf{A}$ and φ:

$$\mathbf{E} = -\frac{1}{c}\dot{\mathbf{A}} - \nabla\varphi. \tag{1.7}$$

Temporarily, let us choose a vanishing scalar potential φ. Then, substituting (1.7) into (1.2) and integrating with the condition $\dot{x} = 0$ at $\mathbf{A} = 0$ (i.e., no motion if no field), we find:

$$\dot{\mathbf{x}} = -\frac{e}{mc}\mathbf{A} \tag{1.8}$$

and thus, by analogy with (1.4),

$$\mathbf{j} = en\dot{\mathbf{x}} = -\frac{e^2 n}{mc}\mathbf{A}. \tag{1.9}$$

Equations (1.8) and (1.9) are the famous "Londons' equations", which were introduced by Fritz and Heinz London brothers in 1935 to explain experimental findings by Meissner and Ochsenfeld, to be discussed below. To be precise, the Londons represented (1.8) and (1.9) in a somewhat different form. Namely, taking the time derivative of (1.8), using (1.7) with the dropped scalar potential and (1.9), one obtains

$$\dot{\mathbf{j}} = \frac{e^2 n}{m}\mathbf{E}. \tag{1.10}$$

Taking curl of (1.9), one can have

$$\mathrm{curl}\,\mathbf{v} = -\frac{e}{mc}\mathbf{B}. \tag{1.11}$$

These (1.10) and (1.11), have been suggested originally by the Londons. The Londons' equations have crucial importance in the physics of superconductivity, since these replace the Ohm's law for superconductors. We will discuss these equations and their generalization further on.

Problem 1. Using (1.9), consider the penetration of a magnetic field into a superconductor.

Tip: Consider a half-space occupied by a superconductor and apply (1.9) together with the Maxwell equations.

Problem 2. Prove that screening of a magnetic field in superconductors takes place at the shortest possible distance.

Tip: Consider the plasma frequency of metals and compare with the London penetration depth.

Problem 3. Estimate the characteristic length of magnetic field penetration into a bulk superconductor.

Tip: Use the result of Problem 2.

Solution to Problem 1

Let us consider a superconductor occupying the half-space $x > 0$; the region $x < 0$ is empty (Fig. 1.1). The surface of the sample is the (y, z)-plane.

Then, both the magnetic field **B** and the current **j** in the superconductor depend only on the x-coordinate. The Maxwell equations required for this task can be written as:

$$\text{curl } \mathbf{B}(x) = \frac{4\pi}{c} \mathbf{j}(x), \tag{1.12}$$

$$\text{div } \mathbf{B}(x) = 0. \tag{1.13}$$

Arbitrary direction of magnetic field can always be decomposed into two vectors: one to be perpendicular to the sample surface $\mathbf{B}(x) = (B_x, 0, 0)$, and the other to be tangential to that surface (see Fig. 1.1). In the first case, from (1.13) it follows that $\partial B/\partial x = 0$, so that B_x is constant. Now, let us take the curl of the Londons' equation (1.9); recalling that $\text{curl } \mathbf{A} = \mathbf{B}$, we get $\text{curl } \mathbf{j} = -[e^2 n/(mc)]\mathbf{B}$. Substituting in it **j** from (1.12), we find

$$\text{curlcurl } \mathbf{B} = -\frac{4\pi e^2 n}{mc^2} \mathbf{B}. \tag{1.14}$$

Clearly, for a constant **B**, the l.h. side of (1.14) is zero, so $B = 0$ (and also, as follows from (1.12), $j = 0$). This signifies that there should be no B-field component perpendicular to the surface, i.e., the magnetic field should always align tangentially

Fig. 1.1 Superconducting
half-space (at $x > 0$) facing
vacuum with the magnetic
field **B**

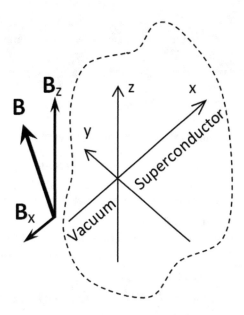

to the surface of superconductor: $\mathbf{B}(x) = (0, B_y, B_z)$. In that case, (1.13) is always
satisfied. To simplify further consideration, we can rotate the local reference frame
aligning its z-axis along **B** so that $\mathbf{B}(x) = (0, 0, B_z)$. Then, only the y-component of
curl **B** is non-zero: curl $\mathbf{B} = (0, \text{curl } B \mid_y, 0)$

$$\text{curl } \mathbf{B} \mid_y = -\frac{\partial B_z}{\partial x}, \tag{1.15}$$

and, according to (1.12), the current is directed along the y-axis: $\mathbf{j} = (0, j_y, 0)$.
Substitution of (1.15) into (1.14) yields

$$\frac{\partial^2 \mathbf{B}}{\partial x^2} = \frac{4\pi n e^2}{m c^2} \mathbf{B}. \tag{1.16}$$

Of the two solutions $B = B_0 \exp\left[\pm\sqrt{4\pi n e^2/(m c^2)}\, x\right]$ of (1.16), we should choose
the exponentially decreasing one since only that solution is finite inside the superconductor. Thus, the externally applied magnetic field exponentially falls to zero inside
of the bulk superconductor; i.e., the superconductor screens off the magnetic field.
The characteristic length λ_L of this screening is defined by the relation:

$$\frac{1}{\lambda_L^2} = \frac{4\pi n e^2}{m c^2}. \tag{1.17}$$

Comment This quantity, λ_L, is called the "London penetration depth", and the expulsion of magnetic field from the depth of a superconductor is called the "Meissner effect".

Solution to Problem 2

The highest frequency which electrons can possess in metals is the plasma frequency: $\omega_{pl} = \sqrt{4\pi ne^2/m}$. One can notice from (1.17) the relation $\lambda_L = c/\omega_{pl}$. The time $\tau \sim 1/\omega_{pl}$ is the shortest possible time for the collective motion of electrons in metals, and the speed of light, c, is the fastest possible in nature. Thus, the magnetic field screening in superconductors takes place at the shortest possible distance.

Solution to Problem 3

Using $\lambda_L = c/\omega_{pl}$ and substituting $c \sim 3 \cdot 10^{10}$ cm/s and the value $\omega_{pl} \sim 10^{16}$ s^{-1}, we find $\lambda_L \sim 3 \cdot 10^{-6}$ cm, i.e., about ~ 30 nm as a typical value.

Comment Obviously, this quantity is material-dependent: both n, the density of electrons, and m, the effective mass of electrons, depend on the material. In high-temperature superconductors, like $YBa_2Cu_3O_{6+\delta}$, n can be smaller than 10^{21} cm^{-3}, while in elemental superconductors, like Al, n can be close to 10^{24} cm^{-3}. The effective mass of electrons can be both smaller or larger than the mass of a bare electron (i.e., the rest mass of electron in vacuum). For example, in the so called "heavy-fermion" superconductors, m is factor of thousand greater than the bare electron mass. Thus, λ_L may vary among metals to a large extent. Note also that c in metals is smaller than in vacuum by a factor of 3.

1.3   Superconducting Disk in a Magnetic Field: COMSOL Example

As the first application of COMSOL Multiphysics, let us consider the expulsion of a magnetic field from a thin superconducting disk in stationary conditions within Londons' approach.

In the case of magnetostatics we have two of Maxwell equations in the form

$$\text{div } \mathbf{B} = 0, \tag{1.18}$$

$$\text{curl } \mathbf{B} = 4\pi \mathbf{j}, \tag{1.19}$$

where $\mathbf{B}$ is the magnetic induction

$$\mathbf{B} = \text{curl}\,\mathbf{A}, \qquad (1.20)$$

and $\mathbf{j}$ is the current density. Substituting (1.20) into (1.19), we have

$$\text{curl}\,\text{curl}\,\mathbf{A} \equiv \text{grad}\,\text{div}\,\mathbf{A} - \nabla^2\mathbf{A} = 4\pi\mathbf{j}. \qquad (1.21)$$

At this point, we will make a choice of the gauge, assuming div $\mathbf{A} = 0$. This is the so-called "London gauge". Substituting (1.9) into (1.21), we get

$$\nabla^2\mathbf{A} = \frac{1}{\lambda_L^2}\mathbf{A}. \qquad (1.22)$$

where the relation (1.17) for λ_L is used.

Remark When deriving (1.22), we used the Londons' relation $\mathbf{j} \propto -\mathbf{A}$. That means that (1.22) is equivalent to

$$\nabla^2\mathbf{j} = \frac{1}{\lambda_L^2}\mathbf{j}. \qquad (1.23)$$

We should note in passing that from (1.19), using (1.18), it also follows that:

$$\nabla^2\mathbf{B} = \frac{1}{\lambda_L^2}\mathbf{B}. \qquad (1.24)$$

Later in this book, we will see that the derivation of (1.23) and (1.24) does not require choosing any particular gauge while that of (1.22) does. That is why they would say that the simple relation $\mathbf{j} \propto -\mathbf{A}$ is valid only in London's gauge.

To take advantage of chosen gauge for solving (1.22), we notice that its free term is proportional to the current $\mathbf{j}$, which can be non-zero only in the superconductor. Thus, (1.22) can be re-written as

$$\nabla^2\mathbf{A} = \frac{I(\mathbf{r})}{\lambda_L^2}\mathbf{A}, \qquad (1.25)$$

where $I(\mathbf{r}) = 1$ inside of superconductor, and $I(\mathbf{r}) = 0$ outside of it. We will consider the external magnetic field to be homogeneous and perpendicular to the disk surface. The problem then is convenient to treat in the cylindrical coordinates: $\mathbf{A} = \mathbf{A}(r, \varphi, z)$ with $\mathbf{H} || \hat{\mathbf{z}}$. Then $\mathbf{A} = A\hat{\varphi}$, where the scalar A depends only on coordinates (r, z). Thus, the vectorial equation (1.25) is reduced to a simpler scalar equation

$$\nabla^2 A = \frac{I(r, z)}{\lambda_L^2} A. \tag{1.26}$$

As soon as the scalar A is known, the magnetic induction can be found as

$$\mathbf{B} = \mathrm{curl}\,\mathbf{A} = -\frac{\partial A}{\partial z}\hat{\mathbf{r}} + \frac{1}{r}\frac{\partial(rA)}{\partial r}\hat{\mathbf{z}}. \tag{1.27}$$

This relation and the symmetry of the problem allow us to formulate boundary conditions for A. Indeed, from the symmetry, the $\mathbf{B}$-field cannot possess an $\hat{\mathbf{r}}$-component on $\hat{z}$-axis. That means, as it follows from (1.27),

$$\left.\frac{\partial A}{\partial z}\right|_{r=0} = 0. \tag{1.28}$$

Far from the disk, the field should remain homogeneous, with $\mathbf{B} = \mathbf{H}$, i.e., B_φ and B_r are absent, and $B_z = H$. Again, from (1.27), one can deduce

$$A|_{r\longrightarrow\infty} = A(r), \quad i.e., \quad A \text{ cannot depend on } z, \tag{1.29}$$

and

$$A|_{z\longrightarrow\pm\infty} = A(r) = \frac{Hr}{2}. \tag{1.30}$$

For a numerical approach to the problem, the infinities in (1.29) and (1.30) should be replaced by some finite values (we will denote them, accordingly, R_0 and H_0, with the assumption that these values should be much larger than the sizes r_0 and h_0 of the disk). As soon as an increase of R_0 and H_0 at fixed values of r_0 and h_0 does not affect the solutions inside and around the disk, the infinities are indeed "infinite".

Let us now implement the finite element modeling of this problem using COMSOL Multiphysics.[1] In this book, we will be using its version 5.4, which was available at the time of the manuscript preparation.

Open COMSOL, and click on Model Wizard. Select 2D Axisymmetric Space Dimension. Next window is Select Physics. Double click on Mathematics, then double click on PDE Interfaces, and then double click on Coefficient Form PDE(c). That choice will become visible in the Added Physics Interfaces window. At this point, it is a good idea to save the file. Then click on Study button. Select Study will come in; choose Stationary by double clicking. Model Builder will open, and you can start building it. Click on Parameters, and start filling in the parameters in Settings window. We will start with the parameters shown in Fig. 1.2.

Now we can choose the geometry. Right click on Geometry, and twice choose rectangles from the pop-up window. Click once on the Rectangle 1 and then in Settings window call it Disk. Do the same with Rectangle 2 and call it Box. Appro-

---

[1] Our treatment follows, with some modifications, the approach suggested by J.-G.Caputo et al., axrXiv:1308.2204v1 [cond-mat.supr-con] 9 Aug 2013.

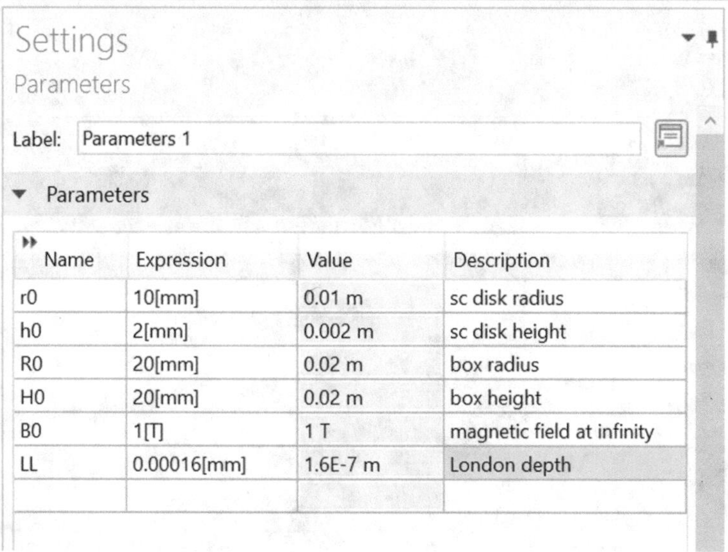

Settings

Parameters

Label: Parameters 1

▼ Parameters

Name	Expression	Value	Description
r0	10[mm]	0.01 m	sc disk radius
h0	2[mm]	0.002 m	sc disk height
R0	20[mm]	0.02 m	box radius
H0	20[mm]	0.02 m	box height
B0	1[T]	1 T	magnetic field at infinity
LL	0.00016[mm]	1.6E-7 m	London depth

Fig. 1.2 Possible choice of parameters

priately, insert the values of r0 and R0 for their Width in Settings window, and h0 and H0 for their heights. Then click on Build All Objects, and on Zoom Extents in Graphics window. You will see the picture shown in Fig. 1.3.

To locate Disk and Box symmetrically relative to the y-axis in Settings window, under Position choose $z = -h0/2$ and $z = -H0/2$ correspondingly. Press Build All Objects and Zoom Extents. Figure 1.4 shows the resultant picture.

Now we can specify the Coefficient Form PDE. Double click on Coefficient Form PDE (c) and click on Coefficient Form PDE 1 in Model Builder. Then click on Equation in Settings window. The equation with coefficients will be displayed as in Fig. 1.5.

We observe that d_a should be turned to zero, as well as the free term f (their default values are 1). The variable u corresponds here to our function A. Coefficients α, β, and γ are zero by default. For this equation to coincide with (1.26), coefficient a should be equal to $I(r, z)/\lambda_L^2$. The function $I(r, z)$ can be defined in COMSOL via Boolean operators:

$$I(r, z) = (r < r_0) * (z^2 < h_0^2/4). \tag{1.31}$$

COMSOL will assign $(r < r_0) = 1$ if $r < r_0$ is true, and zero otherwise. Similarly with the second multiplier in (1.31). Thus the expression for a in Fig. 1.5 should be as shown in Fig. 1.6.

Now we can take care of boundary conditions. The default boundary condition is Zero Flux, which should be replaced, as was discussed above, by the relations which follow from (1.27). Far from the disk, the field should remain homogeneous,

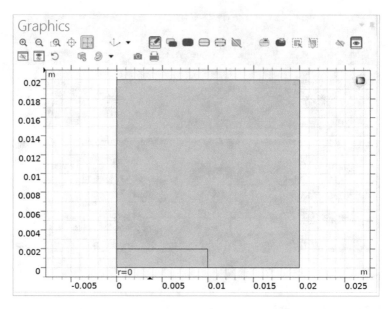

Fig. 1.3 Half disk and half box shown with the symmetry axis (left)

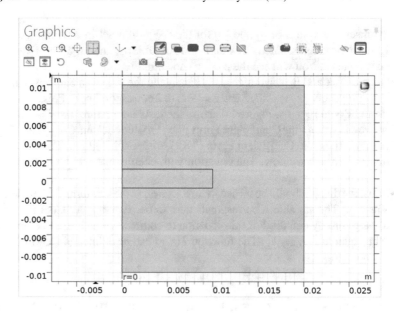

Fig. 1.4 Half disk in the half box—symmetric placement

Fig. 1.5 Equation and coefficients

Fig. 1.6 Expression for the coefficient a

with $\mathbf{B} = \mathbf{H} \equiv \mathbf{B}_0$, i.e., B_φ and B_r are absent, $B_z = B_0$. Again, from (1.27), one can deduce

$$A|_{r=R_0} = HR_0/2, \tag{1.32}$$

$$A|_{z=\pm H_0/2} = \frac{Hr}{2}, \tag{1.33}$$

and

$$A|_{r=0} = 0. \tag{1.34}$$

The last condition is less general than (1.28), which is equivalent to $A|_{r=0} = const$. However, the choice of $const = 0$ matches (1.33). To implement con-

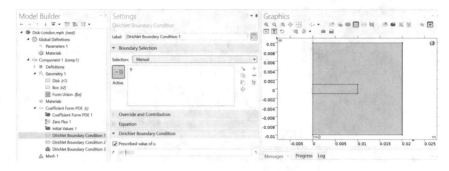

Fig. 1.7 Defining Dirichlet condition for the far vertical boundary

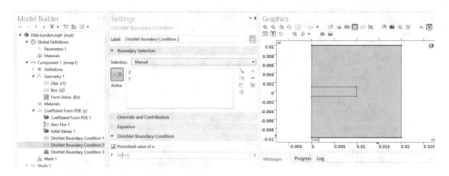

Fig. 1.8 Dirichlet conditions on the far horizontal boundaries

ditions (1.32)–(1.34), right click on Coefficient Form PDE (c) and then click on
Dirichlet Boundary Condition; repeat it three times for (1.32)–(1.34) correspondingly. They will appear in Model Builder window just after Initial Values line. Click
on the first one. Corresponding Settings window will open. We would like the prescribed value of u to be $B0 * R0/2$ on the far vertical boundary. To accomplish this
task, insert this value as shown in Fig. 1.7, then move mouse onto the Graphics panel,
and click onto the vertical line. Hovering mouse on the line will make it red colored,
and after clicking, it will become blue, and the number corresponding to it will appear
in the Boundary selection window, as shown in Fig. 1.7. Second boundary condition,
(1.33), should be inserted for the two long horizontal boundaries in the same manner
(Fig. 1.8). At last, we should implement the boundary condition (1.34) on the $r = 0$
axis. In the Model Builder window, click on Dirichlet Boundary Condition 3, then
hover mouse consecutively on three vertical lines at $r = 0$, and click on them. Their
red color will convert into blue, and the numbers of boundaries, 1, 3 and 5 will appear
in the Settings window. Save the file, and run computation via =Compute button at
Home. The result will look like Fig. (1.9).

Click on Color Black in Settings window to make contours of the disk more
visible. To have more a informative picture, it is worth it to plot not the vector
potential, which is what Fig. 1.9 displays, but rather (1.27), the magnetic field **B**. For

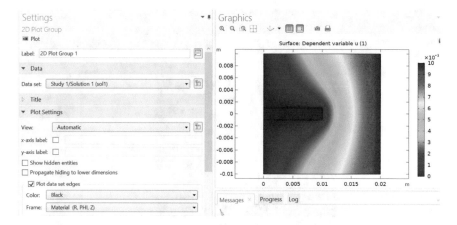

Fig. 1.9 Intermediate result with the scalar A displayed

that in <u>Results</u> of <u>Model Builder</u>, right-click on <u>2D Plot1</u>, and choose in the pop-up window <u>Arrow Surface</u>. In accordance to Fig. 1.9, in the <u>Expression</u> of Settings insert $-uz$ for the R component, and $(u + r * ur)/r$ for the Z component. Then click <u>Plot</u> in <u>Settings</u>. You will see the vector **B**-field on top of the scalar A-field. Scroll down to the scale factor and increase the scale factor to make the arrows more visible. We still need to replace the scalar field A by the value of $B^2 = (\partial u/\partial z)^2 + (u/r + \partial u/\partial r)^2$. For that, click on <u>Surface</u> in <u>2D Plot</u> Group 1 in the <u>Model Builder</u> and replace u by $(uz)2 + (u/(r + 0.0000001) + ur)2$, which is B^2 in COMSOL notations. We added a very small constant in the denominator to avoid division by zero when plotting the figure. The result is shown in Fig. 1.10.

This plotting may be more accurate if a smaller mesh size is chosen: double-click on Component 1, click on <u>Mesh</u>, and in <u>Settings</u> window, switch <u>Element size</u> from <u>Normal</u> to <u>Extra fine</u>.

Another much more essential improvement in plotting may be achieved by representing it in 3D format: the fact that we solved 2D-axisymmetric problem does not mean that we abandoned its 3D nature. All that required here is the postprocessing of the result. For that, double click on <u>Data Sets</u> under <u>Results</u>, and choose <u>Revolution 2D 1</u>. In <u>Revolution Layers</u> choose start angle -180 and Revolution angle 180. Right-click on <u>Results</u> and call-in 3D Plot Group. Then right click on <u>3D Plot Group</u> and call in <u>Volume</u>. In Settings window, in <u>Expression</u> replace u by $sqrt((uz)2 + (u/(r + 0.000001) + ur)2)$, which is the modulus of the magnetic vector. We again added infinitesimal quantity in the denominator so that the plot will be possible to construct. If you would like to visualize the contours

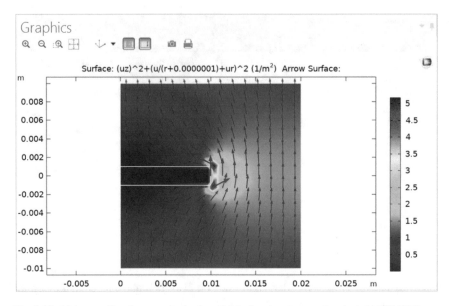

Fig. 1.10 Meissner effect in superconducting disk in Londons' approximation via COMSOL

of the disk better, click on 3D Plot Group and under enabled Plot data set edges, change color from Black to White. We may also want to see the vectors of magnetic field. So again right click on 3D Plot Group, and call in Arrow Volume. Then click on Arrow Volume, and insert the same components as we did for Arrow Surface of 2D Plot Group 1. These are $-uz$ for R component, and $u/r + ur$ for Z component. Do not forget to insert 0 for PHI component, otherwise the code will not run. The arrows corresponding to the vector of magnetic field as of now will be masked off by the color which we assigned to the volume points. To make them visible, you can click on Transparency icon in Graphics window. But even that will not be a good enough visualization (try it!). To improve on this, for the Y grid points in Arrow positioning of Settings window, switch Number of points to Coordinates, and choose 0 there; remove Transparency. You may want to make more data points along the x-axis, which we made equal 20. We left 7 points for Z untouched. The final 3D result is shown in Fig. 1.11.

You can now drag the image with your mouse and rotate to look at it for different angles. You can also try other plotting and modeling options as soon as you learned the rules of the game: the restricting factor is only your imagination. For example, you can try to find configuration of magnetic field when instead of a disk you have a washer. Or two washers of different size, etc.

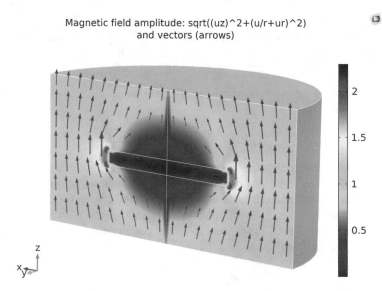

Magnetic field amplitude: sqrt((uz)^2+(u/r+ur)^2)
and vectors (arrows)

Fig. 1.11 Meissner effect in Londons' approximation visualized via COMSOL

1.4 Seminar 1. Rotating Superconductor: London's Moment

The Meissner effect is very famous, and all relevant textbooks contain it: it represents a crucial signature of superconducting state. However, there is another effect, of the same quantum nature, which is called the London moment. Unfortunately, it is not described in textbooks, though it characterizes the superconducting state equally well. We will consider it now.

When dealing with superfluids, an interesting question is: what will happen if we rotate a superconducting ball? The matter is that the external mechanical action is engaging into rotation mainly the solid part of the body, i.e., the ionic lattice. Electrons, which move frictionlessly, should not be directly engaged in motion, so one should expect a large electric current (caused by the charged ions moving through the electronic liquid) and an associated large magnetic moment. In reality, this picture sustains only for a very short amount of time. As soon as the aforementioned current builds up, it generates a magnetic field which grows synchronously with the current. A time-dependent magnetic field, in accordance to Maxwell equations, generates electric field, which accelerates frictionless electrons, and they start moving together with the lattice. Only a very thin layer near the surface of the body, with the thickness of the London penetration depth λ_L is maintaining superconducting current. So the magnetic moment is actually small yet quite detectable, as we will see. The basic cornerstone of our consideration in the assumption that Londons' equations (1.10) and (1.11) are valid in any reference frames, i.e., they are applicable even for moving superconductors. So we will write them for a rotating ball:

$$\operatorname{curl} \mathbf{v} = -\frac{e}{mc}\mathbf{h}, \tag{1.35}$$

$$\dot{\mathbf{v}} = \frac{e}{m}\mathbf{e}, \tag{1.36}$$

Here, the velocity $\mathbf{v}$ describes the motion of superconducting electrons in the ball, $\mathbf{h}$ and $\mathbf{e}$ are local values of the magnetic and electric field vectors. They are unknown and should be determined further on. What is known is that $\mathbf{v}_0 = \mathbf{v}_0(x, y, z, t)$, the local velocity of the body. We will consider a rotating sphere with radius R and angular velocity ω. Consideration is in spherical coordinates r, θ, φ. Rotation is along the polar axis, $\theta = 0$. Then

$$\mathbf{v}_0 = [\omega \times \mathbf{r}]. \tag{1.37}$$

The current is $\mathbf{j} = en_s(\mathbf{v} - \mathbf{v}_0)$, i.e., if all electrons are moving with the ions, there is no current. Substituting $\mathbf{j}$ into the Maxwell equation $\operatorname{curl} \mathbf{h} = (4\pi/c)\mathbf{j}$ where the displacement current is dropped, we have:

$$\operatorname{curl} \mathbf{h} = \frac{4\pi n_s e}{c}(\mathbf{v} - \mathbf{v}_0). \tag{1.38}$$

At the start of rotation, the normal electrons will participate in the relative motion, but when the rotation is stabilized, they move with the body, and so we can neglect their presence. Substituting (1.35) into (1.38), we find:

$$\operatorname{curl}\operatorname{curl} \mathbf{v} = -\frac{4\pi n_s e^2}{mc^2}(\mathbf{v} - \mathbf{v}_0). \tag{1.39}$$

Since $\operatorname{curl} \mathbf{v}_0 = 2\omega$ and $\operatorname{curl}\operatorname{curl} \mathbf{v}_0 = 0$, we can represent (1.39) as

$$\operatorname{curl}\operatorname{curl}(\mathbf{v} - \mathbf{v}_0) = -\frac{1}{\lambda_L^2}(\mathbf{v} - \mathbf{v}_0), \tag{1.40}$$

where we used expression (1.17) for the London penetration length λ_L. We can also substitute (1.38) in (1.35), which yields

$$\operatorname{curl}\operatorname{curl} \mathbf{h} = -\frac{1}{\lambda_L^2}\left(\mathbf{h} + \frac{2mc}{e}\omega\right). \tag{1.41}$$

For the outside of the sphere, $r > R$, the differential equations are

$$\text{div } \mathbf{h} = 0 \tag{1.42}$$

and

$$\text{curl } \mathbf{h} = 0. \tag{1.43}$$

These equations do not preclude having a field outside of the sphere. Symmetry assumes that the field will be dipolar, with the dipole direction along the axis of rotation. Such a field may have the form

$$h_r = \frac{2M}{r^3} \cos\theta,$$
$$h_\theta = \frac{M}{r^3} \sin\theta, \tag{1.44}$$
$$h_\varphi = 0.$$

This dipolar field requires a bit of an explanation. As we know, no magnetic monopoles (i.e., elementary magnetic charges q_m) have been detected. For such a monopole, in analogy with the electric charges and fields, the magnetic field would be

$$\mathbf{H} = \frac{q_m}{r^3}\mathbf{r}. \tag{1.45}$$

Then the dipole field can be imagined as

$$\mathbf{H}_d = \frac{q_m}{r_+^3}\mathbf{r}_+ - \frac{q_m}{r_-^3}\mathbf{r}_- \tag{1.46}$$

where $\mathbf{r}_\pm = \mathbf{r} \mp \mathbf{d}/2$, $\mathbf{r}$ points from the center of the dipole (in our case at $r = 0$) to the point of interest, and $\mathbf{d}$ points from the negative to the positive magnetic charge. In the limit $d/r \to 0$,

$$\mathbf{H}_d = \frac{3(\mathbf{M} \cdot \hat{\mathbf{r}})\hat{\mathbf{r}} - \mathbf{M}}{r^3}, \tag{1.47}$$

where $\mathbf{M} = q_m \mathbf{d}$ is the dipole moment. If in Cartesian coordinates $\mathbf{M} || \hat{\mathbf{z}}$, then in the spherical coordinate system:

$$\mathbf{H}_d|_r = \left[3(\mathbf{M} \cdot \hat{\mathbf{r}})\hat{\mathbf{r}} - \mathbf{M}\right] \cdot \hat{\mathbf{r}}/r^3 = 2M \cos\theta/r^3,$$
$$\mathbf{H}_d|_\theta = \left[3(\mathbf{M} \cdot \hat{\mathbf{r}})\hat{\mathbf{r}} - \mathbf{M}\right] \cdot \hat{\boldsymbol{\theta}}/r^3 = M \sin\theta/r^3, \tag{1.48}$$
$$\mathbf{H}_d|_\varphi = \left[3(\mathbf{M} \cdot \hat{\mathbf{r}})\hat{\mathbf{r}} - \mathbf{M}\right] \cdot \hat{\boldsymbol{\varphi}}/r^3 = 0/r^3,$$

as in (1.44).

The rest is just a technical task. The value of the constant M (i.e., of the magnetic dipole moment) should be determined from the boundary conditions. We need to first determine the field inside the sphere. We will try a current having only a φ-component which is proportional to $\sin\theta$, because $\mathbf{v}_0$ in (1.37) has only a φ-component which is proportional to $\sin\theta$:

$$v_{0\varphi} = \omega r \sin\theta. \tag{1.49}$$

Then one can assume

$$v_\varphi - v_{0\varphi} = f(r)\sin\theta \rightarrow v_\varphi = [\omega r + f(r)]\sin\theta, \tag{1.50}$$

and from (1.40) we have

$$f'' + \frac{2}{r}f' - \left(\frac{2}{r2} + \frac{1}{\lambda_L^2}\right)f = 0. \tag{1.51}$$

Its general solution has a form:

$$f = \frac{C_1}{r^2}\left(\sinh\frac{r}{\lambda_L} - \frac{r}{\lambda_L}\cosh\frac{r}{\lambda_L}\right) + \frac{C_2}{r^2}\left(\cosh\frac{r}{\lambda_L} - \frac{r}{\lambda_L}\sinh\frac{r}{\lambda_L}\right). \tag{1.52}$$

Here C_1 and C_2 are constants. The regular solution (i.e., the one which is finite at $r = 0$) corresponds to $C_2 = 0$. Then

$$v_\varphi = \left[\omega r + \frac{C_1}{r^2}\left(\sinh\frac{r}{\lambda_L} - \frac{r}{\lambda_L}\cosh\frac{r}{\lambda_L}\right)\right]\sin\theta. \tag{1.53}$$

For the next step, we substitute this function into $\mathbf{h} = -(mc/e)$ curl $\mathbf{v}$, which follows from (1.35), and compute components of the curl in spherical coordinates for $r \leq R$:

$$h_r = \frac{mc}{e}\frac{1}{r\sin\theta}\frac{\partial}{\partial\theta}(\sin\theta\, v_\varphi)$$
$$= \frac{mc}{e}\left[2\omega + \frac{2C_1}{r^3}\left(\sinh\frac{r}{\lambda_L} - \frac{r}{\lambda_L}\cosh\frac{r}{\lambda_L}\right)\right]\cos\theta, \tag{1.54}$$

$$h_\theta = -\frac{mc}{e}\frac{1}{r}\frac{\partial}{\partial r}(r\, v_\varphi)$$
$$= \frac{mc}{e}\left\{-2\omega + \frac{C_1}{r^3}\left[\left(1 + \frac{r^2}{\lambda_L^2}\right)\sinh\frac{r}{\lambda_L} - \frac{r}{\lambda_L}\cosh\frac{r}{\lambda_L}\right]\right\}\sin\theta, \tag{1.55}$$

$$h_\varphi = 0. \tag{1.56}$$

Continuity of solutions (1.44) and (1.54)–(1.56) at the boundary $r = R$ delivers two equations for two unknowns M and C_1:

$$2M = \frac{mc}{e}\left[2\omega R^3 + 2C_1\left(\sinh\frac{R}{\lambda_L} - \frac{R}{\lambda_L}\cosh\frac{R}{\lambda_L}\right)\right],$$ (1.57)

$$M = \frac{mc}{e}\left\{-2\omega R^3 + C_1\left[\left(1 + \frac{R^2}{\lambda_L^2}\right)\sinh\frac{R}{\lambda_L} - \frac{R}{\lambda_L}\cosh\frac{R}{\lambda_L}\right]\right\}.$$ (1.58)

The results are:

$$C_1 = \frac{3\omega R\lambda_L^2}{\sinh(R/\lambda_L)}$$ (1.59)

and

$$M = \frac{mc\omega}{e}R^3\left(1 + \frac{3\lambda_L^2}{R^2} - \frac{3\lambda_L}{R}\coth\frac{R}{\lambda_L}\right).$$ (1.60)

Thus the field outside (1.44) and inside (1.54)–(1.56) of sphere is found, as well as the current distribution inside of the sphere (1.53). Let us consider the current first. From (1.53) and (1.59) it follows that $\mathbf{j} = en_s(\mathbf{v} - \mathbf{v}_0)$ has components:

$$j_\theta = j_r = 0,$$ (1.61)

$$
\begin{aligned}
j_\varphi &= en_s(v_\varphi - v_{\varphi 0}) \\
&= \frac{3en_s\omega R\lambda_L^2}{\sinh(R/\lambda_L)}\frac{1}{r^2}\left(\sinh\frac{r}{\lambda_L} - \frac{r}{\lambda_L}\cosh\frac{r}{\lambda_L}\right)\sin\theta \\
&\approx -3en_s\omega\lambda_L\exp\left(-\frac{R-r}{\lambda_L}\right)\sin\theta.
\end{aligned}
$$ (1.62)

In writing the last line in (1.62), we made use of the fact that current flows only at the thin layer $\sim\lambda_L$ of the surface of the sphere. Inside of the sphere, except for this thin surface layer, the magnetic field is homogeneous. Indeed, with exponential accuracy, for $R - r \gg \lambda_L$:

$$h_z = h_r\cos\theta|_{\theta=0} = \frac{2mc\omega}{e}.$$ (1.63)

For $\omega \simeq 10^4\,\mathrm{s}^{-1}$ (rotational speed of typical turbopumps) this field is on the order of a milligauss. Of interest is also the magnetic moment, which in accordance with (1.60) can be represented as

$$M \approx \frac{mc\omega}{e}R^3.$$ (1.64)

Remark 1 Our derivation followed that of Fritz London. He himself took advantage of the previous study by Becker et al. (1933), where all the mathematical expressions were obtained. What was not done, and that was where the genius of London revealed itself, was to understand that the uniqueness of the solutions based on (1.35) and (1.36) *does not depend on the history* of reaching this state. The magnetic momentum of a superconducting ball will appear not only when the ball is first cooled and then rotated, but also when the ball is first rotated (so that all the electrons are moving with the lattice), and then cooled. As soon as the temperature goes below the superconducting transition temperature, the electrons at the surface layer of the rotating sphere will reduce their rotational speed and current will set up, causing the observational magnetic moment. This is quite similar to the Meissner effect, which has the same independence of the history. Both effects have no classical explanation and are due to the quantum nature of the superconducting state.

Remark 2 Very interestingly, experiments with various superconducting materials revealed that the mass m which enters the expressions (1.63) and (1.64) is the *bare electron mass* m_0 as opposed to the effective electron mass m_{eff} one deals with in solid-state physics. These experiments included so-called heavy-fermion superconductors with $m_{eff} \sim 10^3 m_0$, and all these experiments revealed $m = m_0$. Thus, the common belief that *in all experiments*, as soon as an electron is in the crystalline lattice, it becomes a dressed particle and behaves not like a bare electron, should be corrected for inertial experiments. Another example of this type is presented by the well-known Tolman–Stuart experiments, which are also inertial experiments.

Remark 3 Since only the crust of the ball is participating in the build-up of the current, and, accordingly, in the build-up of the magnetic moment, one can replace the bulk superconducting ball by a superconducting shell or by a dielectric ball covered by a thin superconducting layer.

Remark 4 Such dielectric (quartz) balls covered by a very thin Niobium superconductor film indeed were used by NASA in cooperation with Stanford University for the successful space mission *Gravity Probe B* to test predictions of Einstein's General Relativity. Rotating balls were serving as gyroscopes and the orientation of the magnetic moment **M** (which we described above) was monitored by a SQUID sensor. These balls (see Fig. 1.12) when I saw them in person at a local exhibition while visiting the Gravity Probe group at Stanford University a couple of years ago, were the most perfect spheres ever made by mankind.

Fig. 1.12 Superconducting ball of Gravity Probe B mission consisted of fused silica ball (*left*) covered by Nb film (*right*). Four balls were used in the instrument for navigation gyroscopes. Their rotation-caused London magnetic moments were indicating the gyroscope orientation. Manufacturing of balls and development of gyroscope required years of effort. The London moment of rotating superconductor was instrumental for the mission success

1.5 Ginzburg and Landau Approach

Above, we took advantage of the expression (1.9), which has the structure

$$\mathbf{j} \propto -\mathbf{A} \tag{1.65}$$

and explained the Meissner effect, an experimental fact. That means that one can rely on expression (1.65), at least in certain cases. However, in classical physics, observables cannot be proportional to the vector potential. There is a simple reason for that: vector potential is not a gauge-invariant quantity. What does that mean? Suppose we started from Maxwell's equations for some physical problem. We converted these equations using the relations

$$\mathbf{E} = -\dot{\mathbf{A}} - \nabla\varphi \tag{1.66}$$

and

$$\text{curl}\,\mathbf{A} = \mathbf{B} \tag{1.67}$$

to equations for $\mathbf{A}$ and φ and solved them. (An observant reader will notice that we dropped the factor $1/c$ in expression (1.66), *cf.* (1.7). In the context of current discussion this factor is not essential. We will be dropping unessential coefficients without further notice in Part I of this book from time to time, to draw attention to essentials.) Then one can add to the solution for $\mathbf{A}$ the gradient of an arbitrary function $\chi(\mathbf{r}, t)$ without affecting $\mathbf{B}$, since

$$\mathbf{B} = \text{curl}\,\mathbf{A} \equiv \text{curl}(\mathbf{A} + \nabla\chi) \tag{1.68}$$

(the curl of gradient is identically zero). At this process, care should be taken not to change **E** in (1.66). For that purpose, one should add to φ the function $-\dot{\chi}$, so that

$$\mathbf{E} = -\frac{\partial}{\partial t}(\mathbf{A} + \nabla\chi) - \nabla(\varphi - \dot{\chi}) \equiv -\dot{\mathbf{A}} - \nabla\varphi, \qquad (1.69)$$

as in (1.66). This operation that involves simultaneous transformation of the values of **A** and φ is known as the gauge transformation. Thus, the potentials of the electromagnetic field in classical (Maxwellian) physics are defined up to the gauge transformation, i.e., are not defined uniquely. However, the experimentally measurable current (1.65) should be uniquely defined. Yet it is not! This means that we are missing a term in (1.65), which we will call temporarily "*something*". Then the correct expression is

$$\mathbf{j} \propto -\mathbf{A} + something \qquad (1.70)$$

This "*something*" should behave at gauge transformations in a very certain way: the addition of $\nabla\chi$ to **A** should add the same $\nabla\chi$ to this *something*, so that these additions will cancel each other in (1.70). In quantum physics, luckily, there is a quantity which acquires χ at the gauge transformation. That quantity is the phase θ of the wave function $\Psi = |\Psi|\exp(i\theta)$ of the charged particle: $\theta \to \theta + \chi$.

Remark This fact is crucial for further considerations. We will prove it here for curious readers in a simple spatially homogeneous case.

If the potential φ is nonzero, we should incorporate it into the Hamiltonian as a potential energy. Thus, the Schrödinger equation has the form:

$$i\frac{\partial}{\partial t}\Psi = (\hat{H}_0 + \varphi)\Psi \qquad (1.71)$$

Suppose that the solution of this equation at $\varphi = 0$ is $\Psi_0 = |\Psi_0|\exp(i\theta_0)$. We will seek the solution at $\varphi \neq 0$ in the form $\Psi = \Psi_0\exp[i\theta(t)]$. Then

$$i\frac{\partial}{\partial t}\Psi = i\frac{\partial}{\partial t}[\Psi_0\exp(i\theta)] = i\exp(i\theta)\frac{\partial\Psi_0}{\partial t} - \Psi_0\exp(i\theta)\,\dot{\theta}$$

$$\equiv \exp(i\theta)\,i\frac{\partial\Psi_0}{\partial t} - \Psi\dot{\theta} \qquad (1.72)$$

If θ does not depend on coordinate x, then $\hat{H}_0\Psi = \exp(i\theta)\,\hat{H}_0\Psi_0$ and (1.72) yields

$$i\frac{\partial}{\partial t}\Psi = (\hat{H}_0 - \dot{\theta})\Psi. \qquad (1.73)$$

Comparing (1.73) with (1.71) yields:

$$\dot{\theta} = -\varphi, \tag{1.74}$$

or, in the integral form,

$$\theta = -\int \varphi dt + const \equiv \chi. \tag{1.75}$$

Thus, subtraction of $\dot{\chi}$ from φ, (1.69), is associated with adding χ to θ.
This provides us the grounds to express a hypothesis: in (1.70), *something* $\leftrightarrow$ $\nabla \theta$, i.e.,

$$\mathbf{j} \propto -\mathbf{A} + \nabla \theta. \tag{1.76}$$

The expression (1.76) is gauge-invariant: the gauge function gradient added to $\mathbf{A}$ will be compensated by the contribution from $\nabla \theta$. We also have a hint that the missing coefficient in (1.76) should be $|\Psi|^2$. Indeed, in quantum mechanics, the expression for the current density is

$$\mathbf{j} = -\mathbf{A}|\Psi|^2 + \left(\frac{i}{2}\Psi\nabla\Psi^* + c.c.\right) \equiv -(\mathbf{A} - \nabla\theta)|\Psi|^2, \tag{1.77}$$

which will match with (1.76). This Ginzburg-Landau expression for current (1.77), which is more general than (1.9), should thus be used for the explanation of the Meissner effect without any contradiction with the gauge invariance.

This conclusion is very deep and far reaching. It tells us that superconductivity is a quantum phenomenon, and that superconductors are macroscopic quantum objects. Readers may be surprised here: typically we associate the quantum-mechanical wave function with the objects in microworld, such as electrons and photons. Moreover, we learned that macroscopic objects, like pieces of metal, should behave as classical entities in common life. This is not the case with superconductors! Here, Ψ stands for the whole piece of the metal, and that needs an explanation. Below the critical "transition" temperature, the electrons in superconductors become paired into so-called "Cooper pairs". Pairing causes their spins to constitute a whole number (recall that individual electrons have a spin $1/2$), i.e., 0 or 1. In both cases, these Cooper pairs have quantum statistics different from the single-electron statistics: instead of Fermi-statistics, they now acquire Bose-statistics, and can be condensed below a certain "critical" temperature T_c into a condensate that is a relative to the Bose-condensates (though in this case it is called the "Cooper condensate").

Problem 4. Determine the difference between Cooper condensate and Bose-condensate.

Tip: estimate the size of the Cooper pair and compare it with inter-atomic distance.

Solution to Problem 4

Suppose that Cooper pairs have a size ξ. Then, quantum-mechanical uncertainty relates this size with the momentum $\delta p \times \xi \sim h$, where h is Planck's constant. The related energy should be comparable with the transition temperature T_c (since at this temperature, thermal fluctuations are breaking the pairs). Thus:

$$\frac{(\delta p)^2}{2m} \sim \frac{h^2}{2m\xi^2} \sim T_c, \text{ and } \xi \sim h\sqrt{\frac{2m}{T_c}}. \tag{1.78}$$

This can be compared with the inter-atomic distance $a \sim h(2m/\epsilon_F)^{1/2}$, where ϵ_F is the characteristic energy of an electron in the metal. Then

$$\xi \sim a\sqrt{\frac{\epsilon_F}{T_c}}, \tag{1.79}$$

which means, that $\xi/a \gg 1$. Indeed, typical values of ϵ_F are 10^5 K, and at $T_c \sim 10$ K we thus will have $\xi/a \sim 100$. That means that the Cooper pair condensate is constituted of "particles" which have spatial dimensions much larger than the distance between single electrons in metals. We will remind the reader that the density of electrons in metals is approximately equal to the density of ions, and therefore the inter-electron distance coincides with the inter-atomic distance. Thus, Cooper pairs greatly overlap in the the Cooper condensate, while in the Bose-condensate, the size of particles is much smaller than the inter-particle distance, and there is no overlapping. It is worth noting that there could be systems with low density of charge carriers where a crossover between Bose and Cooper pair condensates takes place. Such situations are currently being considered at the frontiers of scientific research.

> **Remark** This is an important difference, but not the only between Cooper and Bose condensates. We refer interested readers to Sect. 3.3 of Part II for more information on this topic.

Problem 5. Consider a hollow superconducting cylinder, and prove that the magnetic flux is quantized in it.

Tip: Take advantage of your knowledge of the Meissner effect.

Solution to Problem 5

We will write (1.77), dropping for a moment the constant $|\Psi|^2$. Then, on the trajectory shown by the dotted line in Fig. 1.13 the value of the current

$$\mathbf{j} = \mathbf{A} - \nabla\theta \tag{1.80}$$

Fig. 1.13 Cross section of
hollow cylinder with its axis
parallel to **H**

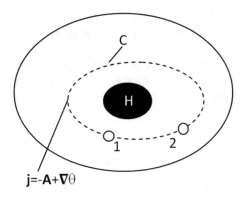

$$\mathbf{j}=-\mathbf{A}+\nabla\theta$$

is zero (it is also zero on any other trajectory C in the bulk of superconductor; the
"bulk" here means away from the boundary by more than λ_L). Then

$$\oint_C \mathbf{j} \cdot \mathbf{dl} = 0 = -\oint_C \mathbf{A} \cdot \mathbf{dl} + \oint_C \nabla\theta \cdot \mathbf{dl}, \tag{1.81}$$

or

$$0 = -\int_S \text{curl } \mathbf{A} \cdot \mathbf{dS} + \oint_C \nabla\theta \cdot \mathbf{dl}. \tag{1.82}$$

Then

$$\int_S \text{curl } \mathbf{A} \cdot \mathbf{dS} = \int_S \mathbf{B} \cdot \mathbf{dS} = \Phi = \oint_C \nabla\theta \cdot \mathbf{dl} = 2\pi n, \tag{1.83}$$

where $n = 0, \pm1, \pm2, etc.$ If proper coefficients are used (see for details Sect. 3.4.5
in Part II), then

$$\Phi = \phi_0 n, \text{ where } n = 0, \pm1, \pm2, etc. \tag{1.84}$$

where

$$\phi_0 = \frac{hc}{2e} \approx 2 \times 10^{-15} Weber.$$

This value is correct for superconductors, and we use it below when considering
SQUIDs. Historically, quantization of flux in superconductors was predicted by Fritz
London, who had no idea about pairing, and thus concluded that one should expect

$$\phi_0 = \frac{hc}{e} \tag{1.85}$$

In reality, we have a double charge $2e$ because of Cooper pairing (1 and 2 in the figure
above denote two electrons in the pair). That $2e$-factor was almost simultaneously

confirmed experimentally by two groups after the appearance of BCS theory. As a matter of fact, the original Ginzburg and Landau theory was developed before BCS theory, and its authors also had no idea about pairing, so the charge doubling in their theory was not taken into account. Based on BCS theory, Lev Gor'kov derived the correct expression for the superconducting current in the Ψ-theory of superconductivity in the form

$$\mathbf{j} = -(2\mathbf{A} - \nabla\theta)|\Psi|^2, \tag{1.86}$$

with explicit doubling of the charge.

Problem 6. Is the flux always quantized?

Tip: Consider a cylinder with a very thin wall.

Solution to Problem 6

Quantization of the magnetic flux requires existence of a trajectory in the bulk of superconductor where the current density is zero (see Problem 5). When the thickness of the cylinder wall is comparable to the London penetration depth, such trajectories are absent, and so is the quantization of the flux.

Remark Suppose that we placed a solenoid in a hollow superconducting cylinder which generates the field $\mathbf{H}$ in its core, as shown in Fig. 1.13. By tuning the current in the solenoid, this H-field can externally introduce any flux, not just a quantized one. However, the bulk superconductor will react with a current generated in its internal wall surface so that the total resultant flux will become quantized. F. London called this total flux a "fluxon". In this language, we should say that "fluxon" (but not flux!) is quantized in superconductors.

1.6 Josephson Effects

Now, let us draw some conclusions from (1.77) for the current in superconductors. The current consists of two contributions: the $\mathbf{A}$-part, and the $\nabla\theta$-part. Three components of the vector potential $\mathbf{A}$ plus the scalar φ are redundant for the definition of three components of the vectors $\mathbf{B}$ and $\mathbf{E}$. This redundancy can be removed by the choice of the gauge. We can do that to eliminate one of the scalars. Three scalar functions are enough for the $3D$-case, two scalars are enough for the $2D$-case, and one scalar is enough for the $1D$-case.

Let us start from the simplest $1D$-case. Suppose we have a superconducting wire along the x-direction. In that case, we can write $j_x = -A_x$, or, in another gauge,

$j = \nabla\theta$. (We drop $|\Psi|^2$ for simplicity, assuming it is constant for the time being.) In the latter case, for the current to be constant, its phase $\theta = const \times x$. Let us understand this better.

If there is no current, then the phase is not x-dependent; it is just a constant, so its gradient (i.e., its derivative over x in this case) is zero. When there is a constant current in the wire (Fig. 1.14), in accordance with the derivative's definition

$$j_x = \lim \frac{\Delta\theta}{\Delta x}|_{\Delta x \to 0} = const \qquad (1.87)$$

the phase increases linearly from point to point along the wire, i.e., the current is associated with the phase-change along the trajectory of the current. Obviously, the current is positive when the phase is increasing, and negative when it is decreasing.

Now consider two pieces of superconducting wire with no current in them (Fig. 1.15). Each piece has its own wave-function:

$$\Psi_{left} = |\Psi| \exp(i\theta_{left}) \text{ and } \Psi_{right} = |\Psi| \exp(i\theta_{right}) . \qquad (1.88)$$

Since there is no current in either one of them, the phases are constant. If the pieces had a superconducting transition independent from one another, their phases would not necessarily be equal. At the same time, we consider $|\Psi|$ to be similar for both pieces—that is possible because in quantum physics $|\Psi|^2$ is the density of particles (in our case, paired electrons) and we can consider both pieces of wire to be made of the same metal. What will happen if we bring them in contact, as shown in Fig. 1.16? Obviously, there will be a phase difference across the boundary line between the pieces. That means there will be a current between these pieces with no voltage applied! The current should be limited in value (any physical border has a finite thickness, so the denominator in (1.87) is finite). It should also be antisymmetric as a function of phase difference $\theta_{left} - \theta_{right} \equiv \theta$. The current should be zero if θ is zero. It should be periodic with argument θ, with a period of 2π. The last statement follows from the fact that $\exp[i(\theta + 2\pi n)] = \exp(i\theta)$ and from the structure of the

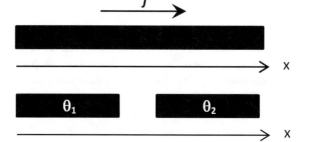

Fig. 1.14 Current in 1D wire. In the 1D case current I is simply proportional to j

Fig. 1.15 Two pieces of superconducting wire

Fig. 1.16 Two
superconducting wires in
contact

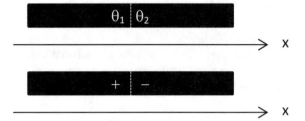

$\theta_1 \mid \theta_2$

$\longrightarrow$ X

Fig. 1.17 Voltage applied to
superconductors junction

$+ \mid -$

$\longrightarrow$ X

functions in (1.88). So one would not be mistaken with a statement that the current
should have the form:

$$j = j_0 \sin \theta, \tag{1.89}$$

and generally should be non-zero when two superconductors are brought in contact.
This statement constitutes the first so-called "stationary" Josephson effect.

   Current between superconductors will flow, of course, until the electric charging of
pieces will stop it. Alternatively, we need an attached source and a sink for electrons.

Problem 7. What will happen if we apply a constant voltage across this superconducting junction (Fig. 1.17)?

   *Hint: consider one of the superconductors to be at $\varphi = 0$ and apply $\varphi = \varphi_0$ to
the other superconductor. Use arguments established via gauge-invariance consideration to find the relation between θ and φ.*

Solution to Problem 7
Non-zero scalar potential in a superconductor can be introduced by performing a
gauge transformation with a function $\dot{\chi} = -\varphi$. That will also add a function χ to the
phase $\theta(t)$, so that $\theta(t) = -\varphi t + \theta_0$ [see (1.74) and (1.75)]. After substitution into
(1.89), we have

$$j = j_0 \sin(\varphi t - \theta_0) \tag{1.90}$$

One can associate φ with V, the potential difference between the superconductors.
Also, the phase $-\theta_0$ at $t = 0$ can be dropped if t is shifted in time. Since the superconducting wave function is associated with Cooper pairs, the pre-factor should be
changed to 2: $V = 2e\varphi$. Restoring all the units to the conventional ones, we can write
the final result as:

$$j = j_0 \sin(\omega t), \text{ where } \omega = 2eV/\hbar. \tag{1.91}$$

Remark This is a very unusual result. Indeed, it tells us that at a constant voltage (or potential difference) between superconductors, the current oscillates in time! It was confirmed experimentally soon after its prediction. It is called the second, or "non-stationary", Josephson effect. Josephson predicted his effects when he was a graduate student and received a Nobel prize for his prediction. It is surprising that these effects were not predicted even before the microscopic theory of superconductivity, say, by Ginzburg and Landau, or even by F. London! Here we have an example of even a genius' imagination being restricted. It is also important to mention that there is always more in the theory than its originators have noticed. We will consider SQUIDs in the next section which will illustrate this statement further.

From the practical point of view, it is very important that the oscillation frequency in (1.91) contains fundamental constants e and $\hbar$. This fact was used to determine these constants with higher accuracy than they were known before. It was also used for time and voltage standards - the junction can be placed in a resonator in which the resonant frequency is known with very high accuracy, and determine at which frequency a resonance is taking place at varying voltage across the junction. Reciprocally, you can keep the voltage constant, and vary the geometry of the resonator to determine the value of voltage from the occurring resonance. Josephson junctions have many other applications, constituting one of the pillars of *superconducting electronics*.

1.7 SQUIDs

We will consider now two Josephson junctions, "**a**" and "**b**" in the loop, as shown in Fig. 1.18. This configuration is called a DC SQUID, the abbreviation standing for "Superconducting Quantum Interference Device". There may be a magnetic flux in the loop, and we are interested in the maximum superconducting current vs. magnetic flux in the loop. The current is supposed to be connected to the loop (current $\mathbf{j}$ is depicted by the white arrows). We will neglect all variation of $|\Psi|$ in the loop and use the formula (1.77) derived above for the current $\mathbf{j} = -(\mathbf{A} - \nabla|\theta|)|\Psi|^2$. Because of the Meissner effect, we expect the current to be zero inside of the bulk superconductor, say, on the dashed line loop. So, let us integrate this zero current density along that line:

$$\oint \mathbf{j} \cdot d\mathbf{l} = 0 = -\oint \mathbf{A}(\mathbf{r}) \cdot d\mathbf{l} + \oint \nabla\theta \cdot d\mathbf{l}, \qquad (1.92)$$

or

Fig. 1.18 Current through DC SQUID

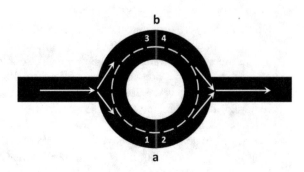

$$\oint \mathbf{A}(\mathbf{r}) \cdot d\mathbf{l} = \oint \nabla \theta \cdot d\mathbf{l}. \tag{1.93}$$

Obviously,

$$\oint \nabla \theta \cdot d\mathbf{l} = \int_1^3 \nabla \theta \cdot d\mathbf{l} + \int_4^2 \nabla \theta \cdot d\mathbf{l}, \tag{1.94}$$

which means that

$$\oint \mathbf{A}(\mathbf{r}) \cdot d\mathbf{l} = \theta_3 - \theta_1 + \theta_2 - \theta_4. \tag{1.95}$$

And the same via Stokes' theorem provides:

$$\oint \mathbf{A}(\mathbf{r}) \cdot d\mathbf{l} = \int_S \operatorname{curl} \mathbf{A} \cdot d\mathbf{S} = \int_S \mathbf{H} \cdot d\mathbf{S} = \Phi, \tag{1.96}$$

where the magnetic flux Φ is:

$$\Phi = \theta_3 - \theta_1 + \theta_2 - \theta_4 = \theta_a - \theta_b \tag{1.97}$$

(the absent coefficients will be restored later on), where we denoted

$$\theta_a = \theta_2 - \theta_1 \text{ and } \theta_b = \theta_4 - \theta_3 \tag{1.98}$$

for the phase jumps on the barriers "a" and "b", Fig. 1.18. In view of our previous knowledge of Josephson junctions, we can expect the current across these two junctions to be:

$$I_a = I_a^0 \sin(\theta_2 - \theta_1) \text{ and } I_b = I_b^0 \sin(\theta_4 - \theta_3) \tag{1.99}$$

Then, the total current is:

Fig. 1.19 Maximum current through DC SQUID vs. magnetic flux is a periodic function of Φ

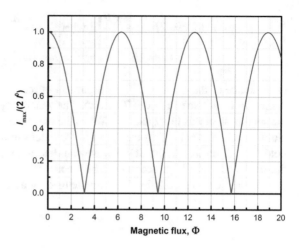

$$I = I_a + I_b = I^0 \left(\sin \theta_a + \sin \theta_b \right)$$
$$= I^0 \left[2 \sin \left(\frac{\theta_a + \theta_b}{2} \right) \cos \left(\frac{\theta_a - \theta_b}{2} \right) \right]. \tag{1.100}$$

Here, we consider symmetric junctions for simplicity:

$$I_a^0 = I_b^0 \equiv I^0. \tag{1.101}$$

Now, we will make simple mathematical transformations and substitutions for Φ using (1.97):

$$I = I^0 \left[2 \cos \left(\frac{\Phi}{2} \right) \sin \left(\frac{\theta_a - \theta_b + 2\theta_b}{2} \right) \right]$$
$$= I^0 \left[2 \cos \left(\frac{\Phi}{2} \right) \sin \left(\frac{\Phi}{2} + \theta_b \right) \right] \tag{1.102}$$

At a given value of Φ, the phase θ_b is still adjustable, so that $\sin (\ldots) = 1$, and thus the maximum current is:

$$I_{\max} = 2I^0 \left| \cos \left(\frac{\Phi}{2} \right) \right|, \tag{1.103}$$

which is shown in Fig. 1.19. To understand the importance of SQUIDs, we will now restore the units in the expression above. To make Φ unitless, we need to divide it by ϕ_0, the flux quantum. Then

$$I_{\max} = 2I_0 \left| \cos \frac{\pi \Phi}{\phi_0} \right|, \text{ where } \phi_0 = \frac{hc}{2e} = \frac{\pi \hbar c}{e}$$

$$\approx 2 \times 10^{-15} Weber = 2 \times 10^{-7} Gauss \times cm^2 \tag{1.104}$$

(appearance of the number π in the argument of the cosine function is related to the definition of ϕ_0; see Problem 7 below). The sensitivity of SQUIDs is related with the smallness of this flux quantum. When the current exceeds $I_{\max}$, there is a voltage along the current path. This can be used for determining $I_{\max}$. At a given magnetic field, or a given flux in the SQUID, we can determine $I_{\max}(\Phi)$, and thus determine the field with very high accuracy. This accuracy stayed unmatched in physics for decades and only recently has been challenged by other quantum phenomena.

1.8 Time-Dependent Ginzburg–Landau (TDGL) Theory

At this point, it became evident that superconductors should be described by the quantum physics wave function, $\Psi = |\Psi| \exp(i\theta)$. The "quantum" part of superconductors, the system of non-dissipative current-carrying electrons, is called Cooper-pair condensate, and it is this condensate that is described by the Ψ-function. However, all conductivity electrons are paired into Cooper-pairs only at absolute zero temperature, $T = 0$. At finite temperatures, $T \neq 0$, a certain amount of unpaired, or "normal", electrons exists. Correspondingly, the current consists of a quantum-mechanical, non-dissipative part and of a normal, dissipative part:

$$\mathbf{j}|_{\text{total}} = -(2\mathbf{A} - \nabla\theta)|\Psi|^2 (non - dissipative\ current)$$

$$+\sigma \left(\frac{\partial \mathbf{A}}{\partial t} + \nabla \varphi \right) (dissipative\ current). \tag{1.105}$$

Because of electric neutrality (or Coulomb interaction), the dissipative and non-dissipative electronic motions strongly interact. The most important corollary of this interaction is that the equation for the Ψ-function *no longer looks* like the Schrödinger equation:

$$i \frac{\partial \Psi}{\partial t} = \hat{H} \Psi. \tag{1.106}$$

It actually turns out that the right equation has a structure similar to (1.106), but with imaginary conjugated Hamiltonian! It took a long time and an intense effort to derive this equation in the theory of nonequilibrium superconductivity for a physically plausible range of parameters. The whole second part of this book is devoted to the derivation of the system of time-dependent Ginzburg–Landau (TDGL) equations, and readers interested in this topic are advised to go through this very detailed background material. For now, in Part I, we will be studying the solutions of these equations, considering the equation for the Ψ-function:

$$-\frac{\pi}{8T_c}\left(\frac{\partial}{\partial t}+2i\varphi\right)\Psi$$

$$+\frac{\pi}{8T_c}\left[D\left(\nabla-2i\mathbf{A}\right)^2\right]\Psi+\left[\frac{T_c-T}{T_c}-\frac{7\zeta(3)}{8(\pi T_c)^2}\mid\Psi\mid^2\right]\Psi=0. \qquad (1.107)$$

as given. This is the simplest, so-called "gapless form" of this equation. In (1.107) $\zeta(3)=\sum_{n=1}^{\infty}n^{-3}\approx 1.2$ is the Riemann zeta function, D is the diffusion coefficient, and T_c is the critical temperature. The most important change here compared to the quantum mechanical equation (1.106) is the disappearance of the imaginary coefficient "i" in front of the time derivative, and its replacement by a real coefficient. While (1.106) is a wave equation, (1.107) is of a relaxation type.

Problem 8. Prove the second part of the statement above.

Tip: Consider spatially homogeneous state with no electric or magnetic fields, and introduce a small deviation to the steady state solution of (1.107).

Solution to Problem 8

In absence of external fields: $\mathbf{A}=0$, $\varphi=0$, and for the homogeneous state (Ψ-real, $\nabla|\Psi|=0$) (1.107) is:

$$-\frac{\pi}{8T_c}\frac{\partial}{\partial t}\Psi+\left[\frac{T_c-T}{T_c}-\frac{7\zeta(3)}{8(\pi T_c)^2}\Psi^2\right]\Psi=0. \qquad (1.108)$$

Its nontrivial steady-state solution ($\Psi_0\neq 0$) follows from the equation:

$$\frac{T_c-T}{T_c}-\frac{7\zeta(3)}{8(\pi T_c)^2}\Psi_0{}^2=0. \qquad (1.109)$$

If $\Psi=\Psi_0+\delta\Psi(t)$, and $|\delta\Psi|\ll|\Psi_0|$, then from (1.108) one can derive

$$-\frac{\pi}{8T_c}\frac{\partial}{\partial t}(\delta\Psi)+\left[\frac{T_c-T}{T_c}-\frac{7\zeta(3)}{8(\pi T_c)^2}(\Psi_0+\delta\Psi)^2\right](\Psi_0+\delta\Psi)=0. \qquad (1.110)$$

Using (1.109) and neglecting smaller $(\delta\Psi)^2$-terms, we represent (1.110) in the form

$$\frac{\pi}{8T_c}\frac{\partial}{\partial t}(\delta\Psi)=-\frac{7\zeta(3)}{8(\pi T_c)^2}(2\Psi_0^2)(\delta\Psi). \qquad (1.111)$$

From here we see that if $\delta\Psi = \Psi - \Psi_0 > 0$ then $\partial(\delta\Psi)/\partial t < 0$. Reciprocally, if $\delta\Psi < 0$, then $\partial(\delta\Psi)/\partial t > 0$. That means that small fluctuations will relax to zero, so that a weakly perturbed Ψ-function will relax to its steady-state value. In more formal language, (1.111) has a relaxational solution $\Psi(t) = \delta\Psi|_{t=0}\exp(-t/t_0)$, with a characteristic relaxation time $t_0 = \pi^3 T_c/[14\zeta(3)\Psi_0^2] \approx 0.5 T_c/\Psi_0^2$.

Remark The relaxation time t_0 is temperature-dependent, since $\Psi_0 = \Psi_0(T)$. Ψ_0 is small near the transition temperature, and larger at $T \ll T_c$. Larger Ψ_0 means shorter relaxation time t_0. This reflects the property of Bose-condensates (which are close relatives of Cooper-condensates): they have ability to successfully fight small fluctuations and heal themselves. That is why the superconducting state is much "quieter" than the normal metal state.

Problem 9. Prove the gauge invariance of TDGL equations.

Tip: Consider (1.105) and (1.107), and take into the account that at gauge transformation of the electric field $\mathbf{E} = -\partial\mathbf{A}/\partial t - \nabla\varphi$ and the magnetic field $\mathbf{H} = \mathrm{curl}\,\mathbf{A}$ with an arbitrary function $\chi(x, y, z, t)$:

$$\mathbf{A}^{new} \to \mathbf{A}^{old} + (1/2)\nabla\chi, \tag{1.112}$$

$$\varphi^{new} \to \varphi^{old} - (1/2)\partial\chi/\partial t, \tag{1.113}$$

$$\theta^{new} \to \theta^{old} + \chi. \tag{1.114}$$

Solution to Problem 9

Let us demonstrate the gauge invariance of the current expression first. For that, it is useful to make a transformation in (1.105):

$$\frac{1}{2i}\Psi\nabla\Psi^* + c.c. = \frac{1}{2i}|\Psi|\exp(i\theta)\nabla[|\Psi|\exp(-i\theta)] + c.c.$$
$$= \frac{\nabla|\Psi|^2}{4i} - \frac{|\Psi|^2\nabla\theta}{2} + c.c = -|\Psi|^2\nabla\theta, \tag{1.115}$$

so that the current density is equal to

$$\mathbf{j} = -(\mathbf{A} - \nabla\theta)|\Psi|^2 + \sigma\left(\frac{\partial\mathbf{A}}{\partial t} + \nabla\varphi\right). \tag{1.116}$$

Substitution of (1.112)–(1.114) into (1.116) confirms that the gauge function χ straightforwardly cancels out:

$$\mathbf{j}^{new} = -(\mathbf{A}^{new} - \nabla\theta^{new})|\Psi|^2 + \sigma\left(\frac{\partial\mathbf{A}^{new}}{\partial t} + \nabla\varphi^{new}\right)$$

$$= -\left(\mathbf{A}^{old} - \nabla\theta^{old}\right)|\Psi|^2 + \sigma\left(\frac{\partial\mathbf{A}^{old}}{\partial t} + \nabla\varphi^{old}\right) \equiv \mathbf{j}^{old}, \quad (1.117)$$

so that the current density (1.105) is gauge-invariant.

We will consider next its counter part, (1.107) for the Ψ-function. Our strategy here is to demonstrate that in (1.107), after application of gauge transformation, the factor $\exp(i\chi)$ comes out as a free-standing multiplier, so that we can divide by that factor and eliminate it. For the time derivative term:

$$\left(\frac{\partial}{\partial t} + 2i\varphi\right)\Psi \rightarrow$$

$$\rightarrow \left[\frac{\partial}{\partial t} + 2i(\varphi - \dot\chi/2)\right]\Psi\exp(i\chi) =$$

$$= \frac{\partial}{\partial t}\left[\Psi\exp(i\chi)\right] + [2i(\varphi - \dot\chi/2)]\Psi\exp(i\chi) =$$

$$= \exp(i\chi)\frac{\partial}{\partial t}\Psi + \Psi\frac{\partial}{\partial t}\{\exp[i(\chi)]\} + [2i(\varphi - \dot\chi/2)]\Psi\exp(i\chi) =$$

$$= \exp(i\chi)\left\{\frac{\partial}{\partial t}\Psi + i\Psi\dot\chi + [2i(\varphi - \dot\chi/2)]\Psi\right\} =$$

$$= \exp(i\chi)\left(\frac{\partial}{\partial t} + 2i\varphi\right)\Psi. \quad (1.118)$$

For the spatial derivative term:

$$\left[(\nabla - 2iA)^2\right]\Psi = (\nabla - 2iA)(\nabla - 2iA)\Psi \rightarrow$$

$$\rightarrow [\nabla - 2i(A + \nabla\chi/2)][\nabla - 2i(A + \nabla\chi/2)]\left[\Psi\exp(i\chi)\right] =$$

$$= [\nabla - 2i(A + \nabla\chi/2)]\{\nabla[\Psi\exp(i\chi)] - 2i[\Psi\exp(i\chi)](A + \nabla\chi/2)\} =$$

$$= [\nabla - 2i(A + \nabla\chi/2)]\exp(i\chi)\{[\nabla\Psi + i\Psi(\nabla\chi)] - 2i\Psi(A + \nabla\chi/2)\} =$$

$$= [\nabla - 2i(A + \nabla\chi/2)]\{\exp(i\chi)[(\nabla - 2iA)\Psi]\} =$$

$$= \exp(i\chi)[i(\nabla\chi)(\nabla - 2iA)\Psi + \nabla(\nabla - 2iA)\Psi - 2i(A + \nabla\chi/2)(\nabla - 2iA)\Psi] =$$

$$= \exp(i\chi)[\nabla(\nabla - 2iA)\Psi - 2iA(\nabla - 2iA)\Psi] =$$

$$= \exp(i\chi)\left[(\nabla - 2iA)^2\Psi\right]. \quad (1.119)$$

Conclusion: indeed, at gauge transformation the $\exp(i\chi)$ appears as a common multiplier in the time and spatial derivative terms of (1.107). Trivially, the same thing happens with the last group of terms (those without derivatives). We also notice that $\exp(i\chi) \neq 0$, so we are not dividing by zero when factoring it out of the equation. Thus, the gauge transformation leaves the system of (1.107) and (1.105) unchanged, which proves the gauge-invariance of TDGL equations.

Remark 1 Gauge transformation simplifies the search for the solutions. For example, one can choose a gauge with $\chi = -\theta$, so that $\theta^{new} \rightarrow \theta^{old} + \chi = 0$ (see (1.114)) and the Ψ-function is real. However, as soon as the gauge is chosen, and the procedure of solving the mathematical problem has started, the gauge can not be changed anymore. So any specific solution is obtained in a specific gauge.

Remark 2 The reader should be alerted that the same Ψ-function enters (1.105) and (1.107) which constitute a system of TDGL equations. That means that the Ψ-function should have the same normalization in both equations which currently is not the case. Normalization does not matter for proving gauge invariance, but it does matter when solving TDGL equations. As shown in Chap. 7, the expression for the current density which has the same normalization for Ψ-function as (1.105) has the form:

$$\mathbf{j} = -\frac{\pi \sigma_n}{4T} |\Psi|^2 \left(2\mathbf{A} - \nabla\theta\right) + \sigma_n \mathbf{E}. \tag{1.120}$$

This form for the current density will be used in the next chapter.

1.9   Seminar 2. When will We Have Superconductors at Ambient Conditions?

Since the discovery of superconductivity, there was an ever growing effort to raise the transition temperature of superconductivity (also called the *critical temperature* T_c) in novel materials. It was not an easy task, especially because the mechanism of superconductivity was stubbornly nondisclosive for almost a half-century after its discovery. Many giant minds, including Einstein, Bohr, Heisenberg, Landau, Feynman (to name a few) failed to decipher the mechanism of superconductivity. However, experimental facts were accumulating, and theoreticians were narrowing the circle of possible options during the years of effort. In 1957, a trio Bardeen, Cooper and Schrieffer brilliantly solved the problem (Nobel Prize, 1972). Their theory explained many experimental facts quantitatively. Curiously, neither one of the *explained* facts *explicitly* included dependence on the so-called "BCS potential". This potential ζ in BCS theory plays a crucial role in determining the value of T_c :

$$T_c \approx 1.13 \omega_D \exp\left[-\frac{1}{|\zeta| N(0)}\right] \tag{1.121}$$

[see Sect. (3.3) for details]. Its negative value corresponds to the indirect attraction between electrons. Why do electrons attract each other? It is not a simple question.

Indeed, electrons interact with each other via the Coulomb force, and this Coulomb interaction *is repulsive* in vacuum. (Landau used to say: "Nobody has abrogated the Coulomb law".) How does the presence of ion lattice change the situation? Let us start with analyzing this crucial point. In vacuum, the Coulomb interaction between two electrons at a distance r is:

$$V(r) = \frac{e^2}{r}. \tag{1.122}$$

In a dielectric medium, the repulsion is screened, as described by the dielectric function $\varepsilon(\mathbf{r}, t)$:

$$V(\mathbf{r}, t) = \frac{1}{\varepsilon} \frac{e^2}{r}, \tag{1.123}$$

so that the repulsion is weaker if $\varepsilon > 1$. As we will see now, the situation is much more complex in metals.

1.9.1 Dielectric Function

For homogeneous solids, it is convenient to work with Fourier transforms, so (1.123) becomes

$$V(\mathbf{q}, \omega) = \frac{4\pi e^2}{\varepsilon(\mathbf{q}, \omega)q^2}. \tag{1.124}$$

The dielectric function is defined according to Maxwell's electrodynamics by the relation:

$$\mathbf{D} = \varepsilon(\mathbf{q}, \omega)\mathbf{E}, \tag{1.125}$$

where

$$\operatorname{div} \mathbf{E} = 4\pi \rho_{\text{total}}, \tag{1.126}$$

and

$$\operatorname{div} \mathbf{D} = 4\pi \rho_{\text{ext}}. \tag{1.127}$$

Also

$$\rho_{\text{total}} = \rho_e + \rho_i + \rho_{\text{ext}} \tag{1.128}$$

where the abbreviations ρ_e, ρ_i, and ρ_{ext} are the charge densities of the electron gas, of the lattice ions, and of the external charge, respectively. "External" in this context means "free to manipulate" while ρ_e and ρ_i are "bound" charges, participating in the build-up of the electron and ion plasma oscillation modes[2] with frequencies:

---

[2]Plasma frequency was already mentioned in relation to the solution of Problem 2.

$$\omega_{ep} = \left(\frac{4\pi n_e^2 e^2}{m} \right)^{1/2}, \tag{1.129}$$

$$\omega_{ip} = \left(\frac{4\pi n_i^2 z^2 e^2}{M} \right)^{1/2}, \tag{1.130}$$

where m and M are electron and ion masses, ze is the effective ion charge, and n_e and n_i are the electron and ion densities. To determine the value of $\varepsilon(\mathbf{q}, \omega)$, we need to calculate

$$\varepsilon(\mathbf{q}, \omega) = \frac{\rho_{\text{ext}}}{\rho_e + \rho_i + \rho_{\text{ext}}} \tag{1.131}$$

as follows from (1.125)–(1.127). For this task, we will note that the motion of ions is governed by the equation

$$\omega^2 \rho_i = \omega_{ip}^2 \rho_{\text{total}}. \tag{1.132}$$

To make sure this is indeed the case, one can start with Newton's law for the ionic motion:

$$M \frac{\partial \mathbf{v}_i}{\partial t} = z \left| e \right| \mathbf{E}, \tag{1.133}$$

and transform it into the equation for current density $\mathbf{j}_i = n_i z \left| e \right| \mathbf{v}_i$:

$$\frac{\partial \mathbf{j}_i}{\partial t} = \frac{n_i z^2 e^2}{M} \mathbf{E}. \tag{1.134}$$

Combining (1.134) with the continuity equation

$$\frac{\partial \rho_i}{\partial t} + \operatorname{div} \mathbf{j}_i = 0 \tag{1.135}$$

and substituting for the wave perturbation a plane-wave in the form

$$\mathbf{E} = \mathbf{E}_0 e^{-i\omega t + i\mathbf{q}\mathbf{r}} \tag{1.136}$$

we justify (1.132) after simple algebra.

We can now look at the electron system. Since $\omega_{ep} \gg \omega_{ip}$ (typically, $\omega_{ep}/\omega_{ip} \sim 10^{16}/10^{13}$), electrons adiabatically follow ionic motion. Electrons have one more characteristic frequency: $\epsilon_F/\hbar$, where ϵ_F is the Fermi-energy which has the same scale as ω_{ep}. The Fermi energy is related to the unperturbed electron density n_e^o via the expression:

$$\epsilon_F = \frac{\hbar^2}{2m} (3\pi^2 n_e^0)^{2/3}. \tag{1.137}$$

From (1.137):

$$n_0^e = \frac{1}{3\pi^2} \left[\frac{2m}{\hbar^2} \epsilon_F \right]^{3/2}. \tag{1.138}$$

Homogeneous perturbation of the electron density: $\delta n_e = n_e - n_e^0$ changes globally the Fermi-energy by $-|e|\,\varphi$, where φ is the (negative, if $\delta n_e > 0$) local electrostatic potential:

$$n_e = n_e^0 + \delta n_e = \frac{1}{3\pi^2} \left[\frac{2m}{\hbar^2} (\epsilon_F - |e|\,\varphi) \right]^{3/2}. \tag{1.139}$$

Taking into account that $|e\varphi| \ll \epsilon_F$, after simple algebra, one can find from (1.133) and (1.134):

$$\delta n_e = -n_e^0 \frac{3}{2} \frac{|e|\,\varphi}{\epsilon_F}, \tag{1.140}$$

or, in terms of $\rho_e = -|e|\,\delta n_e$:

$$\rho_e = -\frac{3}{2} \frac{n_e^0 e^2 \varphi}{\epsilon_F}. \tag{1.141}$$

Now we can use the Poisson equation:

$$\nabla^2 \varphi = -4\pi \rho_{\text{total}} \tag{1.142}$$

and substitute (1.141) into (1.142). For the plane-wave motion Fourier component we get

$$q^2 \varphi = 4\pi \rho_{\text{total}}. \tag{1.143}$$

and, after substitution into (1.141),

$$\rho_e = -\frac{6n_e^0 e^2}{q^2 \epsilon_F} \rho_{\text{total}} = -\frac{k_{TF}^2}{q^2} \rho_{\text{total}} \tag{1.144}$$

Here $k_{TF} \equiv (6\pi^2 n_e^0 e^2 / \epsilon_F)^{1/2}$ is the Thomas-Fermi wave vector. We can now calculate $\varepsilon(\mathbf{q}, \omega)$. Obviously, (1.131) could be presented as

$$\varepsilon(\mathbf{q}, \omega) = \frac{\rho_{\text{total}} - \rho_i - \rho_{\text{ext}}}{\rho_{\text{total}}}, \tag{1.145}$$

and then:

$$\varepsilon(\mathbf{q}, \omega) = 1 - \frac{\omega_{ip}^2}{\omega^2} + \frac{k_{TF}^2}{q^2} = \frac{\omega^2(q^2 + k_{TF}^2) - \omega_{ip}^2 q^2}{\omega^2 q^2}. \tag{1.146}$$

1.9.2 BCS Attraction

Thus, the screened Coulomb interaction (1.123) takes the form:

$$V(\mathbf{q}, \omega) = \frac{4\pi e^2}{q^2 + k_{TF}^2 - \omega_{ip}^2 q^2 / \omega^2}. \qquad (1.147)$$

This is crucial for the explanation of superconductivity. For high frequencies ($\omega > \omega_{ip}$) we have just the screened Coulomb potential:

$$V(\mathbf{q}, \omega) \approx \frac{4\pi e^2}{q^2 + k_{TF}^2}, \qquad (1.148)$$

or, in the coordinate space:

$$V(\mathbf{r}) = \frac{e^2}{r} e^{-k_F r} \qquad (1.149)$$

which is clearly a repulsive potential (sometimes called "Yukawa potential"). Let us consider $q \sim 1/a$, where "a" is the interatomic distance (which is also the interelectronic distance since in a typical metal there is one conducting electron per ion). For $q > k_{TF}$ and $\omega < \omega_{ip}$:

$$V(\mathbf{q}, \omega) \approx -\frac{4\pi e^2 \omega^2}{q^2 \omega_{ip}^2}, \qquad (1.150)$$

so that the attraction dominates in the net interaction for this range of frequencies and wave vectors. This mechanism justifies the BCS model of superconductivity. Indeed, the negative interelecronic potential is the major cornerstone of the BCS-mechanism. We can roughly approximate ω_{ip} by the lattice Debye frequency ω_D, so the attraction takes place at frequencies below ω_D, in accordance to the BCS suggestion.

1.9.3 Phonon Resonance

Small frequencies mean enough lapsed time. That means that inter-electron attraction takes place not immediately, but with a delay thanks to the presence of the lattice. Overall sign change and negativity in the expression (1.147) looks like a resonant effect (see Fig. 1.20), so one may say: "because of the resonance, tiny electrons are able to move heavy ions thus yielding the mechanism of attraction." And on the other hand, at short time scales (high frequencies) the interaction is repulsive (1.148).

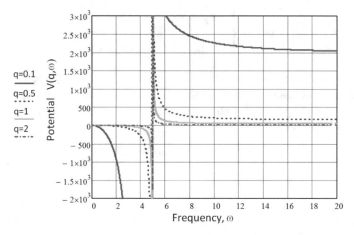

Fig. 1.20 The $V(q, \omega)$ behavior at certain values of q (in units of a^{-1}, where "a" is interatomic distance). We also used $e^2 = 1$, $\omega_D = 5$, and $k_{TF} = 0.1$

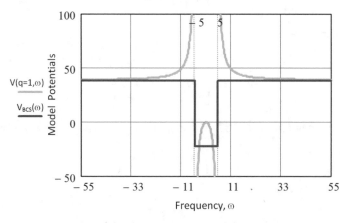

Fig. 1.21 Comparison of the BCS model potential, the Coulomb potential added ("two square well model", rectangular curve), with the resonant model (1.147) (diverging curves)

1.9.4 Coulomb Potential and Tolmachov Logarithm

The choice of q in Fig. 1.20 (from 0.1 to 2) was for illustrative purposes only. For simple metals, interatomic and interelectronic distances have the same mean values, so we will put $q \approx 1$ to move forward with our analysis. Figure 1.21 shows this case. For comparison, we also plotted the BCS approximation for the negative (attractive) potential at $|\omega| < \omega_D$ (taken $\omega_D = 5$) and positive (repulsive) approximation for the Coulomb part of interaction. (Sometimes, it is called the "two square well model.") In analytical form the model potential can be written as:

$$V(\omega) = \begin{cases} -V_p + V_c, & |\omega| < \omega_D \\ +V_c, & \omega_D < |\omega| < \omega_c \\ 0, & |\omega| > \omega_c \end{cases}, \tag{1.151}$$

where ω_c is the cut-off frequency for Coulomb interaction and typically has the order of ω_{ep}. We will see now how this potential works in the BCS model, where the energy gap is determined by the well-known self-consistency equation [see Section (3.3.9) in Part II of this book]:

$$\Delta(\xi) = -\int V(\xi - \xi') \frac{\Delta(\xi')N(\xi')}{2\epsilon} \left[1 - 2f(\epsilon')\right] d\xi', \tag{1.152}$$

where $\epsilon = \sqrt{\xi^2 + \Delta^2}$, $N(\xi)$ is the electron density of states, and $f(\epsilon)$ is the electron distribution function. In the simplest approximation (as we did above), $N(\xi)$ is taken to be constant in the whole range of integration:

$$N(\xi) \approx N(0) \tag{1.153}$$

and $\Delta(\xi)$ is represented as:

$$\Delta(\xi) = \begin{cases} \Delta_1, & \xi \leq \omega_D \\ \Delta_2, & \omega_D \leq \xi \leq \omega_c \\ 0, & \xi \geq \omega_c \end{cases}. \tag{1.154}$$

Accordingly, integration in (1.152) will be split into two parts, yielding two functions:

$$M_1 = -\int_0^{\omega_D} \frac{d\xi'}{\epsilon'} \left[1 - 2f(\epsilon')\right], \tag{1.155}$$

$$M_2 = -\int_{\omega_D}^{\omega_c} \frac{d\xi'}{\epsilon'} \left[1 - 2f(\epsilon')\right]. \tag{1.156}$$

Then (1.152) becomes:

$$\begin{pmatrix} \Delta_1 \\ \Delta_2 \end{pmatrix} = \begin{pmatrix} M_{11} & M_{12} \\ M_{21} & M_{22} \end{pmatrix} \begin{pmatrix} \Delta_1 \\ \Delta_2 \end{pmatrix}. \tag{1.157}$$

Here

$$\begin{aligned} M_{11} &\approx (\mu - \lambda)M_1 \\ M_{12} &\approx \mu M_2 \\ M_{21} &\approx \mu M_1 \\ M_{22} &\approx \mu M_2 \end{aligned}, \tag{1.158}$$

where we denoted $\lambda = N(0)V_p$ and $\mu = N(0)V_c$. For the linear homogeneous equation (1.157) to have non-trivial solutions, the determinant should be equal to zero:

$$\begin{vmatrix} M_{11} - 1 & M_{12} \\ M_{21} & M_{22} - 1 \end{vmatrix} = 0, \tag{1.159}$$

or, using (1.158):

$$- \mu\lambda M_1 M_2 - \mu M_2 - \mu M_1 + \lambda M_1 + 1 = 0. \tag{1.160}$$

Substituting into (1.160)

$$\mu = \frac{\mu^*}{1 + \mu^* M_2}, \tag{1.161}$$

we get:

$$(\lambda - \mu^*)M_1 = -1. \tag{1.162}$$

Using the relation $[1 - 2f(\epsilon)] = \tanh(\epsilon/2T)$, and the fact that at the transition temperature $\Delta_{1,2}/T_c \ll 1$, we have:

$$M_1 \approx - \ln\left(1.13\frac{\omega_D}{T}\right), \tag{1.163}$$

$$M_2 \approx - \ln(\omega_c/\omega_D). \tag{1.164}$$

Then, from (1.162) and (1.163), one can determine

$$T_c = 1.13\omega_D \exp\left[-\left(\frac{1}{\lambda - \mu^*}\right)\right], \tag{1.165}$$

where, as it follows from (1.161) and (1.164),

$$\mu^* = \frac{\mu}{1 - \mu M_2} = \frac{\mu}{1 + \mu \ln(\omega_c/\omega_D)}. \tag{1.166}$$

(This logarithmic factor reducing the strength of the Coulomb repulsion is sometimes called "Tolmachov's logarithm.") Typically, $\mu \cong 1$ and is material dependent, as is λ (see their definitions above and the discussion below). However, since $\omega_c/\omega_D \gg 1$ (say, $\omega_c/\omega_D \approx 10^{2 \div 3}$), $\ln(\omega_c/\omega_D) \sim 4 \div 6$, μ^* is almost independent on μ:

$$\mu^* \approx \ln^{-1}(\omega_c/\omega_D) \sim 0.15 - 0.2. \tag{1.167}$$

This universal value for μ^* is used customarily in scientific literature.

1.9.5   *Superconductivity at* $\mu > \lambda$, *but* $\lambda > \mu^*$.

As follows from (1.165), the Coulomb repulsion reduces the value of T_c compared to the case when it is absent ($\mu = 0$). What is even more important: superconductivity is remarkably present even in cases when the repulsion prevails at all frequencies: it is tolerable to have

$$\mu > \lambda \qquad (1.168)$$

and yet $T_c > 0$ as soon as $\lambda > \mu^*$. This means that the condition $\varepsilon(\mathbf{q}, \omega) < 0$ is not mandatory for superconductivity. However, for some metals the detailed calculations show that it is negative for certain values of q and ω: we will discuss it later. For now, one can consider effective λ defined by the Coulomb renormalization taken into account, i.e., $\lambda_{eff} = \lambda - \mu^*$, so that (1.165) has the same form as (1.121). This redefinition of T_c is valid in the so-called *weak-coupling approximation*: $\lambda_{eff} \ll 1$. It is interesting to explore how high the critical temperature can be in the weak-coupling approximation. The exponential factor in this case is obviously small. High values of T_c can then be delivered only by the large values of the pre-exponential factor. Recalling the elasticity theory, the frequency of ionic oscillation is given by the relation

$$\omega^2 = \frac{k}{M}, \qquad (1.169)$$

where k is the rigidity coefficient related to the elastic forces reversing the ionic excursion from the equilibrium position at oscillations, and M is the ionic mass. The smallest mass among chemical elements corresponds to hydrogen, H. At ambient conditions, as we know, hydrogen is a gas. However, it was predicted by Wigner and Huntington in 1935 that at high pressures, hydrogen will convert into a metal. In accordance with (1.169) ω_D of the metallic hydrogen may be very high, and correspondingly T_c may be high enough to reach room-temperature level. Moreover, there are expectations that the metallic state of hydrogen is metastable. A good insight here is provided by diamond, which is a metastable state of carbon. If you heat a diamond to high enough temperatures, it will come out of the metastable state and readily convert into graphite, which has lower free energy. Reversing this conversion is much harder: you should heat graphite to very high temperatures and squeeze it before cooling it down momentarily to reach and keep the metastable diamond state. One way of doing this is an arc-melting of graphite with droplets falling into water. Another way is the pulsed laser processing of graphite. (These experiments may deliver some small diamond specimens, but you will never get back your original diamond so please do not risk it!)

Experiments with metallic hydrogen have not yet reached much fruition: the required pressures are too high and yet to be conquered.[3] However, squeezed

---

[3] At the time of preparation of the manuscript of the book a report appeared on achieving this milestone at about 4 million atm pressure [P. Loubeyre, F. Occelli, and P. Dumas, Synchrotron infrared spectroscopic evidence of the probable transition to metal hydrogen, *Nature*, **577**, 631–635 (2020)].

hydrogen-rich materials have been tried, and Eremets' group reported in 2015 on $203 K$ superconductivity in $H_2 S$, and then on $250 K$ superconductivity in $La H_{10}$ (in 2019). This class of materials is called Superhydrides. Two International Workshops have been held on these materials: one in Rome, in 2016, and one in Los Angeles, in 2017. There is a remarkably good agreement between the properties of these materials calculated from first principles, and experimental results. This success is pawing the road to room-temperature superconductors. The major problem with them is to recognize which superhydrides may possess metastable superconducting states, and then reach these states experimentally. There is no guarantee that it will be possible; however, there is no show-stopper either. Meanwhile, the most important corollary of this breakthrough to almost room-temperature superconductivity is in demonstration that superconductivity can win the competition over other instabilities, such as charge density waves, spin density waves, etc. These and other instabilities have been detected experimentally, and were considered as major prohibitions for occurring superconductivity at the variation of solid state properties via various methods, or for the success of mechanisms other than the BCS-mechanism, one of which we will consider now.

1.9.6 Little's Model of High-T_c Superconductors and Interplay Between ω, μ and λ

In an attempt to increase the pre-factor in the expression for T_c of weak-coupling superconductors above ω_D, a remarkable idea was expressed by Little almost immediately after the BCS work. He found a way to justify the replacement of the ionic mass M in (1.169) by the (much smaller) electronic mass m, thus reaching much higher values for pre-exponential factor of T_c. To follow Little's idea, one should recall that the above-discussed mechanism of resonant agitation of ions by electrons in physics language means "lattice polarization by electronic motion". However, in certain situations, electronic motion can polarize the environment just electronically. Consider, for example, a metallic nanotube decorated periodically by metallic nano-islands (Fig. 1.22). Electrons of conductivity moving along the nanotube will repulse electrons in the nano-islands, temporarily polarizing them due to their electronic motion. Then, another conductivity electron in the nanotube will be attracted to the first electron via this island polarization, thus creating a Cooper pair in the system.[4] One can expect then much higher values of T_c in accordance with the expression

$$T_c \simeq \bar{\omega} \exp\left(-\frac{1}{\lambda_{eff}}\right). \tag{1.170}$$

---

[4] Actually, Little considered a quasi-1D molecular structure with periodic placement of molecular rings along the tranjectory of the conductivity electron; this nanotube example was brought to my attention by Dr. A. Harutyunyan (Honda Research Institute USA, Inc.).

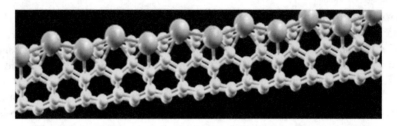

Fig. 1.22 Decorated nanotube as an example illustrating Little's model

where $\bar{\omega}$ is orders of magnitude higher than ω_D in (1.165) [for simplicity, we dropped the numerical coefficient in the pre-factor in (1.170)]. However, one should notice that for the given values of λ and μ, the increase of $\bar{\omega}$ implies the increase of μ^*:

$$\mu^* = \frac{\mu}{1 + \mu \ln(\omega_c/\bar{\omega})}, \tag{1.171}$$

and this dependence of λ_{eff} on $\bar{\omega}$ reduces the T_c for large enough values of $\bar{\omega}$. To obtain the maximum of T_c, one should substitute (1.171) into (1.170) subject to $\lambda_{eff} = \lambda - \mu^*$, and determine its maximum as a function of $\bar{\omega}$. Suppose that $\bar{\omega}$ is as large as ϵ_F. From the condition $dT_c(\bar{\omega})/d\bar{\omega} = 0$, it follows that:

$$\bar{\omega}^{T_c^{\max}} = \epsilon_F \exp\left[-\left(\frac{2}{\lambda} - \frac{1}{\mu}\right)\right]. \tag{1.172}$$

What is very important is that the maximum of $T_c(\bar{\omega})$ does not correspond to the maximum of $\bar{\omega}$, which in this case, is of the order of $\omega_c \approx \epsilon_F$. Rather, it is given by (1.172), which is significantly smaller. Substituting (1.172) into (1.170) and (1.171) subject to $\omega_c \approx \epsilon_F$ yields $\mu^* = \lambda/2$, and we have

$$T_c^{\max} = \epsilon_F \exp\left[-\left(\frac{4}{\lambda} - \frac{1}{\mu}\right)\right]. \tag{1.173}$$

How high can this temperature be? We should keep in mind that (1.173) was derived in the weak coupling limit: $\lambda \ll 1$. Substituting $\lambda \sim 0.3$ and $\mu \sim 0.3$ into (1.173) we find $T_c^{\max} \leq \epsilon_F \exp(-12)$, i.e., if $\epsilon_F \approx 10^5$ K, we get $T_c^{\max} \leq 0.6$ K. This interesting conclusion was derived by Cohen and Anderson (1971). Because of the weak coupling limitation, we cannot choose higher values of λ. If we were able to substitute $\lambda \sim 1$ into (1.173) and still keep $\mu \sim 0.3$, then we would arrive at $T_c^{\max} = \epsilon_F / \exp(-1)$—much higher than room temperature! This indicates how important it is to have a valid expression for T_c beyond the weak coupling approximation. Such a generalization is made based on the so-called Eliashberg model[5] of superconductivity. In this model, as was revealed by McMillan (1968), the most

---

[5]This model in the weak coupling limit will be discussed in Sect. 5.1 of Part II.

essential change occurs in the exponent of the expression for T_c, which, in the first approximation, we can represent as

$$T_c^{McM} = \bar{\omega} \exp\left\{-\left[\frac{1+\lambda}{\lambda - \mu^*(1+\lambda)}\right]\right\}. \tag{1.174}$$

Introducing the renormalized interaction constant $\lambda^* = \lambda/(1+\lambda)$, one can represent (1.174) in the form

$$T_c^{McM} = \bar{\omega} \exp\left[-\left(\frac{1}{\lambda^* - \mu^*}\right)\right]. \tag{1.175}$$

Since λ^* does not depend on $\bar{\omega}$, there is no need to take the derivative again to find the maximum of $T_c^{McM}(\bar{\omega})$; one can just substitute λ^* for λ in (1.172) and (1.173) which yields:

$$\bar{\omega}^{T_c^{max} McM} = \epsilon_F \exp\left[-\left(\frac{2}{\lambda} - \frac{1}{\mu} + 2\right)\right], \tag{1.176}$$

and correspondingly:

$$T_c^{max\, McM} = \epsilon_F \exp\left[-\left(\frac{4}{\lambda} - \frac{1}{\mu} + 4\right)\right]. \tag{1.177}$$

If we substitute now $\lambda = 1$ and $\mu = 0.3$ into (1.177), we find $T_c^{max}/\epsilon_F = \exp(-5) \approx 6.7 \times 10^{-3}$. Thus, at $\epsilon_F = 10^5$ K, we can expect above room-temperature superconductivity!

There is one related aspect we would like to discuss here: the possible range for the parameters λ and μ. For Coulombic interaction, as soon as μ exceeds the value of 1, one can expect ferromagnetism, according to Stoner's criterion (in reality, ferromagnetism may occur even for smaller values of μ). For electron-phonon interaction, λ may be more than 1 for some metals, as shown in the table below.[6]

Metal	Hg	AmPb$_{0.45}$Bi$_{0.55}$	Pb
λ	1.6	2.66	1.55

These metals are not ferromagnets, but rather superconductors, which implies $\mu < 1$. Interestingly, in the physics of metals, there is a relation

$$\mu - \lambda = \int dq \frac{4\pi e^2}{q^2 \varepsilon(\mathbf{q}, 0)}, \tag{1.178}$$

---

[6]L.P. Gor'kov and V.Z. Kresin, Colloquium: High pressure and road to room temperature superconductivity, Rev. Mod. Phys. 90, 011001 (2018) [arXiv:1802.02296].

from which one can deduce that, for these superconductors, the static dielectric function $\varepsilon(\mathbf{q}, \omega = 0)$ should be negative for some values of $\mathbf{q}$. At first glance, one may conclude that "the crystal would be unstable to spatial deformations". However, Kirzhnits proved in 1976 that such negativity does not lead to lattice instability. Moreover, direct calculations by Dolgov and Maksimov (1978) revealed negative values of $\varepsilon(\mathbf{q}, \omega = 0)$ for certain superconducting metals. One can reciprocate this philosophy and look for negative values of static dielectric function in prospective high T_c materials. Ab-initio computations of decorated nanotubes performed by our research group at Chapman U. (2018) can be regarded as first steps in the right direction. We refer also to other studies, in particular by Smolyaninova's group, based on the same philosophy.

One last remark: McMillan's formula of the type (1.174) works fine for $\lambda \lesssim 1.5$. For larger values of λ's microscopic equations yield one more interesting limit for T_c:

$$T_c = 0.18\bar{\omega} \left(\frac{\lambda}{1 + 2.58\mu^*} \right)^{1/2}. \tag{1.179}$$

This expression was derived by Allen and Dynes (1975) via numerical calculations (for $\mu^* = 0$). Analytical treatment was performed by Kresin et al. (1984).

Thus, even leaving aside the opportunity-rich cuprate high-temperature superconductors, which are still "objects for themselves" despite the fact they were discovered decades ago, we conclude that for the superconductivity based on traditional bosonic-exchange mechanism, there is no actual limit on T_c from a theoretical point of view. Experiments are bringing us closer and closer to superconductivity at ambient conditions. I have almost no doubts that current generation of students will work with room-temperature superconductors.

Chapter 2
Exploring Superconductivity with COMSOL via TDGL Equations

In this Chapter, we will apply two very powerful tools to tackle rather serious, advanced problems: COMSOL basic package and TDGL equations. Our first problem will be exploring the Meissner effect in a superconducting disk. Depending on a single parameter in TDGL equations, the dynamics of superconductors in magnetic fields will be quite different: we will realize the existence of two types of superconductors. Abrikosov vortices will come in almost effortlessly at COMSOL modeling. Two different ways of treating 2D-models will be explained in detail. The second way will allow simple-enough generalization to 3D tasks. Using it, we will consider dynamic pattern of penetration of magnetic field into a 3D washer. All these problems will be related to superconducting objects in an externally applied magnetic field. The next group of problems will be related to current carrying superconducting strips. First, we will consider the dynamics of phase slippage in thin superconducting wires: the oscillatory regime at the DC current flow will be discovered and explored. Then, we will treat the flow of current through a thin-finite width superconducting strip: we will observe generation, propagation and annihilation of Abrikosov vortices and anti-vortices. It will be shown that annihilation of vortices generates electric field pulses. The next task will be related with the a close relative of Josephson junctions: we will consider a Superconductor-Normal metal-Superconductor (SNS) junction in a DC-mode and realize a state with periodic single-flux quantum (SFQ) generation. The last task will be cloning of the SFQ pulses: application of an SFQ pulse to SNS junction in the DC-mode will generate a single flux quantum propagating along the junction and generate an SFQ pulse when leaving it.

Electronic supplementary material The online version of this chapter (https://doi.org/10.1007/978-3-030-23486-7_2) contains supplementary material, which is available to authorized users.

2.1 General Notions of TDGL Equations

Let us analyze TDGL equations, (1.107) and (1.120), a little further before using them for practical needs. In solving these equations we can use the most convenient gauge: the choice of any gauge function $\chi(x, y, z, t)$ shall not influence any physical results since the equations are gauge-invariant (see Problem 9). From (1.113) it follows that one can always choose a hypothetical gauge function $\chi(x, y, z, t)$ in such a way that $\varphi \equiv 0$, i.e., φ will drop out from (1.107) and (1.120) and we will end up with a system:

$$\frac{\pi}{8T_c}\frac{\partial \Psi}{\partial t} - \frac{\pi D}{8T_c}(\nabla - 2i\mathbf{A})^2 \Psi - \left[\frac{T_c - T}{T_c} - \frac{7\zeta(3)\,|\Psi|^2}{8(\pi T_c)^2}\right]\Psi = 0, \qquad (2.1)$$

and

$$\mathbf{j} = -\frac{\pi \sigma}{4T_c}|\Psi|^2\,(2\mathbf{A} - \nabla\theta) - \sigma_n\dot{\mathbf{A}}. \qquad (2.2)$$

As always in numerical computations, it is convenient to represent these equations in a dimensionless form. Dividing (2.1) by $\eta = (T_c - T)/T_c$, we have

$$\frac{\pi}{8\eta T_c}\frac{\partial \Psi}{\partial t} - \frac{\pi D}{8\eta T_c}(\nabla - 2i\mathbf{A})^2 \Psi - \left(1 - \frac{|\Psi|^2}{\Psi_0^2}\right)\Psi = 0, \qquad (2.3)$$

where $\Psi_0 = \sqrt{8\pi^2 T_c(T_c - T)/[7\zeta(3)]}$. From here it is obvious that time will become conveniently dimensionless if measured in units $t_0 = \pi/(8\eta T_c)$, and the same will happen with distance if measured in units $\sqrt{\pi D/(8\eta T_c)}$. The latter unit is called coherence length: $\xi \equiv \sqrt{\pi D/[8(T_c - T)]}$, since it characterizes spatial decay of the Ψ-function, which in turn describes coherent behavior of electrons in the Cooper condensate. Thus, we will pass to the dimensionless time $\tau = t/t_0$, but we will not use the coherence length ξ for the similar purpose with the spatial coordinates. Taking this into account, and denoting $\psi \equiv \Psi/\Psi_0$, we can write:

$$\frac{\partial \psi}{\partial \tau} - (\xi\nabla - 2i\xi\mathbf{A})^2 \psi - \left(1 - |\psi|^2\right)\psi = 0. \qquad (2.4)$$

The reason for keeping ξ in (2.4) is that the Ginzburg–Landau set of equations has more than one characteristic scale for spatial variation of parameters characterizing superconducting state. We will use the second such scale, which is the unit characterizing the variation of the vector potential, i.e., the London penetration depth λ_L. This characteristic length will follow from the current density equation (2.2). Using Maxwell's equation (in Gaussian units with $c = 1$): curl $\mathbf{B} = 4\pi\mathbf{j} = \nabla \times \nabla \times \mathbf{A}$, and substituting this into (2.2), we find, temporarily dropping electric field term ($\propto \dot{\mathbf{A}}$) and assuming $\Psi = \Psi_0$ (i.e., $\psi \equiv 1$):

$$\nabla \times \nabla \times \mathbf{A} = -\left(\frac{2\pi^2 \sigma \Psi_0^2}{T_c}\right) \mathbf{A}. \tag{2.5}$$

Comparing (2.5) with (1.22), we can deduce that

$$\frac{1}{\lambda_L^2} = \frac{2\pi^2 \sigma \Psi_0^2}{T_c} = \frac{16\pi^4 \sigma (T_c - T)}{7\zeta(3)}. \tag{2.6}$$

We notice in passing that $\lambda_L(T)$ diverges at $T \to T_c$, i.e., at that point the magnetic field penetrates into the metal unrestrictedly. From now on, we will assume that λ_L is defined by (2.6). [The quantity defined by (1.17) is typically called $\lambda_L(0)$.] We can thus represent (2.2) in the form:

$$\nabla \times \nabla \times \mathbf{A} = -\frac{1}{\lambda_L^2}|\psi|^2 (\mathbf{A} - \nabla\theta/2) - 4\pi\sigma\dot{\mathbf{A}}. \tag{2.7}$$

Introducing dimensionless spatial derivatives via $\tilde{\nabla} = \lambda_L \nabla$, we can represent (2.7) in the form:

$$\tilde{\nabla} \times \tilde{\nabla} \times \mathbf{A} = -|\psi|^2 \left[\mathbf{A} - \tilde{\nabla}\theta/(2\lambda_L)\right] - 4\pi\lambda_L^2 \sigma\dot{\mathbf{A}}. \tag{2.8}$$

Next steps involve denoting $\tilde{\mathbf{A}} \equiv 2\xi\mathbf{A}$, multiplying (2.8) by 2ξ and obtaining

$$\tilde{\nabla} \times \tilde{\nabla} \times \tilde{\mathbf{A}} = -|\psi|^2 \left[\tilde{\mathbf{A}} - \frac{1}{\kappa}\tilde{\nabla}\theta\right] - 4\pi\lambda_L^2 \sigma\frac{\partial\tilde{\mathbf{A}}}{\partial t}, \tag{2.9}$$

where $\kappa \equiv \lambda_L/\xi$ is the so-called Ginzburg–Landau parameter. At this point, we can also introduce the dimensionless time in (2.9), so that

$$4\pi\lambda_L^2 \sigma\frac{\partial\tilde{\mathbf{A}}}{t_0\partial\tau} = 4\pi\frac{7\zeta(3)}{16\pi^4\sigma(T_c - T)}\sigma\frac{8\eta T_c}{\pi}\frac{\partial\tilde{\mathbf{A}}}{\partial\tau}, \tag{2.10}$$

i.e.,

$$4\pi\lambda_L^2 \sigma\frac{\partial\tilde{\mathbf{A}}}{t_0\partial\tau} = \frac{14\zeta(3)}{\pi^4}\frac{\partial\tilde{\mathbf{A}}}{\partial\tau} \equiv \tilde{\sigma}\frac{\partial\tilde{\mathbf{A}}}{\partial\tau}, \tag{2.11}$$

so that (2.9) is representable as

$$\tilde{\nabla} \times \tilde{\nabla} \times \tilde{\mathbf{A}} = -|\psi|^2 \left(\tilde{\mathbf{A}} - \frac{1}{\kappa}\tilde{\nabla}\theta\right) - \tilde{\sigma}\frac{\partial\tilde{\mathbf{A}}}{\partial\tau}. \tag{2.12}$$

Turning back to (2.4), we can represent it now, for the $2D$-case, in the form:

$$\frac{\partial \psi}{\partial \tau} - \left[(\xi/\lambda_L)\tilde{\mathbf{V}} - i\tilde{\mathbf{A}}\right]^2 \psi - \left(1 - |\psi|^2\right)\psi = 0, \tag{2.13}$$

or, equivalently, as

$$\frac{\partial \psi}{\partial \tau} + \left(\frac{i}{\kappa}\tilde{\mathbf{V}} + \tilde{\mathbf{A}}\right)^2 \psi - \left(1 - |\psi|^2\right)\psi = 0. \tag{2.14}$$

Finally, dropping all tildes, and considering all variables and coefficients dimensionless, we can present the system of equations in a form convenient for COMSOL:

$$\frac{\partial \psi}{\partial \tau} = -\left(\frac{i}{\kappa}\mathbf{V} + \mathbf{A}\right)^2 \psi + \left(1 - |\psi|^2\right)\psi, \tag{2.15}$$

and

$$\sigma\frac{\partial \mathbf{A}}{\partial \tau} = -\mathbf{V} \times \mathbf{V} \times \mathbf{A} - |\psi|^2\left(\mathbf{A} - \frac{1}{\kappa}\mathbf{V}\theta\right). \tag{2.16}$$

Since ψ is a complex function, (2.15) consists of two equations for two scalars: $|\psi|$ and θ, or, alternatively, real and imaginary parts of ψ:

$$\psi = \psi_1 + i\psi_2 = \operatorname{Re}\psi + i\operatorname{Im}\psi. \tag{2.17}$$

The latter is more convenient for solving these equations. So we need to separate the real and the imaginary parts of the ψ-function. Then, in the $1D$-case (space $\equiv x$, $A_x \equiv A$), one can represent the system of equations as

$$\dot{\psi}_1 = \frac{1}{\kappa^2}\left(\psi_{1.xx}\right)$$
$$+\frac{2}{\kappa}\left(A\psi_{2.x}\right) + \frac{\psi_2}{\kappa}\left(A_{.x}\right) - \psi_1\left(A^2\right) + \psi_1 - \psi_1\left(\psi_1^2 + \psi_2^2\right), \tag{2.18}$$

$$\dot{\psi}_2 = \frac{1}{\kappa^2}\left(\psi_{2.xx}\right)$$
$$-\frac{2}{\kappa}\left(A\psi_{1.x}\right) - \frac{\psi_1}{\kappa}\left(A_{.x}\right) - \psi_2\left(A^2\right) + \psi_2 - \psi_2\left(\psi_1^2 + \psi_2^2\right), \tag{2.19}$$

$$\sigma\dot{A} = -\frac{1}{\kappa}\left(\psi_2\psi_{1.x} - \psi_1\psi_{2.x}\right) - \left(\psi_1^2 + \psi_2^2\right)A - j_0, \tag{2.20}$$

where j_0 is $4\pi\times$(current density).

For the $2D$-case ($A_x \equiv A_1$, $A_y \equiv A_2$), the equations are:

$$\dot{\psi}_1 = \frac{1}{\kappa^2} \left(\psi_{1.xx} + \psi_{1.yy} \right) + \frac{2}{\kappa} \left(A_1 \psi_{2.x} + A_2 \psi_{2.y} \right) + \frac{\psi_2}{\kappa} \left(A_{1.x} + A_{2.y} \right)$$
$$- \psi_1 \left(A_1^2 + A_2^2 \right) + \psi_1 - \psi_1 \left(\psi_1^2 + \psi_2^2 \right), \tag{2.21}$$

$$\dot{\psi}_2 = \frac{1}{\kappa^2} \left(\psi_{2.xx} + \psi_{2.yy} \right) - \frac{2}{\kappa} \left(A_1 \psi_{1.x} + A_2 \psi_{1.y} \right) - \frac{\psi_1}{\kappa} \left(A_{1.x} + A_{2.y} \right)$$
$$- \psi_2 \left(A_1^2 + A_2^2 \right) + \psi_2 - \psi_2 \left(\psi_1^2 + \psi_2^2 \right), \tag{2.22}$$

$$\sigma \dot{A}_1 = -\frac{1}{\kappa} \left(\psi_2 \psi_{1.x} - \psi_1 \psi_{2.x} \right) - \left(\psi_1^2 + \psi_2^2 \right) A_1 + A_{1.yy} - A_{2.xy}, \tag{2.23}$$

$$\sigma \dot{A}_2 = -\frac{1}{\kappa} \left(\psi_2 \psi_{1.y} - \psi_1 \psi_{2.y} \right) - \left(\psi_1^2 + \psi_2^2 \right) A_2 + A_{2.xx} - A_{1.xy}. \tag{2.24}$$

In the $1D$-case, the current density is a parameter of the problem (it cannot change along the superconducting wire and is given by boundary conditions which can be time-dependent), while in the case of higher spatial dimensionality, the current density can vary in space, so **j** should be determined by the Maxwell equation **j** = curl **B** =curl curl **A**. For example, in the $2D$-case, the x- and y-components of curl curl **A** are substituted into (2.23) and (2.24) instead of **j**$_0$ in (2.20). Naturally, boundary and initial conditions should be specified to solve these equations.

2.2   Disk in a Magnetic Field: Ginzburg–Landau Approach

Let us re-consider the problem which was solved in the previous chapter: a disk in a homogeneous external magnetic field orthogonal to the disk's surface. This time, based on TDGL, we will be able to track its dynamics. It will deliver astonishingly more information compared to London's approach.

Go to Model Wizard and Select Space Dimension 2D. Next, in Select Physics window, double click on Mathematics, double click on PDE Interfaces, and then on General Form PDE(g). You will see the Field name: u in Review Physics Interface. Since we have four equations for four variables, you can choose Number of dependent variables 4. That will lead to a matrix form representation of equations. It is simpler to keep the scalar form of representation, and to keep the default value 1 in the window. Double click on General Form PDE (g) four times. Then click on Study and choose Time Dependent, and click on Done. The Model Builder will come up. Let us introduce Parameters in Global Definitions. Since the (2.21)–(2.24) are dimensionless, we can ignore dimensionality specifications while doing this (that will cause some formulas in COMSOL be colored yellow, which, unlike red colored ones, are tolerable). Insert $R_0 = 5$ as the disk radius, $kappa = 4$ as the Ginzburg–Landau parameter κ, $sigma = 1$ as the conductivity σ of normal electrons, $B_a = 0.9$ as the external field value, and $t_0 = 200$ as the calculation time.

Right click on Geometry, and choose Circle. In the Radius window of Settings insert $R0$. Click Build Selected. The file is now ready for insertion of (2.21)–(2.24). Click on the General Form PDE 1 in General Form PDE (g) area. Then click and open the Equation in Settings window. Recognize that in our notations $\psi_1 = u$, $\psi_2 = u2$, $\overline{A_1} = u3$, and $\overline{A_2} = u4$. Insert (2.21) as shown in Fig. 2.1. Similarly insert equations (2.22)–(2.24), as shown in Figs. 2.2, 2.3, and 2.4. Now, when equations are in order, we need to deal with boundary conditions. Physically, the following conditions should be imposed. There should be no current crossing the boundary of the disk. Not only total current density, but both superconducting $\mathbf{j}_s$ and normal $\mathbf{j}_n$ current densities should not have normal components at the boundary of the disk with vacuum. That means $\mathbf{j}_s \cdot \mathbf{n}|_C = 0$ and $\mathbf{j}_n \cdot \mathbf{n}|_C = 0$ on the boundary C of the disk. One more condition is that magnetic field should be continuous, i.e., curl curl $\mathbf{A}|_z = \mathbf{B}_a$. This last condition is satisfied by our choice of Γ-matrices in equations PDE 3(g3) and PDE 4(g4) and Zero Flux default boundary conditions since z-component of curl curl $\mathbf{A}$ is $\partial A_y/\partial x - \partial A_x/\partial y$. Let us discuss boundary conditions in the first two equations. Conservative flux in PDE (g) and PDE 2 (g2) is equal to $const \cdot \nabla \psi_1$, and $const \cdot \nabla \psi_2$ correspondingly. So in view of Zero Flux default boundary conditions, jointly they will satisfy the equation $\nabla \psi \cdot \mathbf{n} = 0$. Obviously from here one can expect $(\psi^* \nabla \psi - \psi \nabla \psi^*) \cdot \mathbf{n} = 0$. For $\mathbf{j}_s \cdot \mathbf{n}|_C = 0$, as follows from (1.115) and (1.116), one should require also $\mathbf{A} \cdot \mathbf{n}|_C = 0$. That requirement will automatically satisfy the condition $\mathbf{j}_n \cdot \mathbf{n}|_C = 0$. Indeed, in our gauge $\mathbf{E} = -\partial \mathbf{A}/\partial t$. Also, $\mathbf{j}_n = \sigma \mathbf{E}$. Then $\mathbf{A} \cdot \mathbf{n}|_C = 0$ imposes $\partial \mathbf{A}/\partial t \cdot \mathbf{n}|_C = 0$ and $\mathbf{E} \cdot \mathbf{n}|_C = 0$, so that $\mathbf{j}_n \cdot \mathbf{n}|_C = 0$. Thus, implementing $\mathbf{A} \cdot \mathbf{n}|_C = 0$ will yield satisfaction of physically reasonable requirements on superconducting and normal currents. The problem is that we used already all the entries for boundary conditions in our equations. For that purpose a trick was invented by researchers,[1] which we will use now. It involves entering one more equation of the type of the previous four equations, and accordingly, the fifth variable $u5$. To fulfill this goal click on Physics tab, then on Add Physics. On the right side of the screen, Add Physics window will open. Double click on Mathematics, then double click on PDE Interfaces, and then on General Form PDE (g). A fifth equation will appear in the Model Builder. You can close the Add Physics window now. Under General Form PDE 5 (g5) click on General Form PDE 1. In the definition of the Γ-vector insert instead of default $-u5x$ and $-u5y$ correspondingly $u3$ and $u4$. That will enforce the condition $\mathbf{A} \cdot \mathbf{n}|_C = 0$. Insert $u3x + u4y + u5$ for f, and change default value of d_a to zero. It is then a stationary equation of the type $u5 = 0$ for $u5$, with matching initial value 0, so the trivial $u5 \equiv 0$ will always be its solution. At this point the created code should be able to generate solutions. Save the file. Try to press Run. The program will not run! That is because after adding the fifth equation the Study Time Dependent should be inserted again. Click on Study tab, it will open in the right side of the screen, choose Time dependent by double clicking, so that it will appear in the Model Builder. Then the Study option will become available. But the solutions will stay at zero level. That is because all the initial

---

[1] To the best of my knowledge, this approach was first described by T.S. Alstrøm et al. [Acta Appl. Math. 115(1), 63–74, 2010].

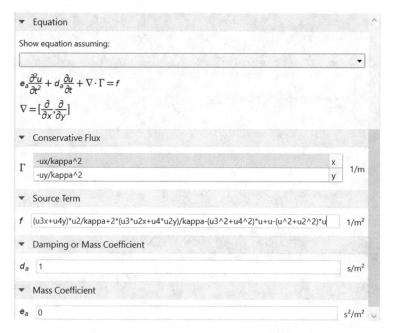

Fig. 2.1 Equation for the real part ($= u$) of the wave function ψ in COMSOL

conditions by default are zero. So let us change one of them, making the real part of ψ in <u>General Form PDE 1</u> equal to one: $\psi_1 = 1$ at $t = 0$. Let us also change one more default: in <u>Study 1</u> in <u>Model Builder</u>, click on <u>Step 1, Time Dependent</u>, and then in <u>Settings</u>, in <u>Times</u>, replace in the <u>range</u> last $\overline{1}$ by $\overline{t0}$, so that it reads now $range(0, 0.1, t0)$. If you run <u>Study</u> now, it will generate nonzero results. However, all default five plot <u>Groups</u> will plot correspondingly functions $u, u2, \ldots, u5$. Rather, we would like to plot density of Cooper pairs, which is $\psi_1^2 + \psi_2^2$. To reach this goal, double click on <u>2D Plot Group 1</u> in the <u>Model Builder</u>, then click on <u>Surface</u>. Find <u>Expression</u> in <u>Settings</u> window, and replace u by $u\hat{}2 + u2\hat{}2$. Click on <u>Home</u> tab, and then on <u>Compute</u> command. After the computation is successful, change time in <u>Settings</u> window. For different values of time, you will see evolving pattern of single flux vortices, so-called Abrikosov vortices.[2] The quality of the solution can be enhanced by choosing a finer mesh. Double click on <u>Mesh 1</u> in <u>Model Builder</u>, and switch <u>Normal</u> to <u>Extra fine</u>. It will be a bit longer run, but most computers can do it fast. To view the result for a desired time, go again to <u>2D Plot Group 1</u>, then in <u>Settings</u> choose the required time (the default will be 0). The plot at <u>Time (s): 100</u> is shown in Fig. 2.5.

---

[2]In 2003, Alexey Abrikosov received a Nobel Prize for this remarkable result which he obtained analytically decades before. With the help of COMSOL you can replicate Nobel Prize results in less than an hour! Interestingly, L. D. Landau, Abrikosov's supervisor, initially did not believe that the solution is physically plausible, and for years the results remained unpublished.

Fig. 2.2 Equation for the imaginary part of the wave function ψ ($=u2$)

Since we have determined the whole set of variables ψ and **A**, we can now plot also the current density vectors. For that right click on 2D Plot Group 1 in Model Builder, and choose Arrow Surface from the pop-up window. Then click on Arrow Surface 2 and start working in Setting window. You need to insert curl curl **A** for the **j**-function. The components view of that is shown in Fig. 2.6. Corresponding plot after zooming in is displayed in Fig. 2.7. At this point, one should pay attention: the current vectors have opposite signs at the circumference of the disk and around the vortices. This is because of a topological difference: on the circumference, they screen the interior of a superconductor from the external field, and in the case of vortices, they screen the superconducting exterior from the internal field.

We can now create an animation which will show the dynamics of the vortex creation, propagation and set-up of the stationary state. Click on 2D Plot Group 1. In the right corner of the screen, click on Animation icon and choose File. Then, in Setting window click on Output and choose AVI. Click on Browse and insert a file name, then save it. Scroll down to the Frames section and change then number of frames to 500. At the very bottom of the Setting window, open Advanced, and uncheck Synchronize scales between frames. After that, click Export at the top part of Settings. That will create a video for you which you can play via Windows player. You can always make a higher quality animation by changing the settings in Settings

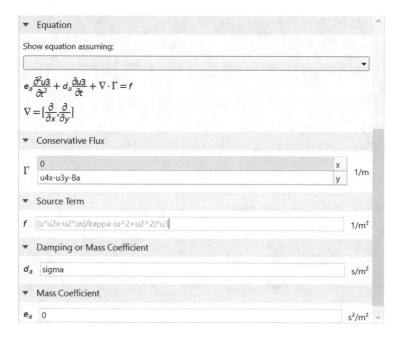

Fig. 2.3 Equation for $\mathbf{A}_x$ $(=u3)$

window—that needs no comments here. This is a very useful tool for exploring the results of your modeling. Alternatively, you can command COMSOL to show you the results of computation during the computation. For that double click on Study 1 in Model Builder, and then click on Time Dependent. Move to Settings window, click on Results While Solving and enable Plot. Click Compute. You will see the evolving solution on your screen. Though it will take longer to accomplish the task. To minimize the time, reset the Times in Settings: replace $range(0, 0.1, t0)$ by $range(0, 1, t0)$. That will have no influence on accuracy of solutions (COMSOL uses its internal time steps for finding solutions and time steps in $range(0, timestep, t0)$ to generate visual frames), but will accelerate the overall execution of modeling.

After the code is tested, you can start exploring superconductivity. The following two problems will be steps in that direction, but you can do much more if you are creative.

Problem 11 Make sure that the vortex feature shown in Fig. 2.5 only occurs when the Ginzburg–Landau parameter *kappa* is large enough. Find the critical value of *kappa* below which vortices do not form. Find the dependence of that critical value on the disk radius.

Tip: for an infinitely large disk, the value of kappa is $1/\sqrt{2}$. See Part II of the book, Sect. 3.2.4.

Fig. 2.4 Equation for $\mathbf{A}_y$ ($=u4$)

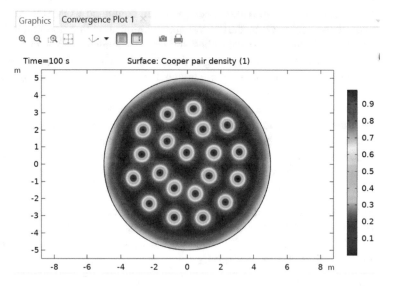

Fig. 2.5 Vortex pattern at $time = 100$

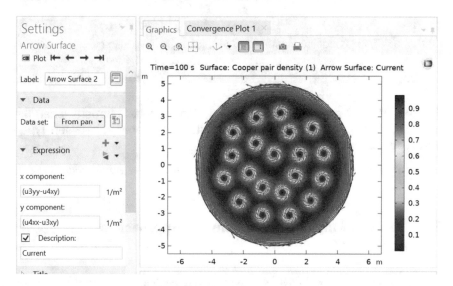

Fig. 2.6 Components of current density $\mathbf{j}_x = u3yy - u4xy$ and $\mathbf{j}_y = u4xx - u3xy$ and resulting plot at $Time = 100\,\mathrm{s}$

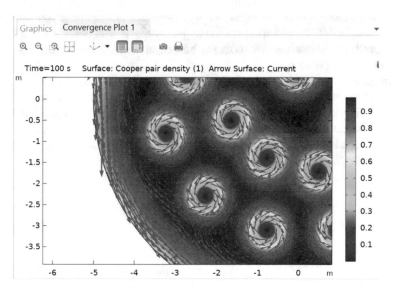

Fig. 2.7 Opposite rotational direction of current density vectors

Remark Superconductors with $\kappa < 1/\sqrt{2}$ are called Type I superconductors. Superconductors with $\kappa > 1/\sqrt{2}$ are called Type II superconductors.

Problem 12 For a given value of *kappa*, find out what happens when the magnetic field B_a is increasingly larger. Is there a difference between Type I and Type II superconductors?

Tip: start with small values of $B_a \ll 1$ and reach values greater than 1.

2.3   Disk in a Magnetic Field: Simpler Approach to Boundary Conditions

The reader probably concluded from the previous section that equations are pretty straightforward to write down for a given problem. At the same time, the boundary conditions are more subtle and require higher level of effort. Using the same simple example of a superconducting disk, we will introduce now another approach with TDGL and COMSOL which evades the major problems with boundary conditions. The equations themselves will take the burden of treating the formalities at the superconducting boundary.

Imagine a superconducting volume surrounded by a non-superconducting, non-magnetic material, which has very poor conductivity. The presence of this surrounding material will have no influence on the physical effects that the magnetic field will generate in the superconductor. We will require that the ψ-function be zero in the non-superconducting material by setting appropriate coefficients in TDGL equations to be zero:

$$\frac{\partial \psi}{\partial \tau} = vol(x, y, z) \left[-\left(\frac{i}{\kappa} \nabla + \mathbf{A} \right)^2 + 1 - |\psi|^2 \right] \psi, \qquad (2.25)$$

where the *vol*-function is defined as

$$vol(x, y, z) = \begin{cases} 1, \ if \ (x, y, z) \in \ \text{superconductor} \\ \quad\ 0 \ \ \text{otherwise} \end{cases} . \qquad (2.26)$$

Boundary conditions for the ψ-function at the outer surface will naturally be set to zero via Dirichlet boundary condition

$$\psi|_{\text{external surface}} = 0. \qquad (2.27)$$

The **A**-function will be non-zero everywhere, and the only boundary condition for it will be the continuity of the magnetic field at the outer surface:

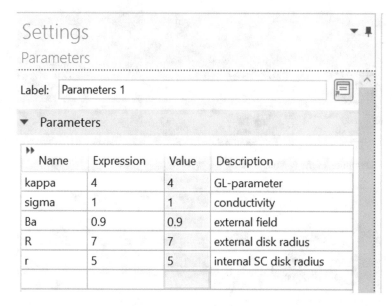

Fig. 2.8 Parameters

$$\text{curl } \mathbf{A}|_{\text{external surface}} = \mathbf{B}_a \qquad (2.28)$$

For simplicity, we will consider a superconducting disk with a radius r, and a surrounding material with a circular symmetry with radius R, though they both may have arbitrary shapes—that is not an issue for COMSOL at all.

Let us now implement this strategy with COMSOL. Again, open and save a COMSOL file with four time-dependent general equations repeating the initial steps described in the previous Section. However, in Parameters, insert both superconducting disk radius r, and external box radius R (Fig. 2.8).

In Model Builder, under Component 1, right click on Definitions, and in the pop-up window, choose Functions and click on Analytic. In the Setting window, Label it Volume Factor for SC and give it the Function name vol (of course, choices here are arbitrary). Then define it via Boolean operator in the Expression window, as shown in Fig. 2.9. You can now use this function in the first two equations, for $u = \text{Re }\psi$ and $u2 = \text{Im }\psi$ (Fig. 2.10).

Next, we should deal with initial and boundary conditions. Initial values should be set to zero for all functions except u, which should be taken equal to $vol(x, y)$. For boundary conditions, we should set up Dirichlet conditions (2.27) for the ψ-function. For that, in Model Builder window right click on General Form PDE (g), and in the pop-up window, choose Dirichlet Boundary Condition. It will come with default value $u = 0$, which is what we need, but you should still specify to which boundaries it should be applied in Settings window. For that, hover the mouse arrow over each quarter of the external circumference in the Graphics window, and as soon

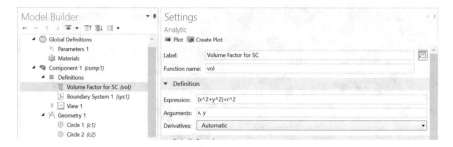

Fig. 2.9 Definition of the *vol*-function

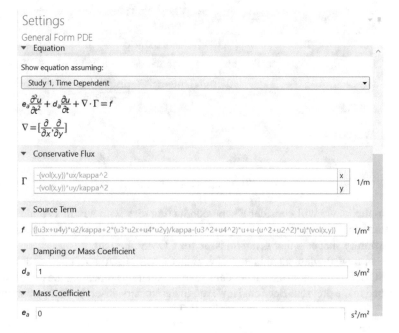

Fig. 2.10 One of TDGL equations with the *vol*-function inserted. The second one is processed similarly

as the color of the line becomes red, click on it. It will change color to blue, and the number will appear in Boundary Selection window in Settings. You should get there boundary numbers 1, 2, 5, and 8. The same procedures should be performed for the second equation, for $u2$. For the last two equations, nothing should be done: default zero Initial Values and Zero Flux conditions at the boundaries 1, 2, 5, and 8 will do the job. Before clicking Compute, do not forget to set up the computation time to 100 or 200 by clicking on Step 1: Time Dependent under Study 1 in Model Builder. Also, in Results, under 2D Plot Group 1, in Surface change the Expression u by $u^2 + u2^2$ to plot the density of Cooper pairs. The resultant pattern is shown in Fig. 2.11 for Physics-Controlled Extra fine mesh.

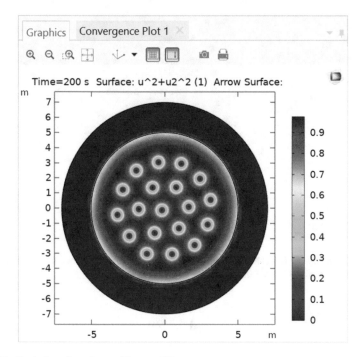

Fig. 2.11 Evolution of vortices at Time $= 200$

In this case also you can make an animation, add current density vectors, etc., in the same manner as was done in the previous Section. This approach is even more useful for higher complexity problems, like penetration of magnetic field into a 3D-washer, which we will consider next.

2.4 Penetration of Vortices into 3D Washer

We will consider now a 3D superconducting washer with height h_0, radius r_0, and a cylindrical cavity with radius r_1, surrounded by a poorly conducting nonmagnetic substance (cylinder with height Z_0 and radius R_0), Fig. 2.12.

Magnetic vector of the external field is along the z-axis and is equal to $\mathbf{B}_z^0$. London penetration depth corresponds to the unit of length: $\lambda_L = 1$. For this task, we will use the same approach outlined in (2.25) and (2.26). Naturally, (2.21)–(2.24) should be "upgraded" to the 3D-case. Upgrading is straightforward:

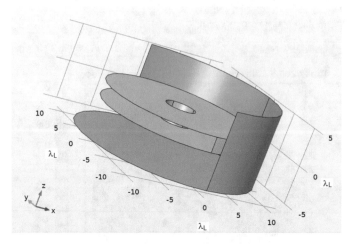

Fig. 2.12 Geometry of 3D washer surrounded by a nonsuperconducting medium. Some boundary surfaces are removed to expose internal details

$$\dot{\psi}_1 = \frac{1}{\kappa^2}\left(\psi_{1.xx} + \psi_{1.yy} + \psi_{1.zz}\right)$$
$$+\frac{2}{\kappa}\left(A_1\psi_{2.x} + A_2\psi_{2.y} + A_3\psi_{2.z}\right) + \frac{\psi_2}{\kappa}\left(A_{1.x} + A_{2.y} + A_{3.z}\right)$$
$$-\psi_1\left(A_1^2 + A_2^2 + A_3^2 - 1 + \psi_1^2 + \psi_2^2\right), \tag{2.29}$$

$$\dot{\psi}_2 = \frac{1}{\kappa^2}\left(\psi_{2.xx} + \psi_{2.yy} + \psi_{2.zz}\right)$$
$$-\frac{2}{\kappa}\left(A_1\psi_{1.x} + A_2\psi_{1.y} + A_3\psi_{1.z}\right) - \frac{\psi_1}{\kappa}\left(A_{1.x} + A_{2.y} + A_{3.z}\right)$$
$$-\psi_2\left(A_1^2 + A_2^2 + A_3^2 - 1 + \psi_1^2 + \psi_2^2\right), \tag{2.30}$$

$$\sigma\dot{A}_1 = -\frac{1}{\kappa}\left(\psi_2\psi_{1.x} - \psi_1\psi_{2.x}\right)$$
$$-\left(\psi_1^2 + \psi_2^2\right)A_1 + A_{1.yy} - A_{2.xy} - A_{3.xz} + A_{1.zz}, \tag{2.31}$$

$$\sigma\dot{A}_2 = -\frac{1}{\kappa}\left(\psi_2\psi_{1.y} - \psi_1\psi_{2.y}\right)$$
$$-\left(\psi_1^2 + \psi_2^2\right)A_2 + A_{2.xx} - A_{1.xy} - A_{3.yz} + A_{2.zz}, \tag{2.32}$$

$$\sigma\dot{A}_3 = -\frac{1}{\kappa}\left(\psi_2\psi_{1.z} - \psi_1\psi_{2.z}\right)$$
$$-\left(\psi_1^2 + \psi_2^2\right)A_3 + A_{3.xx} - A_{1.xz} + A_{3yy} - A_{2.yz}. \tag{2.33}$$

These equations should be used for COMSOL modeling of the problem. We will use the coefficient form PDE system for ψ_1 and ψ_2, and general form PDE system for the components (A_1, A_2, A_3) of the vector potential. The $vol(x, y, z)$-function will be used similarly to the previous task.

Open a new COMSOL file. Open Model Wizard and Select Space Dimension 3D. In Select Physics, double click on Mathematics and then on PDE Interfaces. Then double click on Coefficient Form PDE (c). In Review Physics Interface, set the Number of dependent variables to 2. Then, go back to Select Physics and double click on General Form PDE (g), and in Review Physics Interface change the Number of dependent variables to 3. Then, click on Study, and choose Time Dependent. Click on Done. By clicking in Model Builder on Coefficient Form PDE 1 and on General Form PDE 1, you can make sure that in the Settings window the variables are $u1$ and $u2$ (for ψ_1 and ψ_2) and $u3$, $u4$, and $u5$ (for A_1, A_2, and A_3). Save the file before continuing. It is a good idea to keep saving it periodically for not losing the job by accidental crushing. Next, insert Parameters under Global Definitions, as shown in Fig. 2.13.

We now need to introduce geometrical objects for modeling. Right click on Geometry1 under Component1 and call-in Cylinder three times. Then click on Cylinder1 and insert $r0$ and $h0$ for its radius and height in Settings window. Also replace 0 by $-h0/2$ in the z entry field under Position for the center of the cylinder to be at the center of the coordinate system—not mandatory, but worthwhile. For the Cylinder2 and Cylinder3 insert in a similar way $R0$, $Z0$ and $-Z0/2$ and correspondingly $r1$, $h0$ and $-h0/2$. Next, we need to create the hole to have a washer. For that, right-click on Geometry1, highlight Boolean and Partitions, and click on Difference. This will open two windows in Settings: Objects to add and Objects to subtract. Obviously, we need to subtract $cyl3$ from $cyl1$. However, if you hover the mouse over the drawings in the Graphics window, the boundaries of the external cylinder will shutter the view of the internal cylinders, and you will not be able to choose them. There are different ways to overcome this obstacle. For example, you can click on Select All button in the raw over the graphic, or, alternatively, click on the background of the Graphics window and use the "Ctr+A" command on the keyboard. You will see all three cylinders appear in the Objects to add window. Highlight $cyl2$ and click on "−" in the right column. Do the same with $cyl3$. Next, go to Objects to subtract, and make it active by clicking on the left mouse button. Again, click on the Graphics background, and use "Ctr+A" command. Delete $cyl1$ and $cyl2$. Click Build All Objects: you have now the washer defined as a difference. To see the washer click the Transparency button in the Graphics window (see Fig. 2.14). There is, however, another way which is more effective, and will serve to more goals, so we will introduce it now. In the Graphics window, deselect Transparency, and also click on Select None. Then, in the same window, click on Click and Hide so that it will become active, then click on Select Boundaries. Hover the mouth over the top lid of the cylindrical box (it will change color) and click on it (it will disappear—if not, then click on View Unhidden). You can consecutively click on other boundaries to remove as many as you want, and when done, deselect Click and Hide by clicking on that icon again. After that you can grab the picture by the mouse and rotate it to

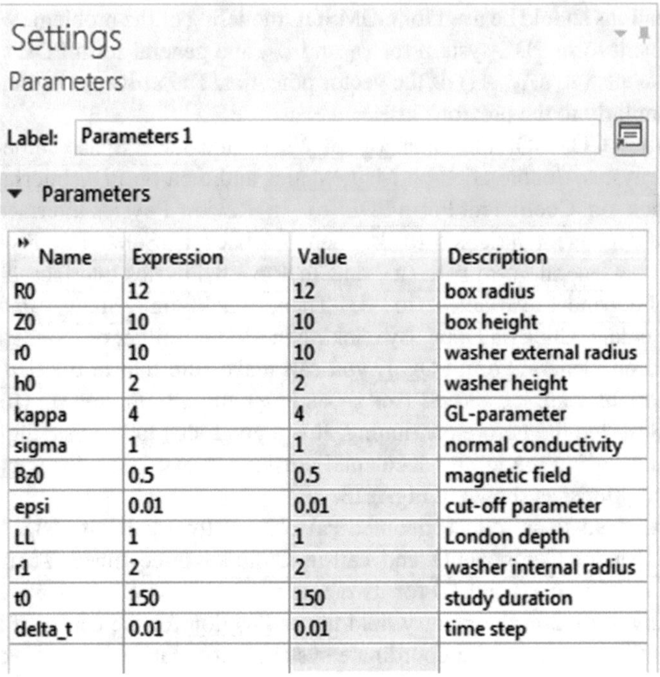

Settings

Parameters

Label: Parameters 1

▼ Parameters

Name	Expression	Value	Description
R0	12	12	box radius
Z0	10	10	box height
r0	10	10	washer external radius
h0	2	2	washer height
kappa	4	4	GL-parameter
sigma	1	1	normal conductivity
Bz0	0.5	0.5	magnetic field
epsi	0.01	0.01	cut-off parameter
LL	1	1	London depth
r1	2	2	washer internal radius
t0	150	150	study duration
delta_t	0.01	0.01	time step

Fig. 2.13 Parameters for modeling

the most informative angle, see, e.g., Fig. 2.15. You have access now to all domains, so that if necessary, you can add Objects to the Difference1 by making Objects to add active and clicking on the plotting.

We are now at the stage where we can start inserting equations. First, we need to introduce the $vol(x, y, z)$-function discussed above. For that right-click on Definitions under Components in Model Builder, select Functions, and click on Analytic. In the Settings window Label it Volume1 and type in the Function name vol. Set it up by a Boolean expression, as shown in Fig. 2.16. Also, type in Arguments x, y, z and set Manual derivatives to zero as shown in the same figure. Now we can insert all the required coefficients in the first and second matrix equations, as shown in Figs. 2.17 and 2.18. One can notice that we added a small number epsi = 0.01 to the time derivative coefficient d_a in the General Form PDE, which will allow computations in regular manner outside of the superconductor volume. The coefficient d_a in the poor conducting material can be arbitrarily small but finite. You can reduce parameter "epsi" (Fig. 2.13) until no influence on solutions is noticeable.

We need to deal now with boundary conditions, which we will do in a way similar to the previously considered disk's case. In the Coefficient Form Equation, we will introduce Dirichlet boundary condition, and require ψ to be zero on the external

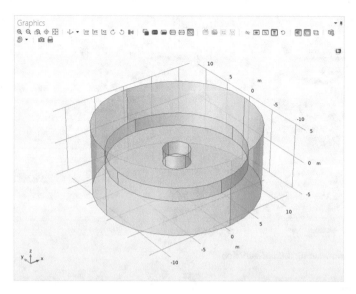

Fig. 2.14 Visualizing superconducting washer inside its inclusion is via transparency mode

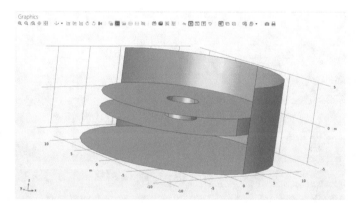

Fig. 2.15 Optimizing the viewing angle

boundary. All we need to do is to right-click on Coefficient Form PDE (c) and then click on Dirichlet Boundary Condition—by default it comes in with zero values at boundaries, but we need to specify these boundaries. Go to the Graphics window and click on visible boundaries first; you will get 3, 11 and 16 in the boundary selection window. Then click on the icon View Hidden Only and you will add hidden boundaries also. You may now click on View Unhidden icon to come to the display shown in Fig. 2.15.

We can proceed now to the second system. Our task is to set up such boundary conditions for the volume that, in absence of the superconducting washer, the field will be homogeneous and directed along z-axis. This can be done with the boundary

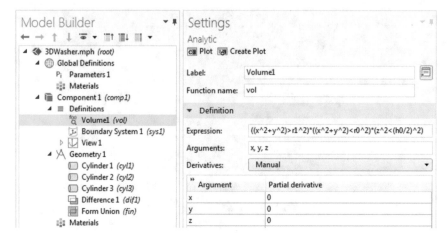

Fig. 2.16 Introducing the vol-function

Fig. 2.17 First two equations in the system

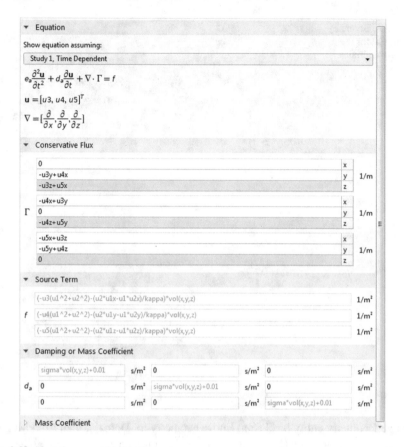

Fig. 2.18 Next three equations in the system

conditions for the **A**-potential: $A_1 = -B_z^0 y/2$, $A_2 = B_z^0 x/2$, $A_3 = 0$. These values should be included into the Dirichlet boundary conditions, which should be called-in as in the previous case, with the same choice of boundaries. The resulting layout in this case is shown in Fig. 2.19.

Initial conditions are simple. As in the case of the 2D-disk, we will choose $\psi_1 = vol(x, y, z)$, and $\psi_2 = 0$. For the **A**-function, all the initial values can be taken to be zero. They are zero by default, so all we need to do is to insert Initial Values in the first system and keep the second one untouched. Let us deal next with Mesh. Find it in Model Builder, right-click on it, and delete. Start building it from scratch: right-click on Component1 and click Add Mesh. Then click on Element Size and switch from Normal to Finer. Then click Build All to see the results. It is preferable to make smaller mesh sizes on the superconducting disk. Right-click on Mesh1, and call-in Free Tetrahedral, then rename it Superconductor for clarity. Right-click on Superconductor and call-in Size which will appear as Size1. Click on Size 1 and switch entire geometry to Domain. We need to choose domain 2. For that

Fig. 2.19 Boundary conditions for the vector-potential **A**

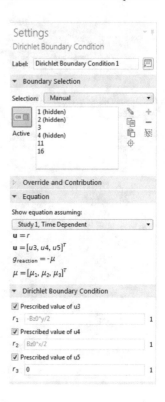

change Geometry entity level from Entire geometry to Domain, then from Manual to All domains, and delete domain 1. Remaining domain 2 will become blue outlined in Graphics window. Now change Normal to Extra fine in the Element Size. Click Build All to see the results (Fig. 2.20). Click on Superconductor to make sure that it has Geometric entity level Remaining. Save the file and run it to make sure if nothing is wrong with it. If no errors are coming up, then you inserted everything correctly, and now can fine-tune your code. Note that when the program ends running, in Results section of Model Builder two groups appear: 3D Plot Group 1 and 3D Plot Group 2. Delete or Disable Group 2, and modify Group 1 to exhibit info which is more informative than the default Slicing. Double click on Group 1 and delete Slice 1. Then right-click on Group 1, and call-in Surface. In Settings window, under Expression, insert $u_1^2 + u_2^2$, i.e., density of Cooper pairs; you may want to insert that title into the Label window instead of Surface1. Then click Plot. You may want also to see the configuration of magnetic induction $\mathbf{B} = \text{curl}\,\mathbf{A}$ on the same plot. Right-click on 3D Plot Group1 and call-in Arrow Volume. Insert expressions for $B_x = u5y - u4z$, $B_y = -u5y + u3z$ and $B_z = u4x - u3y$ into windows for X, Y and Z components. You may also be interested in the current density distribution $\mathbf{j} = \text{curl}\,\text{curl}\,\mathbf{A}$ in superconductor. Right-click on 3D Plot Group 1, and call-in one more Arrow Volume. Insert $j_x = u4xy - u3yy + u5xz - u3zz$,

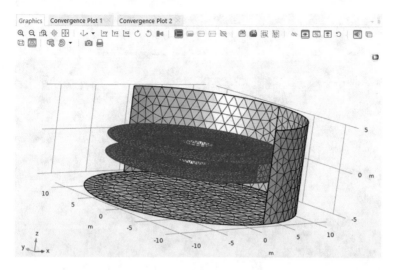

Fig. 2.20 Fine meshing of superconducting washer

$j_y = -u4xx + u3xy + u5yz - u4zz$, $j_z = -u5xx + u3xz - u5yy + u4yz$ for the corresponding components in Settings window. We also recommend you increase the number of grid points to 25, 25, 17. Click <u>Plot</u> to obtain the picture shown in Fig. 2.21. To have titles as in Fig. 2.21, click on <u>Title</u> for each subentry in 3D Plot Group1, switch <u>Automatic</u> to <u>Manual</u>, and enter the relevant text. We can extend now the evolution time to see the vortex dynamics. In Study, click on Step1: Time Dependent, and change *range* in <u>Times</u> to *range*$(0, 0.01, 50)$. After the run is finished (it may take hours), you can create an animation in the regular way described previously. The animation, as well as this COMSOL file, can be downloaded from the Springer website for this book. The website also contains a COMSOL Executable file, which you can run on your computer (Windows or Linux), change parameters, and explore solutions without having COMSOL installed. The final picture corresponding to Time $t = 50$ is shown in Fig. 2.22. Running this file will require at least 16 Gb of RAM.

2.5 Dynamics in Current-Carrying Superconducting Wires

We accumulated enough knowledge in COMSOL and superconductivity to be able to easily consider the current-carrying states in superconducting 1D wires. In the 1D case, the equations, as we mentioned above, are (2.18)–(2.20) for the variables ψ_1, ψ_2, and A, with j_0 as a parameter.

Open <u>Model Wizard</u>, Select Space Dimension 1D, open <u>Mathematics</u>, <u>PDE Interfaces</u>, and double-click on <u>General Form PDE (g)</u> three times. Then click on <u>Study</u>, and choose Time Dependent in <u>Select Study</u>. Click on <u>Done</u> to get

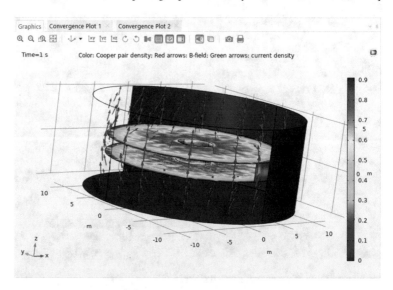

Fig. 2.21 Modeling solution after trial run for time = 1

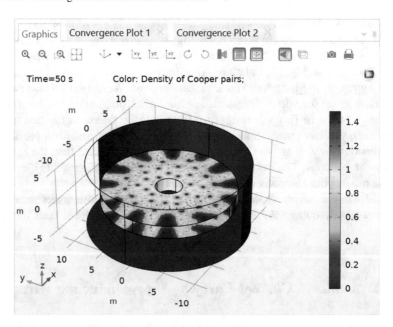

Fig. 2.22 Penetration of magnetic flux in 3D-washer

to Model Builder. Right-click on Geometry and insert an Interval. In Settings for Interval, insert Coordinates $-L_0$, L_0. In Parameters insert $L_0 = 10$. Ignore units everywhere, since our equations are dimensionless. Insert also other parameters, as shown in Fig. 2.23. Start composing equations. In General Form PDE 1 insert $\Gamma = -ux/kappa^2$ and $f = 2*(u3*u2x)/kappa + (u2*u3x)/kappa - u*(u3)^2 + u - u(u^2 + (u2)^2)$. In General Form PDE insert $\Gamma = -u2x/kappa^2$ and $f = -2*(u3*ux)/kappa - (u*u3x)/kappa - u2*(u3)^2 + u2 - u2(u^2 + (u2)^2)$. In the third equation, set $\Gamma = 0$, $f = (-u2*ux + u*u2x)/kappa - (u^2 + u2^2)*u3 - j0$, and $d_a = sigma$. Initial Values may be set to $u = 1$, $u2 = 0$, and $u3 = 0$. That corresponds to the absence of current at the initial moment. The boundary conditions depend on the material of the banks to which the wire is touching with its ends. In the case of massive superconducting banks, we will consider $|\psi| = 1$, and the normal current $\propto \dot{A}$ equals zero at the end-points. Correspondingly, we will choose Dirichlet boundary conditions: $u = 1$ and $u2 = 0$ in the first two equations[3], and $u3 = -j0$ in the third equation. The latter follows from (2.20). To implement these conditions right-click on General form PDE 1 (g) and choose Dirichlet Boundary Condition in the pop-up menu. Then in Graphics window hover the mouse over the end points, and as soon as they change color to red, click on them: they will become blue, and 1 and 2 will appear in the Boundary Selection window at Settings. Do the same in General form PDE 2 (g2). Similarly, insert $-j0$ in the prescribed value of $u3$ in Settings for Dirichlet Boundary Condition for General Form PDE 3 (g3). Next, click on the Mesh and switch Normal mesh to Extra fine. Now, we can test run it, after saving the file. If no mistakes, it will run for a short time (we have not yet set the evolution time to $t0$). After checking for mistakes, click on Study 1, Step 1: Time Dependent, and in Settings, set range to $range(0, 0.1, t0)$. Before running, go to Results, double-click on 1D Plot Group 1, and click on Line Graph 1, then, in Settings window, change Expression to $u^2 + u2^2$, i.e., to Cooper pairs density. The result of computation is shown in Fig. 2.24. This obviously requires some analysis. To start the analysis, let us create an animation. Click on Results tab, then click on Animation and choose File. In Settings window switch GIF to AVI. Then Browse and choose the name of the file and its location. In Frames per second, insert 20. Number of frames set to 500. Finally, click on Export. Be patient: typically, the animation takes longer time than the computation itself. When it is done, watch the video. It immediately becomes clear that a time of 300 is not long enough; make it 1000 and repeat the computation and animation. You do not need to create a new animation, just click on Animation 1 in Results after computation is over. Clearly, the density of Cooper pairs periodically turns to zero at the middle of the wire. Another way to see more oscillations during shorter time-span is to increase the current; no need to increase by much, as changing j_0 from the chosen value 0.4 to 0.41 will make a significant difference.

It makes sense to trace the behavior of superconducting current simultaneously with the dynamics of pair density. Instead of supercurrent, it is easier to deal with

---

[3]Alternatively, and more accurately from a physics standpoint, one should choose $u' = u2' = 0$. Try this option by using the default boundary conditions.

Settings
Parameters

Label: Parameters 1

▼ Parameters

Name	Expression	Value	Description
L0	10	10	half-length of SC wire
sigma	1	1	normal conductivity
kappa	0.4	0.4	GL parameter
j0	0.4	0.4	current parameter
t0	300	300	evoluton time

Fig. 2.23 Initial set of problem parameters

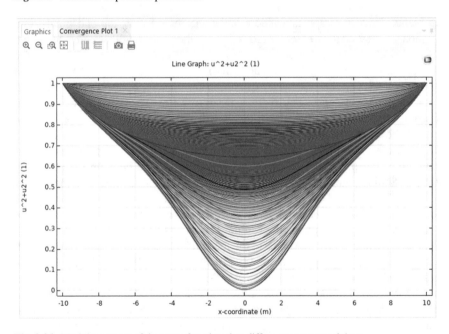

Fig. 2.24 Modulus square of the wave function ψ at different moments of time

the dynamics of normal current $j_n = -\sigma \dot{A}$. Knowing j_n will immediately provide information on j_s, since $j_s = j_0 - j_n$. To implement this, double-click on 1D Plot Group 2, then click on Line Graph 1 under it. In the Expression window in Settings insert $-sigma * u3t$. By default, after computation, COMSOL will plot 1D Plot Group 1. So click on 1D Plot Group 2 to see the normal current plot, Fig. 2.25. The spikes in this figure reveal insufficiency of the mesh size: it should be smaller in size. Click on the Mesh. In Settings window you will see that it is Physics-controlled (by default). Above we switched it to Extra fine which was enough for the Cooper pair density description. However, it will not be enough for a description of the current even if you switch it to Extremely fine size. You can try it: the picture will still have these unphysical spikes. Don't give up. Switch Sequence type in Settings to User-controlled mesh. In Model Builder click on Size under Mesh 1, and then in Settings set the Maximum element size to 0.02. Also, set maximum element growth rate to 1. The result of Computing after that action is shown in Fig. 2.26. What is interesting here is that the values of the normal current exceed 0.4, the value of total current $j0$, Fig. 2.23. As follows from Fig. 2.26, j_n exceeds 0.45 at some moments of time in the center of the wire. That means that, at these moments of time at these spatial points, the superconducting current is negative, i.e., the supercurrent flows in a direction opposite to the direction of total current! It is worthwhile to plot j_s and j_n together, animate them, and display synchronously with $|\psi|^2$. To implement this, right-click on 1D Plot Group 2, and call-in Line Graph. Insert in the Expression $j0 + sigma * u3t$ (which is the j_s), and also in the x-Axis Data change the default Arc length to Expression, and insert x for Expression. Then click Plot. You will get the plot shown in Fig. 2.27.

Animation should be done for the 1D Plot Group 2. You probably noticed that with this mesh, the plotting requires a rather long time. So meshing should be improved so as to have majorly smooth curves with satisfactorily fast speed. You can consecutively increase the mesh size and reveal that even for a mesh size as large as 0.1, the quality of the current curves is still satisfactory. You may want to use that mesh for animation. To fulfill our task of plotting both currents and the Cooper pair density, we need to add one more curve to the plotting, i.e., the curve corresponding to $\psi_1^2 + \psi_2^2$. This could be done in the manner with which we added j_s to 1D Plot Group 2. However, there is an easier way: since 1D Plot Group 1 already contains the desired curve, highlight it with your mouse, then drag it and drop in 1D Plot Group 2. After that, you can even delete the empty 1D Plot Group 1, so that COMSOL will plot 1D Plot Group 2 after the next computation is over. The excerpts from the animation[4] at different moments of time are shown in Fig. 2.28. In Fig. 2.28, we have chosen $j_0 = 0.41$ and $t_0 = 150$; other parameters are similar to those in Fig. 2.23. To see the results while computing, open Study, click on Time Dependent, then, in Settings window, open

---

[4]Sometimes you may have problems with the animation: the software may complain that at least one frame is required for animation and rejects your Export request. When that happens, in Settings window, in Time selection, instead of All there will be From list, without available option of changing it to All. In such cases delete the Animation under Export in Model Builder, and construct animation from scratch again.

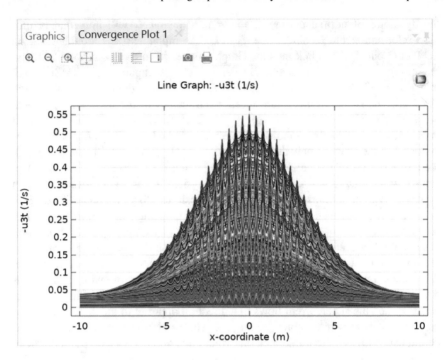

Fig. 2.25 Unphysical spikes indicating insufficient accuracy of modeling. Compare with next figure

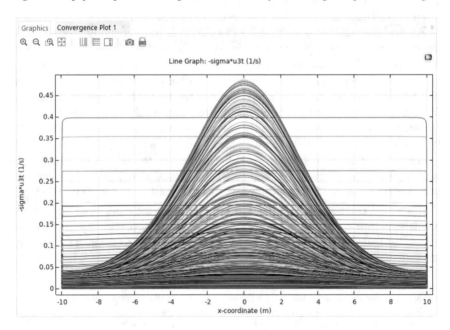

Fig. 2.26 Quality is good for this choice of mesh

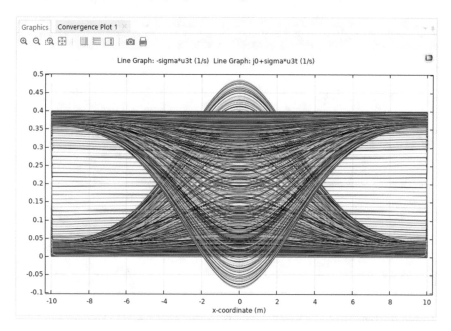

Fig. 2.27 Normal and superconducting currents at different moments of time. Superconducting current becomes negative in the central region of the wire for some moments of time

Results While Solving, and enable Plot. After that, click on the Plot Group which will be visible in the Graphics window, and in the Settings window, under Time selection, switch from All to Last. Then you can compute. One more small secret: to see the time which corresponds to the plots, you need to click on one of the Line Graphs under the 1D Plot Group, then, in Settings window under Legends, make sure Show legends is enabled, Legend itself is Automatic, and Solution is checked in. As soon as the Computation is started, legend will indicate the time variable. Keep in mind that the 1D Plot Group you are interested in plotting should also be chosen in *Study*, Time Dependent: under Results While Solving, choose the appropriate Plot group. You should have no problem with plotting then. You can also choose the place in the graph where the time label appears: click on 1D Plot Group of interest, then in Settings window open Legend, and under Show legends (which should be enabled) choose the desired Position.

Problem 13 Using COMSOL, compute the voltage across the thin superconducting wire in oscillatory regime considered above and plot it as a function of time.

Tip. Use the fact that in our choice of gauge with $\varphi = 0$, the electric field $\mathbf{E}$ is $-\partial A/\partial t$.

Solution to Problem 13
In Results of Model Builder right-click on Derived values, and in the pop-up menu go to Integration, and choose Line Integration. In the Graphics window, hover mouse over the line, and click on it to get number 1 to appear in the Active window of

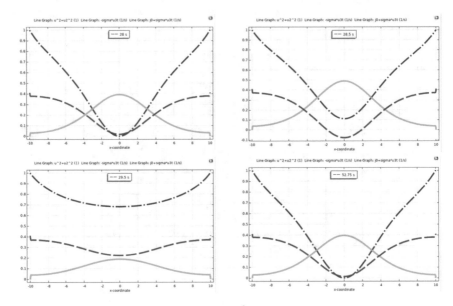

Fig. 2.28 Dynamics of PSC dash-dotted line—$|\Psi|^2$, dashed line—j_s, solid line—j_n

Settings. In the Expression window below it, leave only one entry: make it $-u3t$, which corresponds to $\mathbf{E} = -\partial A/\partial t$. Since the line integral $\int E dl$ corresponds to the voltage across the line, we can now plot it as a function of time to reach our goal. First, click on Evaluate to compute its values. That operation will create Table 1, which in principle you can export and plot, say, by Origin. However you can plot it within COMSOL directly. For that there are two ways. The fastest is to locate the icon Table Graph in the Graphics window. Below it you can see the Table 1 itself. Double-click on that icon, and the desired plot will appear. You will also notice the appearance of a new 1D Plot Group, with Table Graph 1 inside. Alternatively, you could have called it by clicking, say, on 1D Plot Group 3, which by default is for the variable $u3$, then delete Line Graph 1 under it, right-click on it, and call-in a Table Graph from the pop-up menu. You will obtain the same plot as before as soon as you click on Plot in Settings window. Keep in mind that the table should be visible under the Graphics window; if not—you need to first Compute the data. The resultant voltage is shown in Fig. 2.29.

> **Remark 1** We can see from Fig. 2.29 that voltage is an oscillating function of time despite the current is kept constant. Moreover its average value is non-zero—that's why sometimes these states are called "resistive superconducting state". You may be interested to see how it depends on the applied current. You can find the answer using COMSOL.

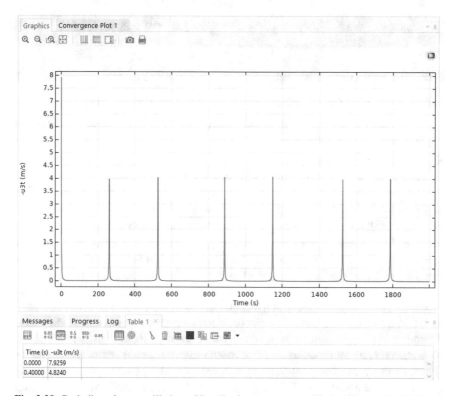

Fig. 2.29 Periodic voltage oscillations. Visually there are two oscillation frequencies: high and low. One can perform Fourier transformation to prove that. Both high and low frequencies were observed experimentally

Remark 2 We used the gauge with $\varphi = 0$. Also A_y and A_z were taken to be zero. Instead of these three gauge conditions, in 1D-case one can use $A_x = A_y = A_z = 0$. In that case, superconducting current $\mathbf{j}_s$ as follows from (2.2) is proportional to $|\Psi|^2 \nabla \theta$. As was clear from previous discussions, the quantity $\partial \theta / \partial t - \varphi$, called the gauge-invariant scalar potential, sometimes denoted μ, is gauge-invariant, and in nonequilibrium situations may be nonzero (see Part II of this book). When a thin wire touches massive superconductive banks, $\mu = 0$ at the boundary. Having average in time $\overline{V} = \overline{\varphi_1 - \varphi_2} > 0$ implies $\overline{\partial / \partial t} (\theta_{1,2}) > 0$. This means that the phase difference between the ends of wire, $\theta_{1,2} = \theta_2 - \theta_1$, is growing unlimited. That will result in unrestricted values for $\nabla \theta$, and therefore for the superconducting current. But in the superconducting state, the current cannot exceed a critical value, or, otherwise, superconductivity will be destroyed. It indeed happens in the middle of the wire, where $|\psi|$ has the smallest value. When the phase difference, and consequently $\nabla \theta$, grow enough that j_s exceeds the critical value, then $|\psi| = 0$, and the phase difference jumps back by the value of 2π. For an instant, this causes an infinitely large phase gradient at the location of this jump, but the current $|\Psi|^2 \nabla \theta$ stays constant since $|\Psi|^2 = 0$. So-called "phase-slippage" takes place.

Problem 14 Find out the phase behavior in our representation.

Tip: COMSOL has built in function $a\tan2(y, x)$ which corresponds to the phase angle of a complex number $x + iy$.

Solution to Problem 14

In the file we were working on, right-click on 1D Plot Group 1, and call-in one more Line Graph. In Settings select All domains, so that 1 will appear in the Active window. In the y-Axis Data insert Expression $(1/(2 * pi)) * a\tan2(u2, u)$. In the x-Axis Data, choose Expression in the Parameter window, and for Expression, insert x. In accordance with Fig. 2.28, the $|\psi|^2$-function first takes on a value of zero at $t \approx 27 - 28$. So click on 1D Plot Group 1, and in Settings window, choose From list in Time selection. Then, scroll down to $27 - 28$ in the list of Time (s), *control click* on certain values, and then click on Plot. The result is shown in Fig. 2.30. From here, one can deduce that at the center of the wire we have phase discontinuity at the time $t = t_{crit} \approx 27.44$. This discontinuity in reality corresponds to the phase difference across the central point of the wire becoming more than 2π: COMSOL plots $a\tan2(x, y)$ by modulus 2π. For continuity, one should replot the curves as shown in Fig. 2.31. The continuity is restored by adding (-2π) phase difference across the central point $x = 0$ for $t \geq t_{crit}$: the phase "slips down" by 2π. The jump is exactly 2π. Quantum states with the phase θ and $\theta + 2\pi$ are indistinguishable: $\psi = |\psi| \exp[i(\theta + 2\pi)] = |\psi| \exp(i\theta)$, since $\exp(2\pi i) \equiv 1$. At the very moment $t = t_{crit}$, $\nabla \theta = \infty$ at $x = 0$. However, this fact is not causing any problem, since the physical quantity, the supercurrent j_s, is proportional to $|\psi|^2 \nabla \theta$, and $|\psi|^2$ itself equals zero at $t = t_{crit}$, so that j_s stays finite. After this "phase slippage" the gradient slowly decreases over the time by absolute value (see Fig. 2.32), until the second phase slippage starts taking place. This second event, which again is related with the touching of zero by $|\psi|$, is even more interesting. After some time, the reduced gradient curves near $x = 0$ will go horizontal, and then will cross the $y = \pm 0.5$ lines both left ($y = -0.5$) and right ($y = 0.5$) of $x = 0$. The function $a\tan2(x, y)$ will immediately split the curve, as shown in Fig. 2.33. This splitting will evolve with some bending of the horizontal parts. Bending will be maximal just before $|\psi|$ touches zero (Fig. 2.34). At $|\psi|^2 = 0$, phase slippage again becomes possible, Fig. 2.35. As we did in the case of the first phase slip (Fig. 2.31 above), we can make again these plots continuous, which will demonstrate the second down-slipping, Fig. 2.36, of the phase at $t_{crit} = 51.6$. One can make sure that the third phase slip will take place quite similarly to the first one, without preceded splitting, and the fourth one will occur similar to the second one, with splitting, and so on with each of the next odd and even events.

Thus we can conclude that at $t = t_{crit}$ when $|\psi| = 0$, the phase "slips" by 2π, so that a qualitative change occurs with the phase pattern, and the phase difference increases by 2π.

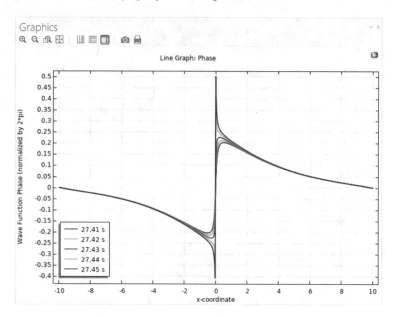

Fig. 2.30   The phase at moment of time preceeding $|\psi| = 0$ (time $= 28$) and succeeding it

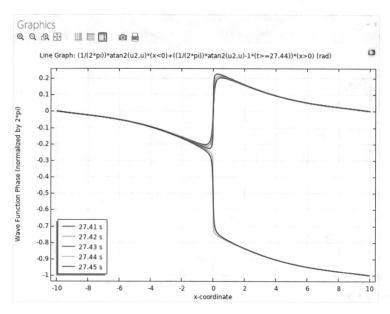

Fig. 2.31   Continuous phase function

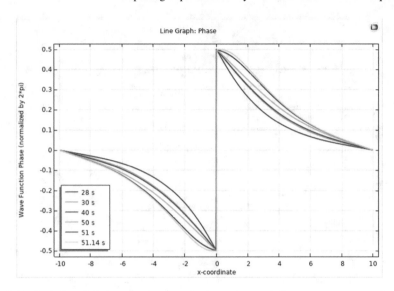

Fig. 2.32 Evolution of phase curves between the first and second slips

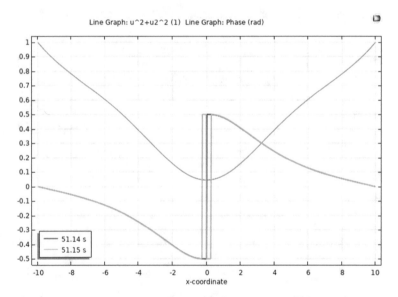

Fig. 2.33 Phase curve splitting at the points when it crosses horizontal lines $y = \pm 0.5$. At this moment of time ($t = 51.15$), $|\psi(x, t)|$ is still far from zero at $x = 0$

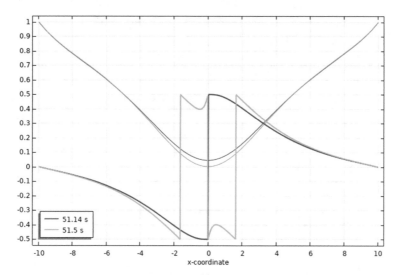

Fig. 2.34   Evolution of split curve

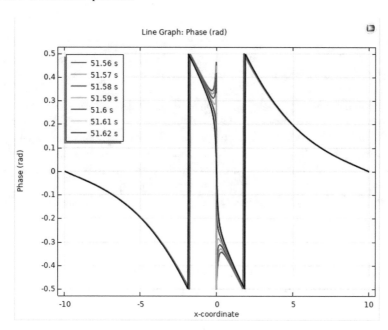

Fig. 2.35   Second phase slippage at $t = 51.6$

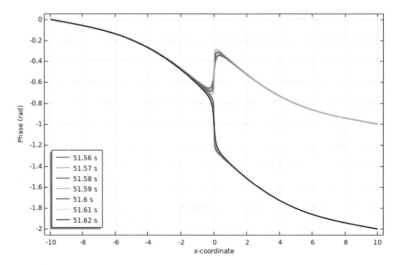

Fig. 2.36 Second jump by 2π at $x = 0$ results in total phase difference 4π between the ends of the wire

Remark The difference between our chosen gauge with $\varphi = 0$ and the case with the gauge $\mathbf{A} = 0$ discussed above is that in our gauge, the slippage adds one full (counterclockwise for positive values of j_0) turn to the wave function while in the other gauge, it reduces phase difference by 2π. In both cases, the supercurrent on average stays constant since in our case, the growth of the phase difference is compensated by the counter growth of the vector potential (readers can explore that themselves!), while in case of $\varphi \neq 0$ gauge, the vector potential is dropped and the phase difference is constant on average. Both gauges are quite legitimate for solving procedures.

2.6 Current Flow in Thin Superconducting Strips: Annihilation of Abrikosov Vortices

We will consider now the flow of an electric current through a thin film of a superconductor. Open Model Wizard, Select Space Dimension 2D. In Select Physics double-click Mathematics, then double click PDE Interfaces, and double click General Form PDE (g) four times. We need four equations: two for the real and imaginary parts of the ψ-function, (2.21) and (2.22), and two for the A_x and A_y components of the vector potential, (2.23) and (2.24). Then click on Study and in Select Study choose Time Dependent. Click Done to get to Model Builder.

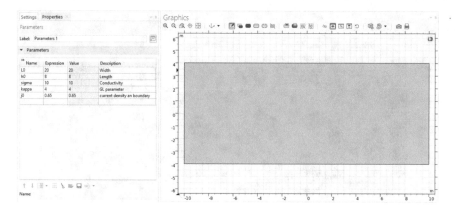

Fig. 2.37 Parameters and film geometry

Right-click on Geometry and choose Rectangle for our thin film. Insert $L0$ and $h0$ as the Width and Height of the rectangle, with Base: Center at $0, 0$. Go to Global Definitions, and in Parameters insert $L0 = 20$ and $h0 = 8$. Also enter $sigma = 10$, $kappa = 4$, and $j0 = 0.65$—these parameters will be used for model building. Click Build All, and then click on Zoom Extents in Graphics window to visualize the film, Fig. 2.37. We can now set up the equations. In the first equation, for the function u, as usual, type in $-ux/kappa^2$ and $-uy/kappa^2$ for x and y components of Conservative Flux Γ. For the Source Term f, type in $2 * (u3 * u2x + u4 * u2y)/kappa + u2 * (u3x+u4y)/kappa - u * (u3^2+u4^2)+u * (1-u^2 - u2^2)$. Leave other coefficients unchanged. The second equation, for the function $u2$, should have $-u2x/kappa^2$ and $-u2y/kappa^2$ for x and y components of Conservative Flux Γ. The Source Term f is $-2 * (u3 * ux + u4 * uy)/kappa - u * (u3x + u4y)/kappa - u2 * (u3^2 + u4^2)+u2 * (1 - u^2-u2^2)$. Leave other coefficients unchanged. In the third equation, for $u3$, insert 0 and $u4x - u3y$ for x and y components of Conservative Flux Γ. Also, change Damping Coefficient d_a to $sigma$. In the fourth equation, for $u4$, insert $-u4x + u3y$ and 0 for x and y components of Conservative Flux Γ, and change Damping Coefficient d_a to $sigma$.

Next, we need to take care of boundary and initial conditions. Let us start with the ψ-function, i.e., with u and $u2$ functions. Since there is no current across the horizontal boundaries of the superconductor, we will impose zero flux on them. On the vertical boundaries, we will impose Dirichlet boundary conditions: $\psi|_{\pm L0/2} \equiv 0$, since we assume the superconductor is contacting the normal metal banks. To implement this plan, right-click on General Form PDE (g) and choose Dirichlet Boundary Condition in the pop-up window. In Settings window, we need to indicate which boundaries this condition refers to. Go to Graphics window, hover the mouse over the vertical boundary, it will change its color to red, click on it, its number will appear in the Selection window of Settings. Do the same thing for the second vertical boundary. Numbers 1 and 4 will appear in Selection window, and the lines will change their color to blue. You can click now on the Zero Flux 1 and make

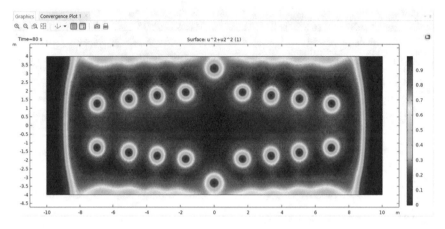

Fig. 2.38 Current flow in superconductors causes vortice pattern

sure that boundaries 1 and 4 are <u>overridden</u> in <u>Selection</u>. The same action should
be carried out for the second equation, for $u2$. <u>Initial Values</u> for these two functions
should be compatible with the boundary conditions. For example, u and $u2$ equal to
0 are compatible with them, but if we start with such a zero state we risk the solutions
to stay at 0 forever, which would correspond to normal state (non-superconducting)
solution. So instead, we will start with $u = \cos(pi * x/L0)$, and $u2 = 0$ initial values
which will guarantee non-zero solutions. We need now to insert boundary and initial
values for the vector potential. By default, we have <u>Zero Flux</u> conditions $-\mathbf{n} \cdot \Gamma = 0$
for $u3$ and $u4$ on all boundaries. That should be changed. In the equation for $u3$,
right-click on <u>General Form PDE 3(g)</u> and insert <u>Flux/Source:</u> $-n \cdot \Gamma = g - qu3$,
and substitute $-Ba$ for g. Leave $q = 0$. This condition is related to horizontal boundaries. So hover mouse on them, and click them in as we did in case of the ψ-equations
above. In the equation for $u4$, right-click on <u>General Form PDE 4(g)</u> and insert
<u>Dirichlet Boundary Condition:</u> $u4 = r$, $r = 0$. This condition is related to vertical
boundaries. So hover mouse on them, and click them in as above. Save the program
and run it. If no errors, then go to <u>Study</u>, time dependent and change computation time
to 100: $range(0, 0.1, 100)$. Click <u>Compute</u>. After <u>Computation</u> is complete, you
will find four 2D Plot Groups in <u>Results</u> of <u>Model Builder</u>. Double-click on the first
one, and click on <u>Surface 1</u>. In <u>Settings</u> window, change the <u>Expression</u> from u to
$u2 + u2^2$, i.e., Cooper condensate density. To have higher quality output change
mesh to <u>Extremely fine</u>. You will see the vortex structure shown in Fig. 2.38. Prior to
continuing our analysis we should explain the logic behind the boundary conditions
which are used above.

The total current I through the strip is governed by the external source, and
it is constant through any cross section of the strip: $I = I_0$ (in view of div $j = 0$
in good metals, including superconductors). Flow of current through the metallic
lead generates a magnetic field with circular field lines around it at far distance.
If the current lead has a circular cross section, the field lines are circular at any

distance from the lead. If the lead is flat, then the field line geometry deviates from circularity to ellipticity near the lead, and its proper form may be defined by proper consideration, for example, using the COMSOL DC Module. However, the most important aspect for our modeling feature can be deduced without computation, just from the symmetry of the problem: for thin films the magnetic field lines at the edges of the film should be orthogonal to the film surface. Moreover, they should have opposite directions at opposite longitudinal edges of the film strip. There is a one-to-one correspondence between the parameter I_0 and the value of B at the longitudinal boundaries of the strip. Thus, instead of setting the value of I_0 via boundary conditions at the vertical facets of the strip, one can set the value of B on the lateral boundaries. This simplifies the finding of numerical solutions, and was suggested and repeatedly used by researchers. From this condition

$$\mathbf{B}_z = \nabla \times \mathbf{A}|_z = \frac{\partial A_y}{\partial x} - \frac{\partial A_x}{\partial y} \equiv u4x - u3y \tag{2.34}$$

and for the symmetric distribution of current density relative to the x- axis (which indeed follows from calculations), we can write $u4x - u3y = Ba$ on the top boundary and $u4x - u3y = -Ba$ on the bottom boundary. The value of $B(\equiv Ba)$ is proportional to the total current through the cross section of the strip: $B = \beta I$, where

$$I = \int_{-h_0/2}^{h_0/2} \mathbf{j} \cdot \mathbf{dh}. \tag{2.35}$$

The function β should be determined self-consistently from the current constancy: $I = I_0$, if I is calculated via (2.35) with $\mathbf{j}$ following from the boundary condition (2.34). In the case of a homogeneous normal metal strip $\beta = 1/2$, so that $B = Ba = I_0/2$. For homogeneous superconductors, as in Fig. 2.37, the current distribution inside the strip is inhomogeneous due to the Meissner effect and the strip's finite length (Fig. 2.38). However, if the current density distribution is still symmetric relative to the x-axis, the same value of β holds for the superconducting case.

Prior to formulating the full set of boundary conditions in the case of a superconducting strip, let us first consider a homogeneous stationary (DC) current flow *in a normal metal strip*. Bearing in mind the standard relation $j = \sigma E = \sigma \nabla \varphi$, one scalar function (φ) is enough to determine this current. Since we chose the gauge $\varphi = 0$, the electric field is determined by the relation $E = -\partial A/\partial t$. Thus, one component (A_x) of the vector potential suffices for this task. The second component (A_y) will effectively be zeroed if we choose zero initial conditions, and zero boundary values for it. Then (2.34), where the $\partial A_y/\partial x$ term is dropped, will serve as the lateral boundary condition for A_x. On the vertical facets of the strip, one can use the Dirichlet boundary condition: $A_x = 0$. Also, the initial condition $A_x = 0$ can be applied without loss of generality.

For a superconducting strip in contact with the normal leads, the boundary conditions and the initial condition for the function A_1 are the same as above. However, the current density $\mathbf{j}$ is not distributed homogeneously inside of the strip, and the function A_y does not vanish here. We can still apply the Dirichlet condition $A_y = 0$ both on the vertical and the horizontal facets. This condition on the vertical facets follows from the relation at the boundary with the normal metal: $\mathbf{j}|_x = \pm L_0/2 = \mathbf{j}_n$, where $\mathbf{j}_n$ is orthogonal to the facet. The condition on the horizontal facets follows from the fact that the y-component of supercurrent $\mathbf{j}_s|_{y=\pm h_0/2} = (\psi^* \nabla \psi - \psi \nabla \psi^*)/(2i\kappa) - |\psi|^2 \mathbf{A}$ is absent. Taken together with the Neumann boundary condition for the ψ-function: $\hat{\mathbf{y}} \cdot \nabla \psi|_{y=\pm h_0/2} = 0$, it yields $A_y|_{y=\pm h_0/2} = 0$. The boundary condition for the ψ-function on the vertical facets is of Dirichlet type: $\psi|_{x=\pm L_0/2} = 0$. This condition neglects the proximity effect (see Part II of this book), which is unimportant for the effects described in this section. This is because the details regarding the current density distribution near these vertical facets of the strip are essential at distances $\leq \xi$ and are not critical to solutions for strips with a length greatly exceeding this distance.

Remark Recall that we explore the case $\kappa = \lambda_L/\xi \gg 1$, and the length is in λ_L-units.

Problem 15 Using COMSOL, demonstrate that the annihilation of vortex-antivortex pairs yields spikes of an electric field, i.e., causes electromagnetic radiation. Also plot the current vectors and show that vortex and antivortex have opposite clockwiseness.

Tip: use the tip to Problem 13.

Solution to Problem 15
As soon as the functions $u3$ and $u4$ are known, the electric field can be straightforwardly plotted as $E_x = -u3t$ and $E_y = -u4t$. In Results, right-click on 2D Plot Group 1 and call-in Arrow Surface from the pop-up window. Replace ux by $-u3t$ and uy by $u4t$ in the Expression window for X and Y components. Change Color to Green, and click Plot. You may also want to adjust the scale factor to see the arrows better. Compute the simulation with Plot checked on in Results While Solving at Step 1: Time Dependent in Study to see the effect while computing goes on. Typical spike generation is shown in Fig. 2.39.

For visualization of current vectors, we should again do the same procedures as above; however, ux should be replaced by $-(u3yy - u4xy)$, and uy by $-(u4xx - u3xy)$. We can disable the electric vector plot to have a more distinct pattern. However, as the readers can check themselves, the picture of the current plots is not what we expect it to be. The reason is in insufficient accuracy of solutions. Solutions are enough accurate for the functions themselves (Fig. 2.38), and even for derivatives (Fig. 2.39) but not for higher order derivatives, which describe the current density vectors. We need to enhance the mesh. Click on Mesh in Model Builder, and in Settings window, Sequence type, switch Physics-controlled mesh to User-controlled mesh. Then go back to Model Builder, and click on Size under Mesh 1. In Settings window, select Custom instead of Predefined. This will allow you to change mesh sizes.

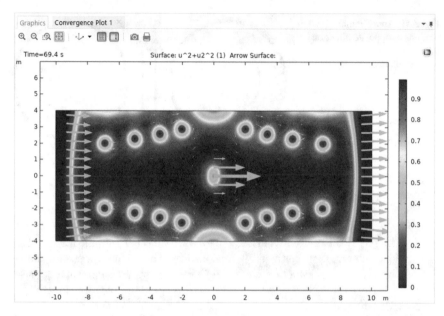

Fig. 2.39 Annihilation of vortex-antivortex pair results in a spike of an electric field (green arrows)

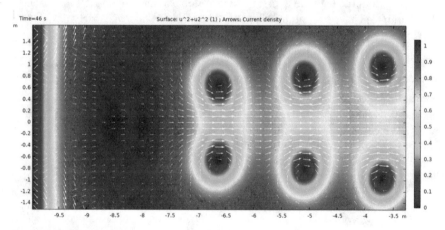

Fig. 2.40 Current distribution in current-carrying superconducting strip with vortices and antivortices as second derivative of vector potential

Reduce <u>Maximum element size</u> from 0.2 to 0.05, and <u>maximum element growth</u> rate to 1. The result after computing, which naturally will take a longer time is shown in Fig. 2.40. This looks much more convincing!

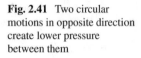

Fig. 2.41 Two circular motions in opposite direction create lower pressure between them

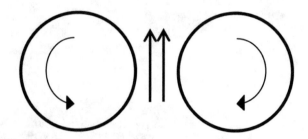

Remark 1 To better understand the physics of vortices, it is useful to make an analogy to tornados. The air flow may be laminar at small velocities. For a higher speed of air, turbulence prevails. You probably have noticed how leaves are engaged by an autumn wind and move fast rotationally around a vertical axis. This motion is viscous naturally, and the leaves drop quickly because of a dissipation of rotational motion energy. However, if energy of rotational motion is very large, like in tornadoes, the dissipation is negligible, and tornadoes can move distances much larger than their sizes without any noticeable change. This is like the superfluid motion of Cooper pair condensate in vortices. The analogy to tornadoes can explain why a vortex and an antivortex attract each other. If you have two circular motions of air mass in opposite directions, and you will bring them close, they will facilitate air motion between them, since both are driving air between them in the same direction (Fig. 2.41).

Thus, one can expect lower pressure between tornado-antitornado pair in accordance with the Bernoulli law. In case of realization of this event in nature catastrophic consequences will far exceed the consequences of a single tornado action, since the pair will dissipate locally all its energy, as vortex-antivortex do. Typical in literature explanation of vortex-antivortex attraction is by the Lorentz force; however the Bernoulli force is sometimes also mentioned, and it is more vivid.

Remark 2 To continue analogy with the circular air motion, let us recall that turbulence is easier to observe when you have some obstacle to the blowing wind. For example, at the corner of a building, the wind is much more turbulent than in an open space. This means that dents in the superconducting strip may cause generation of vortices.

Problem 16 Consider a strip with two triangular dents in the middle, both at the top and bottom (Fig. 2.42) and analyze the vortex-antivortex generation by them at DC current flow.

Tip: At the edge of the dent, magnetic field is stronger than in an undented area— neglect this small effect, and just take into account the geometry.

Fig. 2.42 The proposed
geometry of the strip for
initiating "turbulent"
superfluid motion

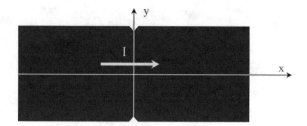

Solution to Problem 16

Let us start with modifying the geometry of the strip in the file which we were
using above. Save it under a different name. Right-click on Geometry, and click-in
Square 1, then, in similar way, Square 2 . For both of them, change in Settings Position
Base to Center. Then change their y-coordinates to $h0/2$ and $-h0/2$ respectively.
For both of them, change in Rotational Angle to Rotation 45 degree. Then right-
click on Geometry, choose Booleans and Partitions, and click on Difference. Setting
window will open, where you should insert Rectangle 1 ($r1$) in the Objects to add
window, and Square 1 ($sq1$) and Square 2 ($sq2$) in the Objects to subtract window.
By default, Objects to add is Active. Hover the mouse over the strip, it will change
color, click on it, and $r1$ will appear in the Objects to add. Then click and make
active Objects to subtract. Then hover the mouse over the rotated squares, click
on them, so that $sq1$ and $sq2$ will appear in Objects to subtract window. Click on
Build All Objects, and you will see the final geometry as shown in Fig. 2.43. The
equations stay the same, but we need to modify the boundary conditions. Magnetic field at the area of dents will be somewhat stronger than average value,
but we will neglect that fact for now, and will come back to it in Remark 1
below. What is more important is that the **n**-vector in the dented area of the film
edges has both n_x and n_y components, and they are less than 1. Thus, we need
to take care of keeping the value of $curl\mathbf{A} = \mathbf{B}$ on dents which can be done
in the following manner. In the Flux/Source 1 entry in General Form PDE 3 (g3),
replace $-Ba$ by $-Ba(sign(y) * ny) * (t > t0)$. This will take care of changed
values of n_y on the boundary. Also, we would like to switch on the current with
some delay, so we added a Boolean condition, and the stationary superconducting state will set up in the film prior to switching on the current at $t = t0$. In the
General Form PDE 4 (g4), we need to take into account that at the dented area the
n-vector has x-component. Thus, right click on General Form PDE 4 (g4) and click-
in Flux/Source, then hover the mouse over the dents segments and click on them,
so that their numbers 4, 5, 6 and 7 appear in the Settings window. Then, in the
Boundary Flux/Source, insert $-(-u4x + u3y) * nx$ for g. These boundary elements will be overridden in Zero Flux window. Since we used $t0$ for the switching of the current, in the Parameters you need to insert $t1$, which should then
be introduced in Study, Time Dependent, Times range $(0, 0.1, t1)$ as the computation time limit of Settings window. You can start studies with $t1 = 100$ to grasp
what is going on, Fig. 2.44. At calculations, User Defined Mesh was used, with
Maximum element size 0.1, minimum element size 0.01, and maximum growth rate
1 in Custom Element Size Parameters.

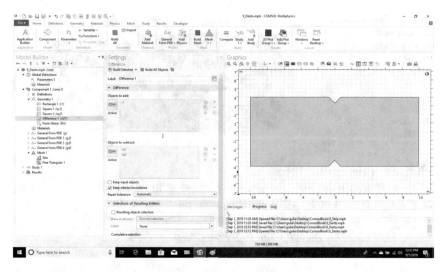

Fig. 2.43 Setting up the geometry of superconducting film with dents

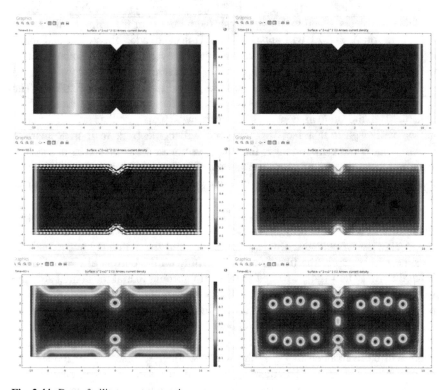

Fig. 2.44 Dents facilitate vortex creation

Remark 1 Above, we neglected the increase of Ba near the dent edge. If that effect had been taken into account, the disturbance created by the dent, and consequently, the origination of vortices would be further intensified.

Remark 2 The disturbing role of the edge roughness which we modeled via a pair of dents can play very negative role in superconducting devices. Indeed, as we now know, the pair of vortices can annihilate and create an electric pulse which will mimic SFQ pulse (considered below), which is a cornerstone of certain class of superconducting electronics. The considered example demonstrates how important the quality of lithographic patterning is in these devices.

2.7 Generation of SFQ Pulses in SNS Junctions

We will consider now an SNS junction. It consists of two superconductors with a normal metal layer in between them (Fig. 2.45). This junction is similar to the Josephson junction, where a very thin tunnel barrier is replaced by a much thicker normal metal layer. When the superconductor and normal metal are brought into contact with each other, Cooper pairs can propagate a finite distance into the normal metal. Thus, a thin layer of the normal metal that is in contact with the superconductor behaves like a weaker superconductor. This is called the "proximity" effect, which we neglected in the previous section. In the current case, it plays an important role. "Weaker" in the context of the proximity effect means the ψ-function modulus in the normal layer is somewhat smaller than that of a genuine superconductor.

If biased as shown in Fig. 2.45, the supercurrent, while moving along the junction, squeezes through the proximitized normal metal and gradually reduces its amplitude in the direction from the left of Fig. 2.45 to the right. Motion of charge carriers should create magnetic field around the SNS-junction. It is easy to conclude that the distribution of the magnetic field should qualitatively correspond to the picture shown in Fig. 2.45. For the parameters of a homogeneous junction we can assume

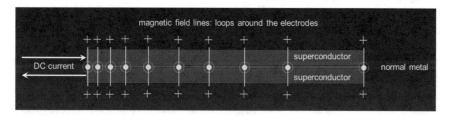

Fig. 2.45 Superconductor-Normal metal-Superconductor junction

linear behavior of magnetic field: $B_z = B_0(L_0/2 - x)/L_0$, where L_0 is the length of the junction, and the coordinate system is located at junction's center, with x-axis along it, y-axis is directed vertically, and z-axis is towards us. As in the previous section, the current will be determined via the boundary conditions involving B_z.

We will consider the junction as a whole piece of the superconductor, which, in absence of the current, has ψ-function weaker around the horizontal median line. That could be done with the help of the p-function introduced in Part II. Then (2.15) will take a form:

$$\frac{\partial \psi}{\partial \tau} = -\left(\frac{i}{\kappa}\nabla + \mathbf{A}\right)^2 \psi + \left(1 - |\psi|^2 + p\right)\psi, \tag{2.36}$$

so that (2.21) and (2.22) will correspondingly be modified as:

$$\dot{\psi}_1 = \frac{1}{\kappa^2}\left(\psi_{1.xx} + \psi_{1.yy}\right) + \frac{2}{\kappa}\left(A_1\psi_{2.x} + A_2\psi_{2.y}\right) + \frac{\psi_2}{\kappa}\left(A_{1.x} + A_{2.y}\right)$$
$$-\psi_1\left(A_1^2 + A_2^2\right) + \psi_1(1 + p) - \psi_1\left(\psi_1^2 + \psi_2^2\right), \tag{2.37}$$

$$\dot{\psi}_2 = \frac{1}{\kappa^2}\left(\psi_{2.xx} + \psi_{2.yy}\right) - \frac{2}{\kappa}\left(A_1\psi_{1.x} + A_2\psi_{1.y}\right) - \frac{\psi_1}{\kappa}\left(A_{1.x} + A_{2.y}\right)$$
$$-\psi_2\left(A_1^2 + A_2^2\right) + \psi_2(1 + p) - \psi_2\left(\psi_1^2 + \psi_2^2\right), \tag{2.38}$$

As follows from (2.36), negative values of $p = p(y)$ reduce locally critical temperature mimicking the proximitized N-layer (reasonable value $p = -0.3$ is used in the example below). To understand why this happens one can notice that in the spatially homogeneous stationary case, the non-zero values of $|\psi|$ are equal to $1 + p$ in accordance with (2.36), i.e., are weaker than in the superconductor layers.

Problem 17 Using COMSOL, consider a Superconductor-Normal Metal-Superconductor (SNS) junction in DC-mode. Realize a state with periodic single-flux quantum (SFQ) generation.

Tip: Represent the p-function via the Boolean condition: $p(y) \equiv p * (y/2)^2 < 0.02$, which corresponds to normal metal layer of thickness $\approx \lambda_L$.

Solution to Problem 17
The character of this problem has analogy with the disk in a magnetic field that we solved in Sect. 2.2. So it will save us time to save the resultant COMSOL file under a different name and start modifying it. First, delete in Geometry Circle 1 by right-clicking on it, then insert Rectangle by right-clicking on Geometry. Set up Width $L0$ and Height $h0$, with Base: Center. They will come out red, because we have not yet entered them in parameters. Click on Parameters 1 under Global Definitions and delete R; insert $L0 = 5$ and $h0 = 4$. You can see now that in Geometry 1, Rectangle 1 the red color is gone. Click on Build Selected to visualize the junction. Then click on Zoom Extents icon in Graphics window to optimize its plotted size.

Next, let us modify the equations. Double-click on General Form PDE (g), then on   General Form PDE 1   and transform   the f-function to $(u3x + u4y) *$ $u2/kappa + 2 * (u3 * u2x + u4 * u2y)/kappa - (u3\text{^}2 + u4\text{^}2) *   u + (1 + p *$ $((y/2)\text{^}2 < 0.02) - u\text{^}2 - u2\text{^}2) * u$. Then go to the second equation and transform the f-function to $-(u3x + u4y) * u/kappa - 2 * (u3 * ux + u4 * uy)/kappa -$ $(u3\text{^}2 + u4\text{^}2) * u2 + (1 + p * ((y/2)\text{^}2 < 0.02) - u\text{^}2 - u2\text{^}2) * u2$. In the third equation, the expression for Γ should be modified. As was discussed above the magnetic field should linearly decrease to zero at the right end of the junction, being maximal at the left end. So the y-component of Γ should be written as: $u4x - u3y - Ba * (L0/2 - x)/L0$. The equation for the junction domain will not change, but the boundary condition will correspond to a linear decrease of B_z with respect to the x-coordinate. Similarly, the x-component of Γ in the fourth equa-tion should become $-u4x + u3y + Ba * (L0/2 - x)/L0$. This generates the cor-rect boundary condition, but also changes the equation in the domain, unless we add a term $-Ba/L0$ to the f-term. For all the equations, we will keep the boundary and initial conditions. Before saving the file, insert also the parameter $p = -0.3$ into the Parameters1 of Global definitions. You may also want to visualize the results during the computation; to do so, click on Step 1: Time Dependent, and then in Settings window, click on Results While Solving, and check in Plot. It is a good idea to start at relatively small values of Ba, say, 0.45, and increase it step by step.

Figure 2.46 shows what happens when the current through the junction is above the certain critical value.

Problem 18 Explore the generation of electric field at SFQ leaving the SNS junction.

Tip: Plot the vectors of electric field using the solution obtained for Problem 17.

Solution to Problem 18
The electric field is given by the relation $\mathbf{E} = -\partial \mathbf{A}/\partial t$. Thus, right-click on 2D Plot Group 1, and call-in Arrow Surface. In Settings, insert $-u3t$ for x component and $-u4t$ for y component. Click Plot. Regulate the length of $\mathbf{E}$-vectors by Scale factor: you can highlight the number and type in your suggested value, then click Plot. You can plot the Graphs at different moments of time. At leaving the junction, the amplitude of electric field is maximum (see Fig. 2.47), which corresponds to the generation of an electric field pulse. On the left-hand side of the junction we have electrodes supplying a DC current. If we attach electrodes on the right-hand side of the junction plates, we will detect periodic voltage pulses!

2.8   Cloning of SFQ Pulses

We introduced above the SFQ pulses generated by the SNS junction. For the needs of superconducting electronics, sometimes it is required to clone SFQ pulses. That is, to regenerate a single flux quantum using a propagating voltage pulse. That could

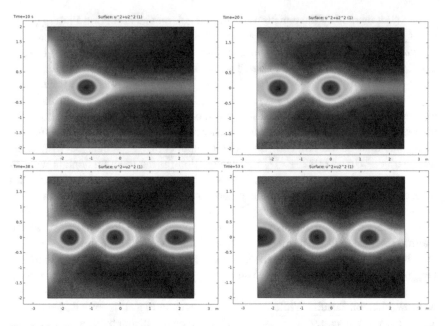

Fig. 2.46 When current exceeds some critical value a single vortex enters the junction ($t = 10$). While it propagates along the junction, the second vortex is in ($t = 20$). At $t = 38$ we have three vortices, and the initial one begins to leave the junction. A little later, it is gone, and a new one is entering the junction ($t = 53$)

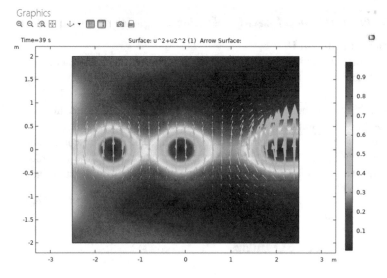

Fig. 2.47 At the moment of leaving the junction, SFQ generates an electric pulse

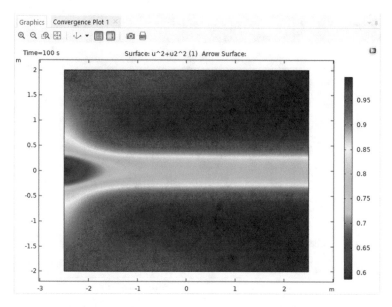

Fig. 2.48 If current is below certain critical value, the SQF pulse will not be generated. In the case shown here, the amplitude of B-field is $Ba = 0.45$, at $t = 10$ the junction state is all set, and will stay indefinitely long if no external parameter changes

be achieved by applying an SFQ pulse to a DC-biased SNS junction. Let us model such a process. For accomplishing this task let us recall that demonstrated above periodic generation of SFQ pulses at DC bias took place for overthreshold applied current. If current is below the threshold, then there is no SFQ generation, Fig. 2.48. If the junction is in that mode, application of a voltage pulse to it will instantaneously increase the current, and for a short period of time the conditions for generation of a vortex will be satisfied, so we will see a propagating vortex, and when leaving the junction it will generate an SFQ pulse itself.

Problem 19 Using COMSOL, consider SNS junction in a stationary DC-mode. Apply an SFQ pulse to generate an SFQ propagating along the junction.

Tip: Pulsing of current will create pulsing of magnetic field surrounding the junction. Thus, the appropriate boundary conditions should become time-dependent.

Solution to Problem 19

Let us modify the code which was created for Problem 15. Save that file under a different name. Increase of the magnetic field should start when the stationary condition has been well-established. So Compute it and make sure that at given values of parameters at $t = 20$ it is stationary. Let us effectively double the amplitude Ba of magnetic field using mathematical expression $(1 + exp(-(t - 20)^2/25)) * Ba$ (as before, we use here COMSOL notations). This modification should be performed in

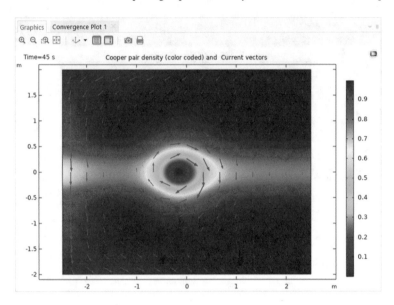

Fig. 2.49 Regeneration (cloning) of the SFQ via applied SFQ pulse to the DC-biased SNS junction

the expression for Γ both in the third and fourth equations. Don't forget to renormalize by this multiplier Ba in the equation for the Source Term f for the fourth
equation. Start Compute with the value of $Ba = 0.45$. In five units of time, it will
be doubled at around $t = 20$. As you may see, with rather simple means we were
able to visualize a rather complex physical phenomenon: cloning of the SFQ pulses,
Fig. 2.49.

2.9 Discovering New Effects with COMSOL-TDGL

In this final example, we will demonstrate how to discover new features of known
phenomena using COMSOL and TDGL equations. We will consider a very old task,
well-described in textbooks: a superconducting ring in a magnetic field. The statement made in textbooks is that the flux through the superconducting ring should stay
constant while the external magnetic field is changing. The proof of this fact is pretty
simple (see, for example, the section *Superconducting Current* in **Electrodynamics
of Continuous Media** by Landau and Lifshitz). Consider a ring in a magnetic field
and apply Maxwell's equation

$$\frac{\partial \mathbf{B}}{\partial t} = -\text{curl}\,\mathbf{E} \tag{2.39}$$

for the description of a temporal evolution of the flux through it. Integrating (2.39)
on the equatorial plane of the ring, we have:

$$\frac{\partial}{\partial t} \int \mathbf{B} \cdot d\mathbf{s} = - \int \text{curl } \mathbf{E} \cdot d\mathbf{s}, \tag{2.40}$$

where $\mathbf{s}$ is normal to the surface. Limiting the integration inside of the ring opening, we have:

$$\frac{\partial}{\partial t} \Phi_{\text{ring}} = - \int \mathbf{E} \cdot d\mathbf{l}, \tag{2.41}$$

where $\mathbf{l}$ is tangential to the internal line along which the plane is crossing the ring. The *r.h.s.* of (2.41) in textbooks is assumed to be zero based on arguments that the tangential component of $\mathbf{E}$ on the surface of conductors should be a continuous function, and inside the superconductors $\mathbf{E}$ is zero. This delivers the relation

$$\frac{\partial}{\partial t} \Phi_{\text{ring}} = 0, \tag{2.42}$$

i.e., flux through the ring remains constant regardless of any changes in the external magnetic field. In particular, if the ring is cooled into the superconducting state in the absence of a magnetic field, the appearance of the magnetic field should not change this zero value. (To avoid any confusion, we should stress that we consider a weak enough magnetic field to avoid destruction of superconductivity or appearance of vortices in it.) Let us now determine how correct is this statement using COMSOL modeling based on TDGL equations. Be ready to discover some new features!

Let us create the code Ring from scratch, though we can also modify the Washer-3D file. Open Model Wizard, and Select Space Dimension 3D. In Select Physics go to Mathematics and open PDE Interfaces. Double-click on Coefficient Form PDE(c). In Number of dependent variables insert 2 (for real and imaginary parts of the Ψ-function). Then double-click on General Form PDE (g) and in Number of dependent variables insert 3 (for three components of the vector potential). Then click on Study and choose Time Dependent. Then click on Done. Next, let us construct a cylindric box and a ring in it. Right-click on Geometry 1 and choose Cylinder. Then right-click again, open More Primitives and choose Torus. We should now define the sizes of the introduced domains. We recall that in our TDGL computational scheme we set up the unit size to be equal to λ_L, the London penetration depth. The rings can be of size much larger than that. However, in superconducting nanoelectronics sizes may be comparable to λ_L. We will take sizes larger than λ_L, but not much larger to make the computations affordable for ordinary computers. Let us choose the major radius of the torus $r = 10$ (in λ_L units), the minor radius of the torus $r1 = 2$, the cylinder radius $R0 = 15$, and its height $Z0 = 10$. Insert them into Parameters under Global Definition. Then click on Cylinder 1 in Model Builder, and in Settings insert $R0$ for Radius, and $Z0$ for Height. Do the appropriate similar actions on Torus under Geometry 1. Then click on Build All Objects in Settings. Objects will become invisible in the Graphics window. So click there on Zoom Extents icon and the visibility will be restored. However, the ring is still hidden under the walls of the cylinder. To visualize the ring, we can do what we did previously for the Washer-3D problem: activate Select Boundaries, then activate Click and Hide icon. Hover the mouse over those cylinder edges which you would like to hide. They will change color. Click on them, and they will disappear. The picture you see will look like in Fig. 2.50. Obviously, you should move the external cylinder down by its half-height. Click on Cylinder 1 under Geometry 1, and in Position in Settings window insert $-Z0/2$ for z. Click Build All Objects to normalize the relative arrangement. We can now insert the equations. They are similar to the ones in the Washer-3D case, so we will omit a detailed description. The only difference is that now, the function $vol(x, y, z)$ should correspond to a torus, not to a washer. The boolean expression which follows from the analytical geometry for the torus, defining the torus volume with chosen parameters, is $(r0 - (x^2 + y^2)^{(1/2)} + z^2) < (r1)^2$. Right-click on Definitions under Component 1, select Functions and click on Analytic. Label it Volume 1 in Settings, and give it Function name: vol, then insert the above-mentioned Boolean expression into Expression window, and set Arguments: x, y, z. You have now the function $vol(x, y, z)$ ready for use. And this works fine when the geometrical objects are easy for analytical description. However, there is a more universal and powerful way of dealing with this task which we will describe as an alternative. The issue is that the Boolean function $vol(x, y, z)$ has two values: 0 and 1 depending on whether or not its arguments are outside of the domain. Action of the same type can be performed by a dual-value variable, which will be equal to 1 inside the domain and 0 outside. To use this opportunity, first disable the function $vol(x, y, z)$ under Definitions. Then right-click on Definitions, and call-in Variables. Variable 1 will arrive. Go to Settings and change Geometry Entity Selection to Domain. We need to choose a domain that corresponds to the torus. In Selection switch to All domains. 1 and 2 will appear in the window. To understand which domain corresponds to the torus, click on these numbers. From the color change, you will easily recognize that it is domain 2. Leave both domains in the window—that is the operational space of Variable 1. In the Name under Variables insert vol, in the next Expression insert $1 * (dom == 2)$ – the coefficient here can have any value. This will act now as a step function. The resultant picture is in Fig. 2.51. To use this function instead of $vol(x, y, z)$ just go to the equations and convert everywhere (including boundary and initial conditions) $vol(x, y, z)$ into vol. Compared to the case of the Washer-3D problem, we would like to introduce one more change: to switch the magnetic field gradually after the superconductivity is settled in the ring, say at $t = t0$, and then gradually decrease it to zero at $t = t1$. Hyperbolic tangent may work nicely for this task. In Dirichlet Boundary Condition for General Form PDE 1, in Prescribed value of $u3$ and $u4$ append the multiplier $((t > t0) * \tanh(omega1 * (t - t0)) - (t > t1) * \tanh(omega2 * (t - t1)))$ to $-Bz0 * y/2$ and to $Bz0 * x/2$ correspondingly, as shown in Fig. 2.52. Insert next the computational time $t2$ into Parameters 1 under Global Definitions. That time should also be inserted into Step 1 under Study 1 in Settings Times window to have range(0, 0.1, t2). Insert also other parameters of the problem, as shown in Fig. 2.53. Save it and run it to make sure everything was done correctly. You can customize the Mesh similarly to how it was done for the Washer-3D case—no need for additional instructions. We would rather focus on the exploration of modeling results. Let us visualize the density of Cooper pairs, the magnetic field lines, and the current vectors. Go to the 3D Plot Group 1 in Results and delete Slice 1. Also delete 3D Plot Group 2. Right-click on 3D Plot Group 1 and call-in Surface. Rename the Surface 1 label to Density of Cooper Pairs. Correspondingly, in Expression window of Settings, replace $u1$ by $u1\textasciicircum2 + u2\textasciicircum2$. Plot it. Next, right-click on 3D Plot Group 1 and two times click-in Arrow Volume. Rename one of them to Magnetic field, and the other one to Current density. Magnetic field is calculated as curl $\mathbf{A}$, and current density as curl curl $\mathbf{A}$. These plots are shown in Fig. 2.54. We can next compute the net current through the ring cross section as a function of time. For that right-click on Data Sets under Results in Model Builder and click-in Cut Plane. Cut Plane 1 will appear. We will later need one more, so repeat and get Cut Plane 2. Click on Cut Plane 1, go to Settings and define it as ZX-planes. Next click Plot, and make sure that current through the cross section is along the y-direction. Now, right-click on Derived Values and in Integration choose Surface Integration. In the Expression for Surface Integration 1 there are three entry lines. Delete two of them, and keep only one. You need the y-component of the current. Go to Current density under 3D Plot Group 1, and copy and paste $(-u4xx + u3xy + u5yz - u4zz) * vol$ from the expression in Settings window. This expression contains vol-variable, which restricts integration within the ring cross section. It is interesting to know how much does the artificial conductance of the environment contribute to the screening current. For that we can insert one more similar integration, but without vol variable. Before pressing the Evaluation icon in Settings window, we need to add one more factor to the integrand Expressions. The matter is that currents in these two cross sections are in opposite directions, so we need to multiply the integrands by $sign(x)$ and divide by 2. Next, you should indicate where this integration should take place. For that, click and highlight each of Surface Integration 1 and 2, go to Settings window, and in Data set choose Cut Plane 1. Finally, you can click on Evaluate for each of them. Evaluation creates tables under Tables icon, which you can see by double-clicking on it. You can plot these data in COMSOL, or export them using bottom icons in the Graphics window. However, plotting can be done better via Group plotting. Right-click on Results, and click-in 1D Plot Group. Then click on it, and in Settings window choose Cut Plane 1 in Data set. Then right-click on 1D Plot Group 2, and call-in twice Table Graph. Then click on Table Graph 1, and in Settings window choose Table 1. Do the same with Table Graph 2, and choose Table 2. That will allow you to see the time evolution of the current, as well as to notice the parasitic contribution of the artificial low-conductance environment, Fig. 2.55. An important feature of the superconducting state is noticeable from this graph: as soon as the magnetic field is off, the current is also gone. This illustrates a well-known fact in the theory of superconductivity: by simply introducing and removing the magnetic field, it is not possible to generate a sustainable current in superconducting closed contours. More sophisticated actions are required: you can find their description elsewhere.

We can now undertake the final steps in exploring the behavior of the flux through the ring. For this, click on the prepared Cut Plane 2. In Settings window, choose XY-planes. By default it will be at $z = 0$, and serve as an equatorial plane for the ring. Now, insert as above another Surface Integration, click on Surface Integration 3,

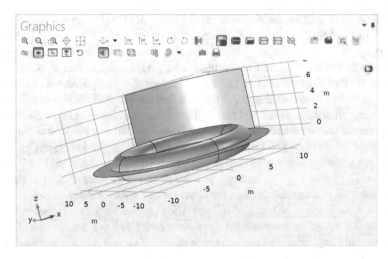

Fig. 2.50 Positioning of external cylinder: it should be shifted down via changing its z-coordinate from default 0 to $-Z0/2$

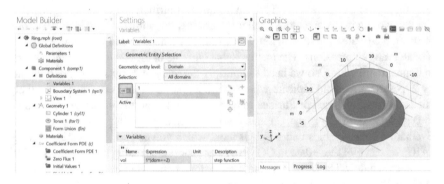

Fig. 2.51 Introducing the vol-variable

Fig. 2.52 Using hyperbolic tangent function for switching magnetic field on at $t = t0$ and off at $t = t1$

Parameters

Name	Expression	Value	Description
R0	12	12	cylinder radius
Z0	8	8	cylinder height
r0	8	8	torus radius major
r1	1.5	1.5	torus radius minor
sigma	1	1	normal conductivity
sigmadiel	0.00000000001	1E-11	artificial conductivity
t0	2	2	field on begins
t1	10	10	field off begins
omega1	2	2	increase speed
omega2	1	1	decrease speed
t2	15	15	evolution time
Bz0	0.01	0.01	external magnetic field
kappa	0.4	0.4	G-L parameter

Fig. 2.53 Parameters which correspond to the the performed calculations and plotted figures

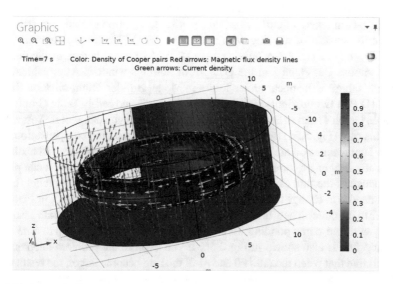

Fig. 2.54 Magnetic flux density lines (red arrows) and superconducting current (green lines) induced by the external magnetic field

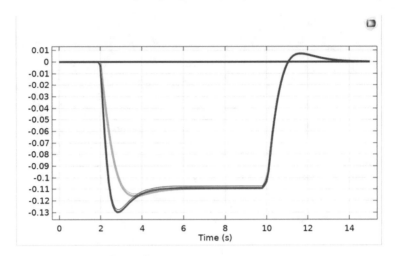

Fig. 2.55 Temporal behavior of the total current through the ring cross section. Its initial and final values are zero in accordance with experiments and theoretical predictions: the external field sets it up and removes it when it is off at arbitrary speeds. Two values of *omega*1 are used (1-green curves and 2-red curves). *Omega*2 is the same in both cases. Thick curves correspond to the case when the artificial conducting environment's contribution to current is included: this proves that it is negligible

and in Settings delete two lines under the Expression. In the remaining line, we need to insert z-component of curl **A**, i.e., $u4x - u3y$. Integration should be restricted by the interior surface of the ring with radius $(r0 - r1)$. Thus, the integrand should be $(u4x - u3y) * ((x^2 + y^2) < (r0 - r1)^2)$. As above, before evaluation, one should choose Cut Plane 2 in the Data set of Settings window. After evaluation is finished, Table 3 will appear under Tables in Model Builder. Right-click on Results, call-in 1D Plot Group, right-click on 1D Plot Group 3, and call-in Table Graph. Click on Table Graph 1, and in Settings, choose Table 3. Then you can plot it. You will see from this plot (Fig. 2.56) that the flux is not a flat line, i.e., the compensation is not 100%. That means you are able to discover results that are overlooked in textbooks! I leave any further exploration of the properties of this state to you. You can plot the dependence of the flux as a function of distance from the equatorial plane: for that, all you need is to change the parameter Z-coordinate in the horizontal Cut Plane 2. It is better to introduce a new one, which is called Cut Plane 3, and generate associated Table 4 with the corresponding Table Graph 2 under 1D Plot Group 3. The corresponding flux is also shown in Fig. 2.56. If you explore the problem deep enough, you will find that when the radii $r0$ and $r1$ of the ring become larger, the result (2.42) becomes more and more accurate. You may ask: what was wrong with its textbook derivation? Obviously, Maxwell's equation (2.39) is not something to question. The Stokes' theorem which allowed us to perform the transformation from the surface integral to a linear one is also beyond the doubts. The reason lies in the used boundary condition. When the external magnetic field is changing, the flux through the ring changes, and in accordance with Faraday's induction law, an electromotive force is

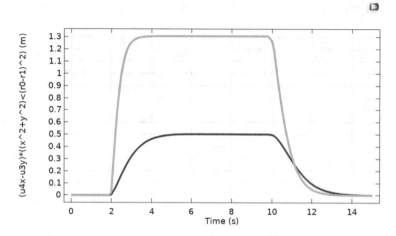

Fig. 2.56 Incomplete zeroing of the flux through the opening of a microsize ring (blue curve). Green curve indicates temporal behavior of the flux away from the ring (computed at the coordinate close to $Z0/2$). With the increase of the ring sizes ($r0$ and $r1$), zeroing is coming in

generated in the ring. The integral $\int \mathbf{E}\cdot\mathbf{dl}$ is this electromotive force, which during the flux variation is non-zero, contrary to the assumption made at the derivation of the relation (2.42). Indeed, for some moments in time, the electric field in the superconductor is not zero.[5] For macroscopic sizes, the overall effect is negligible, but it turns out to be important for sizes in the realm of nanoelectronics!

2.10 Final Remarks on COMSOL Modeling

At this point, we conclude the COMSOL examples in this book. Actually, one can generate endless set of such examples based on TDGL equations. We should emphasize, however, that we used in this Chapter TDGL equations corresponding to the so-called gapless superconductors. More exact correspondence with experimental results can be obtained with the TDGL equations valid for finite gap superconductors. These equations have a similar structure, but are a bit more mathematically complicated. Chapter 7 of Part II of this book explains them. These equations can be combined with COMSOL in the same manner as we described above. At this point, you have gained enough expertise for doing that yourself and, combining creativity and patience, you can obtain many useful enjoyable results. For your convenience,

---

[5]Interested readers can be directed to Sect. 7.4 "Longitudinal Electric Field in Superconductors" of Part II.

and to accelerate your progress, all the COMSOL files described in this book, except the very last one, can be downloaded from the Springer website of this book.

In total, there are 10 uploaded examples. Each of them is in its own folder. The folder contains original COMSOL code (with the extension .mph), as well as the animation in the .avi format (except example 1_DiskLondon which is static). Exe-files for Windows and sh-files for Linux are located in the dedicated folders. These online files generally correspond to the material considered in the book. However, they may slightly differ from the exact files described and constructed in the text. The .mph files require COMSOL Multiphysics version 5.4 or higher to run them. The files in Windows and Linux folders are executable files, which can operate on corresponding platforms. They include both the codes and COMSOL runtime files and thus do not require COMSOL itself to be installed on your computer. When running them, you will have the opportunity to change certain characteristic parameters and obtain your own solutions and create animations. For example, you will be able to switch between superconductors of Type I and Type II. You also will be able to share these files with any other computer users.

Examples of animations are included as well. You can increase the study time and obtain animations which run for a much longer time. One should keep in mind two issues related with the dynamic output of the modeling. (1) Initial pattern as a rule is not yet accurate since the initial conditions are in most of cases impossible to introduce correctly. Thus, the system adjusts itself to correct values based on the equations themselves, and it takes a certain amount of time for that process to finish. Thus, typically, there is an initial disturbance in the system which is unphysical. (2) During computation, the processor cannot compute with an ideal accuracy. There is a small error in solutions, and with time that error accumulates. If you are negligent, the results may become artifacts. So an increase of time should be accompanied with additional steps that account for this fact, for example, refining the mesh.

I will provide some advice for each of the exe-files now.

1_DiskLondon. Default values of geometrical parameters, when changed, sometimes may cause homogeneous blue color of the background. That is related with plotting procedure (occasional division by zero at $x = 0$). That is a minor drawback which I had no opportunity to fix. Keep changing parameters, and the proper coloring scheme will come back.

2_MeissnerTDGL. Click Compute after downloading. When computation is over, you can also generate and export an animation. You can choose the number of frames, and Frames per second. Do not forget to arrange a path for recording in the Filename window. You can change the value of kappa and enjoy the difference between Type I and Type II superconductors.

3_Meissner4Eqs. This is the same problem, with different mathematical approach to the solving procedure as described in the text.

4_Washer-3D. This example may require more computer resources. However, 16Gb of RAM should be enough. Also, computation will take a longer time.

5_PSC. This is demonstration of phase-slip oscillations in a 1D wire. Please note that there is an artificial weak point (called "inhomogeneity") in the code. Its negative values (-p) correspond to weakening of superconductivity. If you choose its amplitude to be 0 instead of the default -0.3, the oscillations will take place at the middle point of the wire. Otherwise, they are shifted towards the inhomogeneity. Have fun!

6_CurrentStrip. Current flows through the strip and generates vortex-antivortex pairs, which annihilate. You can change almost all the relevant parameters and watch the consequences.

7_PhononBasic. This illustrates my statement that the initial pattern often contains an unphysical disturbance which will soon disappear. You will see then how the annihilation of the vortex-antivortex pair creates a spike of the electric field, i.e., generates electromagnetic radiation. Default evolution time is a bit short: increase it to above 150 to see the annihilation event. Keep in mind that the normalized color scale depicts the modulus square of the wave-function, i.e., the Cooper pair density. The file name contains the word "phonon" because there is a phonon flux in the middle of strip orthogonal to the flow of current, weakening superconductivity as in the case of Example #5. You can regulate its effect by changing its amplitude p.

8_StripWithDents. This example illustrates how the imperfect edges of the strip, say, defects created by inaccurate nanolithography, may cause problems in superconducting electronics. Increase the evolution time, and generate your own avi-file to see the annihilation of vortex-antivortex pair. Recall that in accordance with the previous example, such processes generate (unexpected for the electronics engineer!) spikes of voltages.

9_SFQgenerator. Passage of DC current through the SNS junction creates a vortex, and the vortex moves and leaves the junction. SNS junction itself is created by the same "phonon weakening" as above. Leaving the junction creates a voltage spike during this event. SFQ pulse generation is at the heart of superconducting electronics. Unlike previous examples, this one can be regulated. You may change the applied current amplitude and find the threshold needed to move the vortex. That tells us that the pulses can be generated on demand by the applied current.

10_SFQcloning. In this case, the initial current whose presence is mimicked by the magnetic field, is not enough for the excitation of a moving vortex. At some moment in time, at the "Arrival time of SFQ" an additional current pulse is applied. That creates overthreshold conditions and the vortex starts moving, eventually leaving the junction, creating a voltage spike, and thus cloning the SFQ which triggered its motion. You can increase the Arrival time to make the process more distinct.

All these resources are available to everyone, even those who do not possess the COMSOL package. The websites for downloading are mentioned at the bottom of the first pages of Chap. 1 and this Chapter. Needless to say, having basic COMSOL package will allow you to do much more; you can download and use these mph-files, and create your own, perhaps even better, examples. Good luck!

Part II
Derivation of Time-Dependent Ginzburg Landau Equations

Chapter 3
Stationary Ginzburg–Landau Equations

Starting with this Chapter, we will consecutively introduce the basic framework which will eventually allow us to present the derivation of time-dependent Ginzburg–Landau equations. This chapter will deal with static phenomena. After preliminary notions, some of which are repeated for self-consistency from the material already introduced in Part I, we will first derive the set of stationary Ginzburg–Landau equations following the original phenomenological approach. Then, BCS theory will be introduced, using Gor'kov's Green's function formulation. On the basis of this theory, the phenomenological Ginzburg–Landau equations will be justified, and their phenomenological parameters will be computed from the microscopic theory.

3.1 Introductory Concepts

The first main property of superconductors—the flow of electric current without resistance—was discovered by Kamerlingh-Onnes [1] in 1911. For more than 20 years this phenomenon was interpreted as superfluidity, or lossless motion of the electron liquid in metals. The second main property—the total expulsion of the magnetic field out of the bulk superconductor's interior—was discovered by Meissner and Ochsenfeld in 1933. Their half-page report [2] immediately led to very deep insights into the nature of the phenomenon of superconductivity, which eventually allowed bridging the properties of superconductors with quantum mechanics.

We start with a brief discussion of the main superconducting properties. In this section, we sacrifice the historical chronology of superconductivity in favor of introducing some major concepts which will be used in later sections of the book without special explanation.

© Springer Nature Switzerland AG 2020
A. Gulian, *Shortcut to Superconductivity*,
https://doi.org/10.1007/978-3-030-23486-7_3

3.1.1 Infinite Conductivity

In an attempt to understand ideal (infinite) conductivity, one can first refer to Ohm's law $\mathbf{j} = \sigma\mathbf{E}$, and try to handle the infinitely large values of σ. To reach this goal, it is necessary to go back to Ohm's law and eliminate the dissipative term at the initial stage of its derivation. Then one would get "Newton's law" describing the lossless motion of the charge carrier: $m\dot{\mathbf{v}} = e\mathbf{E}$. Combining it with the relation $\mathbf{E} = -(1/c)\,(\partial\mathbf{A}/\partial t)$, and performing the integration, we find:

$$m\mathbf{v} = -\frac{e}{c}\mathbf{A} \tag{3.1}$$

(here the constant of integration is chosen to be zero owing to the appropriate initial condition; also full time derivative in Newton's law is replaced by partial time derivative). To avoid for the moment the difficulties related to the arbitrariness of the gauge for $\mathbf{A}$ in (3.1), one should take a curl of (3.1), bearing in mind the relations curl $\mathbf{A} = \mathbf{B}$, and $\mathbf{j} = en\mathbf{v}$, where n is the carrier density. From this follow the gauge-invariant relations ($\Lambda = m/e^2 n$):

$$\frac{\partial\Lambda\mathbf{j}}{\partial t} = \mathbf{E}, \text{ and } c \text{ curl } \Lambda\mathbf{j} = -\mathbf{B}, \tag{3.2}$$

first introduced by London and London [3], which should replace Ohm's law for superconductors.

3.1.2 Ideal Diamagnetism

Let us now move to the second main property of the superconducting state, namely, to ideal diamagnetism, which is frequently called the Meissner effect. At first glance there is nothing especially surprising in this phenomenon: we know, that application of magnetic field causes the screening currents, which shunt the interior of the conductor and can persist infinitely, if the conductivity is ideal. However, the experiment with magnetic field repulsion may be performed in a different way: the magnetic field is applied initially at sufficiently high temperatures (in the normal state) and after the screening currents have died out, the temperature is lowered below the superconducting transition point $T = T_c$. Such experiments have shown that in superconductors the screening currents arise again after cooling down through T_c and this distinguishes superconductors from ideal conductors. Thus these currents cannot be explained on the basis of classical concepts because the static magnetic field of classical electrodynamics cannot perform work, and consequently cannot produce the circulating screening currents. Formally the Meissner effect can be explained by the relations (3.2)—we already discussed that in Part I of the book. As we concluded there, both ideal conductivity and the Meissner effect are related to the proportionality of the current **j** to the vector potential **A**. This contradicts the classical electrodynamics,[1] in which the current is proportional to the electric field **E**, and provides fair grounds to assign quantum properties to superconductors.

3.1.3 Energy Gap

In 1935 London published insightful arguments elucidating how the Meissner effect is coupled with the possible existence of the gap in the energy spectrum of the charge carriers [7]. Namely, within the quantum mechanical description the current

$$\mathbf{j} = \frac{\hbar e}{2im}(\psi \nabla \psi^* - \psi^* \nabla \psi) - \frac{e^2}{mc}\mathbf{A}|\psi|^2 \tag{3.3}$$

(here $\hbar$ is Planck's constant h divided by 2π) consists of two components: the term, containing $\nabla \psi$ (the "paramagnetic" term) and the term, explicitly proportional to **A** (the "diamagnetic" term). If **A** $\neq 0$, the ψ-function in normal metals acquires dependence on **A** as well, so these two components are of the same order and usually cancel each other to a large extent, so that a weak dia- or paramagnetism occurs, depending on the details of the electronic structure. If one assumes that there is a gap in the energy spectrum associated with the transition to the superconducting state, then in the magnetic field the electronic spectrum of the system will not be changed; the wave function ψ of the state will behave as "rigid": $\psi = \psi_o = \psi(\mathbf{A} = 0)$, so that the paramagnetic term should continue to be zero (as in the case of $\mathbf{A} = 0$), while the diamagnetic term should provide the main response.

The presence of a gap in the energy spectrum will make the creation of single-particle excitations impossible, providing the non-dissipative motion alike of described by (3.1).

In normal metals the spectrum of elementary excitations of electrons with momenta in the vicinity of p_F has the form

$$\xi_\mathbf{p} = v_F(p - p_F), \tag{3.4}$$

which follows straightforwardly in the parabolic band approximation $\xi_\mathbf{p} = p^2/2m - \epsilon_F$, where ϵ_F is the Fermi energy: $\epsilon_F = p_F^2/2m = mv_F^2/2$ (the same type of relation as (3.4) can be justified in more general cases). Evidently, for normal Fermi-liquids there is no gap in the energy spectrum. Let us suppose that in superconducting state

---

[1]Classical electrodynamics is based on Faraday's concept of local influence of the electromagnetic fields on charges. Meanwhile, for a long enough solenoid (one can even realize "infinitely long" option in the toroidal geometry—the magnetic field **H** outside of the solenoid is absent though in a wire looping the solenoid a current will start flowing when the loop will be cooled down to the superconducting state! As was pointed out by Aharonov and Bohm [4] (see also [5]), the quantum objects can "sense" the field potentials **A** and φ while the values of **E** and **H** are zero. That fact allowed Feynman to declare that the real physical fields are not **E** and **H**, but rather **A** and φ [6].

the excitation spectrum possesses a gap-like peculiarity:

$$\varepsilon_{\mathbf{p}} = \sqrt{\xi_{\mathbf{p}}^2 + |\Delta|^2}. \tag{3.5}$$

In the presence of this gap $|\Delta|$ the birth of single excitations with small energies is impeded. In Sect. 3.3 we will see that the microscopic theory indeed leads to the spectrum of the type (3.5). Here we discuss some of the consequences that follow from (3.5).

3.1.4 Bogolyubov–De Gennes Equations: Analogy with Relativistic Quantum Theory

The spectrum (3.5) has an analog in the relativistic quantum theory [8], where the electron energy is $E_p = \sqrt{\mathbf{p}^2 + m_0^2}$ (m_0 is the electron rest mass, $c = 1$). One can try to reconstruct the wave function of the particle having the spectrum (3.5), by writing down the stationary Schrödinger equation (using the equivalent Hamiltonian method; this is sometimes called the equivalent mass approach) [9]:

$$\hat{\epsilon}(-i\hbar\nabla)\psi = \varepsilon\psi. \tag{3.6}$$

As was done by Dirac in the relativistic quantum theory (see, e.g., [10]), one can extract the square root from the operator by linearizing it (here and below $\hbar = 1$):

$$\hat{\epsilon}(-i\nabla) = \widehat{\alpha}\left(-\frac{1}{2m}\nabla^2 - \epsilon_F\right) + \widehat{\beta}|\Delta| \tag{3.7}$$

where $\widehat{\alpha}$ and $\widehat{\beta}$ are some mathematical objects to be identified by squaring the operator (3.7). In the absence of external fields we should obtain the spectrum (3.5). This leads to the relations

$$\widehat{\alpha}^2 = 1, \quad \widehat{\beta}^2 = 1, \tag{3.8}$$
$$\widehat{\alpha}\widehat{\beta} + \widehat{\beta}\widehat{\alpha} = 0. \tag{3.9}$$

In the simplest case $\widehat{\alpha}$ and $\widehat{\beta}$ are not numbers, but they may be expressed as superpositions of the Pauli matrices [the unity matrix $\widehat{1}$ does not participate in these linear combinations, as is seen from (3.9)]:

$$\widehat{\alpha} = {\sum}' a_i\widehat{\sigma}_i; \quad \widehat{\beta} = {\sum}' b_k\widehat{\sigma}_k, \tag{3.10}$$

where the prime denotes $i \neq k$ according to (3.9), and (3.8) gives

$$\sum a_i^2 = \sum b_k^2 = 1. \tag{3.11}$$

Thus the wave function in (3.6) is a one-column matrix:

$$\widehat{\psi} = \begin{pmatrix} \mathcal{U} \\ \mathcal{V} \end{pmatrix}. \tag{3.12}$$

One can demand now, that in the case of normal metal ($\Delta = 0$) the components $\mathcal{U}$ and $\mathcal{V}$ should become decoupled. This means that in the composition (3.10) for $\widehat{\alpha}$ only the coefficient at the matrix

$$\widehat{\sigma}_z = \begin{pmatrix} 1 & 0 \\ 0 & -1 \end{pmatrix} \tag{3.13}$$

will not vanish. In view of relations (3.10) and (3.11), this leads to

$$\widehat{\beta} = \begin{pmatrix} 0 & e^{i\theta} \\ e^{-i\theta} & 0 \end{pmatrix}, \tag{3.14}$$

where θ is the (real) phase factor. Introducing the notation

$$\Delta = |\Delta| e^{i\theta} \tag{3.15}$$

we represent (3.6) in the form

$$\varepsilon \mathcal{U} = -\left(\frac{\nabla^2}{2m} + \epsilon_F \right) \mathcal{U} + \Delta \mathcal{V}, \tag{3.16}$$

$$\varepsilon \mathcal{V} = \left(\frac{\nabla^2}{2m} + \epsilon_F \right) \mathcal{V} + \Delta^* \mathcal{U}. \tag{3.17}$$

In the presence of a magnetic field[2] these equations become

$$\varepsilon \mathcal{U}(\mathbf{r}) = \left[\frac{1}{2m} \left(-i \frac{\partial}{\partial \mathbf{r}} - \frac{e}{c} \mathbf{A} \right)^2 - \epsilon_F \right] \mathcal{U}(\mathbf{r}) + \Delta \mathcal{V}(\mathbf{r}), \tag{3.18}$$

$$\varepsilon \mathcal{V}(\mathbf{r}) = -\left[\frac{1}{2m} \left(-i \frac{\partial}{\partial \mathbf{r}} + \frac{e}{c} \mathbf{A} \right)^2 - \epsilon_F \right] \mathcal{V}(\mathbf{r}) + \Delta^* \mathcal{U}(\mathbf{r}). \tag{3.19}$$

---

[2]The magnetic field may be introduced by generalization of the standard method [11]. One can start from the Lagrangian $\widehat{L}(\mathbf{r}, \dot{\mathbf{r}}) = \widehat{\alpha} m \dot{\mathbf{r}}^2 / 2 - \widehat{\beta} |\Delta| + \widehat{\alpha} e \dot{\mathbf{r}} \cdot \mathbf{A}/c$. The scalar potential φ can be included in ϵ_F. This Lagrangian gives the correct expression for the Lorenz force acting on the electron in normal metal (when $\Delta = 0$). Thus in presence of a magnetic field we have the Hamiltonian $\widehat{H} = \mathbf{p} \cdot \dot{\mathbf{r}} - \widehat{L}$, where $\mathbf{p} = \partial \widehat{L} / \partial \dot{\mathbf{r}} = \widehat{\alpha} (m \dot{\mathbf{r}} + e\mathbf{A}/c)$, i.e., $\widehat{H} = \widehat{\alpha} (\widehat{\alpha} \mathbf{p} - \widehat{1} e\mathbf{A}/c)^2 / 2m - \widehat{\alpha} \epsilon_F + \widehat{\beta} |\Delta|$, which yields (3.18) and (3.19).

This system of (3.18) and (3.19) is called the Bogolyubov–De Gennes equations [12].

3.1.5 Andreev Reflection

For stationary states we consider the wave function in the Schrödinger representation, which includes the factor $\exp(-i\varepsilon t)$. Separating also the fast-oscillating factors in the wave function:

$$\widehat{\psi}(\mathbf{r}, t) = \exp(-i\varepsilon t)\widehat{\psi}_{p_F}(\mathbf{r})\exp(i\mathbf{p}_F \cdot \mathbf{r}) = \widehat{\psi}_{p_F}(\mathbf{r}, t)\exp(i\mathbf{p}_F \cdot \mathbf{r}) \qquad (3.20)$$

and restoring the time-differentiation operators, one can transform (3.18), (3.19) to the form:

$$\left(i\frac{\partial}{\partial t} - \widehat{\alpha}e\varphi\right)\widehat{\psi}_{p_F}(\mathbf{r}, t)$$
$$= \left(-i\widehat{\alpha}\mathbf{v}\cdot\frac{\partial}{\partial \mathbf{r}} + \frac{e}{c}\mathbf{v}\cdot\mathbf{A}\right)\widehat{\psi}_{p_F}(\mathbf{r}, t) + |\Delta|\widehat{\beta}\widehat{\psi}_{p_F}(\mathbf{r}, t). \qquad (3.21)$$

Thus the analogy between the particle spectra has led to an apparent parallel between (3.18), (3.19) and the Dirac equation [8, 10]. There are numerous consequences from (3.21) [or, equivalently, from (3.18) and (3.19)] that are analogous to relativistic quantum effects.[3] We consider the most prominent of them: the so-called Andreev reflection.

In the absence of external fields, the $\widehat{\psi}$-function may be taken as real (without loss of generality), and using the normalization

$$\text{Tr }\widehat{\psi}^2 = \mathcal{U}^2 + \mathcal{V}^2 = 1, \qquad (3.22)$$

one finds from (3.22), (3.16), and (3.17):

$$\mathcal{U}^2 = \frac{1}{2}\left(1 + \frac{\xi}{\varepsilon}\right), \qquad \mathcal{V}^2 = \frac{1}{2}\left(1 - \frac{\xi}{\varepsilon}\right). \qquad (3.23)$$

According to (3.22), the charge carrier with spectrum (3.5) is in a superposition of states, having the probability amplitudes $\mathcal{U}$ and $\mathcal{V}$. As follows from (3.23), this superposition is essential in the energy range $\xi \sim |\Delta|$. For $\xi \gg |\Delta|$ we have from (3.22) and (3.23): $\psi_{p>p_F}^2 \approx \mathcal{U}^2$; this is an electron-like excitation. For $\xi \ll -|\Delta|$

---

[3] We note that there is not a one to one correspondence between effects in superconductors described by the Bogolyubov–De Gennes equations and the physics of the Dirac equation. Even formally there are pronounced differences between these equations. Due to them, for example, the quasiparticle has no magnetic moment associated with "zitterbewegung" [13] of electrons in superconductors.

we have $\psi^2_{p<p_F} \approx \mathcal{V}^2$; this is a hole-like excitation. Because the charges of these excitations have the opposite signs, one has for the charge of quasiparticles, described by the $\widehat{\psi}$-function (3.12), the expression [14]:

$$q_* = \mathcal{U}^2 - \mathcal{V}^2 = \xi/\varepsilon \qquad (3.24)$$

(the electron charge is unity). It follows that quasiparticle charge, as well as its group velocity

$$\mathbf{v}_g = \frac{\partial \varepsilon}{\partial \mathbf{p}} = \frac{\xi}{\varepsilon} \frac{\mathbf{p}}{m} = q_* \mathbf{v}, \qquad (3.25)$$

vanishes (and reverses its sign) at $p = p_F$.

Consider now the propagation of such a particle in a medium, where $|\Delta(\mathbf{r})|$ is a spatially inhomogeneous function. For instance, $|\Delta(\mathbf{r})|$ may increase smoothly from zero to Δ_0 at the boundary between normal and superconducting phases. Let the particle be moving from a normal to a superconducting region, with its energy ξ_p obeying the relation $0 < \varepsilon_p = \xi_p < \Delta_0$ in the normal region. In the superconducting region, as can be seen from (3.5), somewhat smaller values of $\xi_{p'}$ and consequently smaller values of p', correspond to the same energy ε_p. At the point where $|\Delta(\mathbf{r})| = \varepsilon_p$, we have $p = p_F$, and according to (3.24) and (3.25), the particle should stop and be reflected. The group velocity in this case reverses its sign, but the momentum retains. This means that the reflected particle reverses its charge [see (3.24)], i.e., the reflected electron excitation becomes a hole. In the relativistic theory, this phenomenon is known as the Klein paradox [15]. In the superconductivity theory, it corresponds to the Andreev reflection [16]. The Andreev reflection takes place when a current flows across the boundary between a normal metal and a superconductor. This process was demonstrated experimentally by the radio-frequency size effect [17]. As can be seen from Fig. 3.1, the specifics of such a reflection allow one to fit the electron's trajectory (having a diameter D in presence of a magnetic field H_0) within the normal metal layer of thickness $d = D/2$. Evidently if the film borders a vacuum, the minimal value of the field is $H = H_0$, but for a film deposited on a superconductor (see Fig. 3.1) the closed trajectory is achievable in the field $H = H_0/2$ [17]. This example demonstrates one of the remarkable kinetic properties caused by the nature of superconductivity.[4]

3.1.6 Electron Density of States

There are two additional points. The first relates to the density of the energy levels of quasiparticles having the spectrum of (3.5). In a description of normal metals, the

---

[4]It is recognized nowadays that the Andreev reflection plays a major role in "our ability to insert current into a superconductor" [18]. The related physics is very important for various fundamental and applied problems of superconductivity [19–28].

Fig. 3.1 Experimental
evidence for Andreev
reflection: scattering on the
normal-metal
(N)–superconductor (S)
boundary transforms
"electron" into "hole" and
vice versa. Closed
trajectories are possible at
half amplitudes of the
applied magnetic field **H**
compared with the case of a
free-standing single film

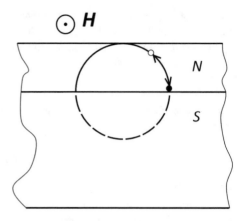

density of the levels can be obtained when one passes from the momentum summation
to the energy integration:

$$\Sigma(\ldots) \to \int (\ldots) \frac{d^3\mathbf{p}}{(2\pi)^3} \approx \frac{mp_F}{2\pi^2} \int (\ldots) d\xi \equiv N(0) \int (\ldots) d\xi, \qquad (3.26)$$

so for the momenta $p \approx p_F$, the levels' density is a constant $N(0) = mp_F/2\pi^2$.
However, in superconductors, the levels' density is a singular function

$$N(0) \int (\ldots) d\xi \to N(0) \int (\ldots) \frac{\partial \xi}{\partial \varepsilon} d\varepsilon = N(0) \int (\ldots) \frac{\varepsilon \theta(\varepsilon^2 - |\Delta|^2)}{\sqrt{\varepsilon^2 - |\Delta|^2}} d\varepsilon. \quad (3.27)$$

Thus, when there is a gap in the excitation spectrum, the energy levels (or states) of
quasiparticles are pushed out from the intra-gap into the gap-edge region of energies
$|\varepsilon| \gtrsim |\Delta|$. This singularity of superconducting energy levels will play an important
role in further discussions.

3.1.7 Coherence Factors

The second point relates to the coherence factors. The form of the wave function
(3.12) indicates that in calculations of the matrix elements connected with the transition of quasiparticles from level ε to level ε', combinations of the type

$$\mathcal{U}(\varepsilon)\mathcal{V}(\varepsilon') \pm \mathcal{V}(\varepsilon)\mathcal{U}(\varepsilon'), \quad \mathcal{U}(\varepsilon)\mathcal{U}(\varepsilon') \pm \mathcal{V}(\varepsilon)\mathcal{V}(\varepsilon') \qquad (3.28)$$

would appear, depending on the form of the interaction operator. The squared quantities (which define the corresponding transition probabilities) would be, for example,
$(\mathcal{U}\mathcal{U}' + \mathcal{V}\mathcal{V}')^2$. For the last quantity one obtains, after a simple calculation taking

into account (3.23):

$$[\mathcal{U}(\varepsilon)\mathcal{U}(\varepsilon') + \mathcal{V}(\varepsilon)\mathcal{V}(\varepsilon')]^2 = \frac{1}{2}\left(1 + \frac{\xi\xi'}{\varepsilon\varepsilon'} + \frac{|\Delta|^2}{\varepsilon\varepsilon'}\right). \qquad (3.29)$$

Other combinations in (3.28) lead to analogous relations, differing only in the signs in the parentheses in (3.29). These factors are called "coherence factors". They renormalize the transition matrix elements in superconductors relative to those in normal metals.

For the processes that are symmetric over the electron-hole excitation branches, the odd terms in $\xi\xi'$ in (3.29) disappear and the coherence factors are of only two types:

$$(1 + |\Delta|^2/\varepsilon\varepsilon') \quad \text{and} \quad (1 - |\Delta|^2/\varepsilon\varepsilon'). \qquad (3.30)$$

Note that for $\varepsilon, \varepsilon' \sim |\Delta|$, the transition probability doubles for the first factor and vanishes for the second one. This is important, because the states with $\varepsilon \sim |\Delta|$ play an essential role in kinetic processes, as may be expected from the peculiarity in the density of states (3.27).

3.2 Phenomenological GL Theory: Triumph and Limits of Human Imagination

The Ginzburg–Landau (GL) theory permits a deep insight into the phenomenon of superconductivity in the case of thermodynamic equilibrium and provides the most transparent technique for investigating this phenomenon. Developed before the microscopic theory of superconductivity, its predictive power is tremendous and can be qualified as the triumph of human imagination. However, it is puzzling that the authors did not make one more step, and discover the Josephson effects in the way we did in Part I. Even a genius' imagination is sometimes limited! We continue our discussion with the presentation of this theory [29].

In formulating this theory, the experimental knowledge that the superconducting transition is a second order phase transition was used. We will remind that the first order phase transitions (e.g., melting of crystals or evaporation of liquids) are related with the discontinuities of thermodynamic potentials (such as the free energy). In case of the first order phase transitions the derivatives of thermodynamic potentials are divergent at the transition temperature. In case of second order phase transitions the thermodynamic functions are continuous while their derivatives are discontinuous. Accordingly, GL assumed that below the phase transition temperature T_c all the electrons of the superconducting metal can be characterized by a superconducting order parameter $\Psi \neq 0$ at $T < T_c$ and $\Psi = 0$ at $T > T_c$. On intuitive grounds, the order parameter Ψ was considered as the "effective wave function of superconducting electrons".

3.2.1 Free Energy Functional

According to the theory of second order phase transitions [30] the free energy of superconducting state in the vicinity of T_c may be presented as a functional of the complex variable Ψ, permitting the expansion (we make for a moment $\mathbf{H} = 0$):

$$F_s^0 = F_n^0 + \alpha |\Psi|^2 + \frac{\beta}{2} |\Psi|^4 + \cdots \qquad (3.31)$$

(the terms proportional to Ψ and Ψ^* do not enter this expansion in view of the gauge invariance of the free energy). In expression (3.31) F_n^0 is the free energy of the normal phase. At fixed temperature $T < T_c$, the free energy (3.31) is minimized by the value of $|\Psi(T)|$, which can be found from equation

$$\frac{\partial F_s^0}{\partial |\Psi|^2} = 0, \qquad (3.32)$$

subject to the condition

$$\frac{\partial^2 F_s^0}{(\partial |\Psi|^2)^2} > 0. \qquad (3.33)$$

From (3.32) and (3.31) one finds

$$|\Psi|^2 \equiv |\Psi_0|^2 = -\frac{\alpha}{\beta}, \qquad (3.34)$$

i.e., the factors α and β have opposite signs at $T < T_c$. From the condition (3.33) it follows that

$$\beta_c = \beta(T_c) > 0 \qquad (3.35)$$

and combining this with (3.34), we find

$$\alpha_c = \alpha(T_c) = 0. \qquad (3.36)$$

According to this for temperatures near T_c it could be written

$$\alpha(T) = \left(\frac{d\alpha}{dT}\right)_{T_c} (T - T_c), \qquad (3.37)$$

$$\beta(T) = \beta_c. \qquad (3.38)$$

Based on these expressions one can conclude that in the case of thermodynamic equilibrium at $T \leq T_c$

$$|\Psi_0|^2 = -\frac{1}{\beta_c}\left(\frac{d\alpha}{dT}\right)_{T_c}(T - T_c),\tag{3.39}$$

$$F_s^0 = F_n^0 - \frac{\alpha^2}{2\beta} = F_n^0 - \frac{1}{2\beta_c}\left(\frac{d\alpha}{dT}\right)_{T_c}^2(T - T_c)^2.\tag{3.40}$$

Let us now consider a superconductor that is placed in a static magnetic field $\mathbf{H}(\mathbf{r})$. The free energy F_s^H must have an additional term that is equal to the field energy $\mathbf{B}^2/8\pi$. Accordingly, the critical magnetic field in a spatially homogeneous case may be found from the equation

$$\frac{H_c^2}{8\pi} = F_n^0 - F_s^0 = \frac{\alpha^2}{2\beta}.\tag{3.41}$$

Generally, one must also take into account the energy, which is proportional to the inhomogeneity of the wave function $|\nabla\Psi|^2$. Thus at small gradients,

$$F_s^H = F_s^0 + \frac{\mathbf{B}^2}{8\pi} + \text{const }|\nabla\Psi|^2.\tag{3.42}$$

The last term in (3.42) corresponds to quantum-mechanical kinetic energy. Hence there are reasons to represent it in the form

$$\frac{\hbar^2}{2m_*}|\nabla\Psi|^2 = \frac{1}{2m_*}|-i\hbar\nabla\Psi|^2,\tag{3.43}$$

where m_* is some coefficient having the dimensionality of a mass. To include the magnetic field in the scheme, it is necessary to make in (3.43) the usual quantum-mechanical substitution

$$-i\hbar\nabla \longrightarrow -i\hbar\nabla - \frac{e_*}{c}\mathbf{A},\tag{3.44}$$

which enables one to obtain the gauge invariant equations. Here $\mathbf{A}$ is the magnetic field's vector-potential, and e_* is the charge of the carrier, represented by the wave function Ψ. Thus, the free energy density may be written in the form

$$F_s^H = F_s^0 + \frac{\mathbf{B}^2}{8\pi} + \frac{1}{2m_*}\left|-i\hbar\nabla\Psi - \frac{e_*}{c}\mathbf{A}\Psi\right|^2.\tag{3.45}$$

Demanding the minimum for the total free energy

$$\mathcal{F}^H = \int_{V_0} F_s^H d^3\mathbf{r}\tag{3.46}$$

(V_0 is the system's volume), one can obtain the equation for Ψ. Varying (3.46) by Ψ^* gives

$$\delta \mathcal{F}^H = \int_V \left\{ \frac{1}{2m_*} \left(-i\hbar\nabla - \frac{e_*}{c}\mathbf{A} \right)^2 \Psi + \alpha\Psi + \beta|\Psi|^2\Psi \right\} \delta\Psi^* d^3\mathbf{r}$$

$$+ \frac{\hbar^2}{2m_*} \int_S \delta\Psi^* \left(\nabla\Psi - \frac{ie_*}{\hbar c}\mathbf{A}\Psi \right) \cdot d\mathbf{s} \qquad (3.47)$$

(S is the metal's surface). Because $\delta\Psi^*$ is arbitrary, we find from (3.47) an equation for the order parameter

$$\frac{1}{2m_*} \left(-i\hbar\nabla - \frac{e_*}{c}\mathbf{A} \right)^2 \Psi + \alpha\Psi + \beta|\Psi|^2\Psi = 0 \qquad (3.48)$$

and also the boundary condition

$$\mathbf{n} \left(-i\hbar\nabla\Psi - \frac{e_*}{c}\mathbf{A}\Psi \right) = 0 \qquad (3.49)$$

(here $\mathbf{n}$ is a vector normal to the metal surface). The variation of (3.46) by $\mathbf{A}$ yields the Maxwell equation

$$\operatorname{curl} \operatorname{curl} \mathbf{A} = \frac{4\pi}{c}\mathbf{j}, \qquad (3.50)$$

where the current

$$\mathbf{j} = -\frac{ie_*\hbar}{2m_*}(\Psi^*\nabla\Psi - \Psi\nabla\Psi^*) - \frac{e_*^2}{m_*c}|\Psi|^2\mathbf{A} \qquad (3.51)$$

has a typical quantum-mechanical form (3.3).

Expressions (3.48) and (3.51) comprise the Ginzburg–Landau system of equations describing the behavior of superconductors in a static magnetic field. Presenting the complex function Ψ in the form

$$\Psi = |\Psi|e^{i\theta}, \qquad (3.52)$$

one can rewrite the expression for the current:

$$\mathbf{j} = \frac{\hbar e_*}{m_*}|\Psi|^2 \left(\nabla\theta - \frac{e_*}{\hbar c}\mathbf{A} \right) = e_*N_s\mathbf{v}_s. \qquad (3.53)$$

Here $|\Psi|^2$ is the "density of superconducting electrons" in the Ginzburg–Landau normalization, $|\Psi|^2 \equiv N_s$, and the superconducting velocity $\mathbf{v}_s$ is equal to

$$\mathbf{v}_s = \frac{\hbar}{m_*}\left(\nabla\theta - \frac{e_*}{\hbar c}\mathbf{A}\right). \tag{3.54}$$

Putting $\mathbf{B} = \text{curl }\mathbf{A}$, one can easily prove that (3.53) coincides with the London equation[3]

$$\text{curl }\mathbf{j} = -\frac{e_*^2 N_s}{m_* c}\mathbf{B}. \tag{3.55}$$

Substituting into (3.55) the Maxwell equation

$$\text{curl }\mathbf{B} = \frac{4\pi}{c}\mathbf{j} \tag{3.56}$$

and taking into account that

$$\text{div }\mathbf{B} = 0 \tag{3.57}$$

we obtain the equation

$$\nabla^2\mathbf{B} = \frac{4\pi e_*^2 N_s}{m_* c^2}\mathbf{B}. \tag{3.58}$$

3.2.2 London Penetration Depth

Equation (3.58) subject to condition (3.57) describes the expulsion of a magnetic field from superconductor's interior (the Meissner effect). Let us consider the distribution of a magnetic field in a superconductor near its surface, assuming the latter to be a plane. The characteristic parameter, which has the dimension of a length, in this situation is

$$\lambda_L = \left(\frac{4\pi e_*^2 N_s}{m_* c^2}\right)^{-1/2}, \tag{3.59}$$

as may be seen from expression (3.58). In the case considered here, the field distribution depends on one (say, x) coordinate only. Then

$$\frac{d^2\mathbf{B}}{dx^2} = \frac{\mathbf{B}}{\lambda_L^2} \tag{3.60}$$

with the boundary condition

$$\frac{d\mathbf{B}_x}{dx} = 0, \tag{3.61}$$

which follows from (3.57). From the expressions (3.61) and (3.57) one can conclude that the vector of induction $\mathbf{B}$ in the depth of the superconductor has the form

$$\mathbf{B}(x) = \mathbf{B}_0 e^{-x/\lambda_L}, \tag{3.62}$$

where $\mathbf{B}_0$ is a tangential component of the external field. The characteristic length λ_L (3.59) is called the "London penetration depth".

3.2.3 Coherence Length

The Ginzburg–Landau set of equations has one more characteristic scale, which has the dimensionality of the length. Its value [usually marked $\xi(T)$], as may be seen from the (3.48), characterizes the scale of spatial evolution of Ψ-function and is given by

$$\xi^{-2}(T) = \frac{2m_*|\alpha(T)|}{\hbar^2}. \tag{3.63}$$

The temperature dependence of $\xi(T)$ in the vicinity of T_c is found using the formula (3.37):

$$\xi(T) = \hbar \left\{ 2m_* \left(\frac{d\alpha}{dT} \right)_{T_c} (T_c - T) \right\}^{-1/2}. \tag{3.64}$$

We are able also to obtain the temperature dependence of $\lambda_L = \lambda_L(T)$. Indeed, in the vicinity of T_c one can substitute (3.39) into (3.59) with the result:

$$\lambda_L(T) = \left\{ \frac{m_* c^2 \beta_c}{4\pi e_*^2 (d\alpha/dT)_{T_c}(T_c - T)} \right\}^{1/2}. \tag{3.65}$$

The ratio of these two characteristic lengths

$$\kappa = \frac{\lambda_L(T)}{\xi(T)} = \frac{m_* c}{\hbar e_*} \frac{\beta_c^{1/2}}{(2\pi)^{1/2}} \tag{3.66}$$

is temperature independent and, as we have seen in Part I, is an important parameter of a superconductor, defining its behavior in a magnetic field. We will understand the reasons now.

3.2.4 Sign of Surface Energy

Let us consider the surface energy of the flat boundary between normal and super-conducting phases, which may exist in a magnetic field. In the normal phase, the free energy density, including the field energy, is equal to $F_n^0 + H_c^2/(8\pi)$. In the region where $\Psi \neq 0$, the free energy density is F_s^H (3.42). Near the boundary one must take into account the energy associated with the magnetization of a superconductor

$$\mathbf{M} = \frac{\mathbf{B} - \mathbf{H}}{4\pi}. \tag{3.67}$$

In the depth of the normal phase, the equation $\mathbf{B} = \mathbf{H} = \mathbf{H}_c$ holds (the second of these equations is also valid in the *depth* of bulk superconducting region, because of $c\,\mathrm{curl}\,\mathbf{M} = \mathbf{j}$). Thus the surface energy

$$\sigma_{ns} = \int \left\{ F_s^H(z) - \frac{B(z) - H_c}{4\pi} H_c - F_n^0 - \frac{B^2}{8\pi} \right\} dz \tag{3.68}$$

taking into account (3.45), is equal to

$$\sigma_{ns} = \int \left\{ \alpha \Psi^2 + \frac{\beta \Psi^4}{2} + \frac{\alpha^2}{2\beta} + \frac{\hbar^2}{2m_*} \left(\frac{\partial \Psi}{\partial z} \right)^2 \right.$$
$$\left. + \frac{e_*^2}{2m_* c^2} A^2 \Psi^2 + \frac{B^2}{8\pi} - \frac{H_c B}{4\pi} \right\} dz. \tag{3.69}$$

[according to (3.41), $\alpha^2/2\beta = H_c^2/(8\pi)$]. In (3.69), the Ψ-function was assumed to be real, because the term $i\mathbf{A} \cdot \nabla \Psi$ vanishes owing to the condition $\mathbf{A}_z = 0$. The analogous term also disappears from (3.48) for the order parameter, so (3.53) for the current acquires the form

$$\mathbf{j} = -\frac{e_*^2}{m_* c} \Psi^2 \mathbf{A}. \tag{3.70}$$

It is expedient to use the variables

$$\bar{z} = \frac{z}{\lambda_L}, \quad \bar{\Psi} = \Psi \left(\frac{\beta}{|\alpha|} \right)^{1/2}, \quad \bar{A} = \frac{A}{H_c \lambda_L}, \quad \bar{B} = \frac{d\bar{A}}{d\bar{z}} = \frac{B}{H_c}. \tag{3.71}$$

Removing the bars above the symbols to simplify the notation, we present (3.48) in the form

$$\Psi'' = \kappa^2 \left[\left(\frac{1}{2} A^2 - 1 \right) \Psi + \Psi^3 \right] \tag{3.72}$$

and write expression (3.69) as

$$\sigma_{ns} = \frac{\lambda_L H_c^2}{8\pi} \int_{-\infty}^{\infty} \left[\frac{2}{\kappa^2} (\Psi')^2 + (A^2 - 2)\Psi^2 + \Psi^4 + (A' - 1)^2 \right] dz. \tag{3.73}$$

Integrating the first term in (3.73) by parts, using the condition $\Psi'(\pm\infty) = 0$, one can reduce (3.73) to the form

$$\sigma_{ns} = \frac{\lambda_L H_c^2}{8\pi} \int_{-\infty}^{\infty} [(A' - 1)^2 - \Psi^4] dz. \tag{3.74}$$

Expression (3.74) vanishes if the integrand is identical to zero: $A' - 1 = \pm \Psi^2$. Since $B = A'$ and B must decrease with increasing z, then

$$A' - 1 = -\Psi^2. \tag{3.75}$$

From (3.50) and (3.70) it follows that

$$A'' = A\Psi^2. \tag{3.76}$$

Taking the derivative of (3.75) and introducing it into (3.76), we find

$$\Psi' = -\frac{1}{2}A\Psi. \tag{3.77}$$

Substituting Ψ' (3.77) and A' (3.75) into (3.72) shows that (3.72) is fulfilled identically at $\kappa = 1/\sqrt{2}$. One can also see, that the condition (3.75) which was used earlier does not contradict the boundary conditions: $A' = 1$ at $z = -\infty$ and $A = 0$ at $z = \infty$. Thus we arrive at a very important conclusion: at $\kappa = 1/\sqrt{2}$ the solution of order parameter equation causes the surface energy to vanish (this criterion was established numerically by Ginzburg and Landau [29] and proven analytically [12] by the Sarma method; the alternative method we used here is by Lifshitz and Pitaevskii [31]). Generally, σ_{ns} may have an arbitrary sign [32]. To see this we will once again use (3.73), as well as the first integral of (3.72) subject to (3.76), which has the form[5]

$$\frac{2(\Psi')^2}{\kappa^2} + (2 - A^2)\Psi^2 - \Psi^4 + (A')^2 = \text{const} = 1. \tag{3.78}$$

As a result we find

$$\sigma_{ns} = \frac{\lambda_L H_c^2}{4\pi} \int_{-\infty}^{\infty} \left[\frac{2}{\kappa^2}(\Psi')^2 + A'(A' - 1) \right] dz. \tag{3.79}$$

Note that the second term in (3.79) is always negative, because the field $B = A'$ which penetrates into the superconductor, is always smaller than the critical one: $A' \leq 1$ (or, otherwise, superconductivity will vanish). The first term in (3.79) is always positive, but its value and consequently the sign of σ_{ns} are determined by the magnitude of κ [see (3.66)]. Superconductors, in which

$$\lambda_L < \xi/\sqrt{2} \tag{3.80}$$

are called type-I superconductors, or Pippard superconductors. In these superconductors, as we have seen, $\sigma_{ns} > 0$. Superconductors, in which

---

[5]The last identity here follows from the equation (3.76) and boundary conditions $\Psi(-\infty) = 0$, $A'(-\infty) = 1$; $\Psi(\infty) = 1$, $A'(\infty) = 0$.

$$\lambda_L > \xi/\sqrt{2} \tag{3.81}$$

are called type-II superconductors, or London superconductors. In these superconductors surface energy is negative: $\sigma_{ns} < 0$. This explains our modeling results in Part I. Appearance of vortices with normal cores introduces internal $ns-$boundaries thus reducing the free energy of superconductors at values of $\kappa > 1\sqrt{2}$.

3.3 BCS-Gor'kov Theory

The basic cornerstone of the microscopic theory of superconductivity was laid down by Cooper [33] in 1956. Cooper considered the indirect (mediated by the phonon exchange) interaction between electrons in metals; this is a process of the second order in electron-phonon interaction. As is known from perturbation theory, the second order correction to the energy of the ground state is always negative (see, e.g., [34]), i.e., the Cooper interaction is attractive. Because the Fermi sphere at low temperatures is almost completely occupied, the motion of conducting electrons in the momentum space is quasi-two-dimensional. This means that any weak attraction between electrons produces the bound state, or leads to electrons pairing. In the absence of total current, the electrons with opposite momenta have the largest pairing probabilities. The paired electrons become bosons, with the spin equal to 0 or 1. The electron system must have rearranged itself (because the paired state is energetically preferable), forming the Bose condensate of paired electrons (the Cooper", or "pair" condensate). The properties of the Cooper condensate are typical for all the Bose condensates. In particular, at temperatures lower than the condensation temperature T_c, the occupation number of paired states with zero momentum is macroscopically large. This means that in presence of the pair condensate, the anomalous components of Green's functions should be introduced into the theoretical description. Such a generalization of the theoretical scheme was made by Belyayev [35] in the theory of superfluidity, and the concept of off-diagonal long-range order was developed even earlier (see discussion in [36]). In the theory of superconductivity, this generalization was introduced by Gor'kov [37] and in a slightly different way by Nambu (see, e.g., [38]). The microscopic description of the superconductor in terms of these formulas is fully adequate to BCS theory and, being considerably simpler, allows one to avoid all the problems connected with the gauge-invariance of the theoretical scheme.

3.3.1 Equations for Ψ-Operators

We start with the BCS-Gor'kov model, considering first the case of $T = 0$. The model Hamiltonian has the form ($c = \hbar = 1$):

$$\widehat{H} = \sum_{\alpha\beta} \int \left\{ -\left[\Psi_\alpha^\dagger(\mathbf{r}) \left(\frac{\nabla^2}{2m} + \epsilon_F \right) \Psi_\alpha(\mathbf{r}) \right] \right.$$

$$\left. + \frac{\zeta}{2} \left[\Psi_\alpha^\dagger(\mathbf{r}) \Psi_\beta^\dagger(\mathbf{r}) \Psi_\beta(\mathbf{r}) \Psi_\alpha(\mathbf{r}) \right] \right\} d^3\mathbf{r}, \tag{3.82}$$

where $\Psi(\mathbf{r})$ and $\Psi^\dagger(\mathbf{r})$ are the field operators in the Schrödinger representation (from now on the repeated spin indices imply summation and the symbol $\sum$ will be omitted). The BCS-potential ζ corresponds to the indirect interaction of electrons. Let us move to the Heisenberg representation, where operators Ψ and $\Psi^\dagger$ are the functions of $x \equiv (\mathbf{r}, t)$ and obey the equations

$$\left\{ i\frac{\partial}{\partial t} + \frac{\nabla^2}{2m} + \epsilon_F \right\} \Psi_\alpha(x) - \zeta \left(\Psi_\beta^\dagger(x) \Psi_\beta(x) \right) \Psi_\alpha(x) = 0, \tag{3.83}$$

$$\left\{ i\frac{\partial}{\partial t} - \frac{\nabla^2}{2m} - \epsilon_F \right\} \Psi_\alpha^\dagger(x) + \zeta \Psi_\alpha^\dagger(x) \left(\Psi_\beta^\dagger(x) \Psi_\beta(x) \right) = 0. \tag{3.84}$$

Here we have used the usual equations for the field operators

$$-i\frac{\partial \Psi(x)}{\partial t} = \left[\widehat{H}, \Psi(x) \right]_-, \tag{3.85}$$

and the commutation rules for the Schrödinger operators

$$\left[\Psi_\alpha(\mathbf{r}), \Psi_\beta^\dagger(\mathbf{r}') \right]_+ = \delta_{\alpha\beta} \delta(\mathbf{r} - \mathbf{r}'), \tag{3.86}$$

$$\left[\Psi_\alpha(\mathbf{r}), \Psi_\beta(\mathbf{r}') \right]_+ = \left[\Psi_\alpha^\dagger(\mathbf{r}), \Psi_\beta^\dagger(\mathbf{r}') \right]_+ = 0. \tag{3.87}$$

Green's functions for superconductor are defined by familiar [39] expressions

$$G_{\alpha\beta}(x, x') = -i \langle T \left(\Psi_\alpha(x) \Psi_\beta^\dagger(x') \right) \rangle, \tag{3.88}$$

and from (3.169) and (3.84), and (3.86) and (3.87), we get

$$\left\{ i\frac{\partial}{\partial t} + \frac{\nabla^2}{2m} + \epsilon_F \right\} G_{\alpha\beta}(x, x')$$

$$+ i\zeta \left\langle T \left(\left(\Psi_\gamma^\dagger(x) \Psi_\gamma(x) \right) \Psi_\alpha(x) \Psi_\beta^\dagger(x') \right) \right\rangle = \delta(x - x') \delta_{\alpha\beta}. \tag{3.89}$$

According to Wick's theorem, the T-product of Ψ-operators may be presented as the averaged product of binary field operators. Owing to the presence of the pair condensate in the system, the T-product may be written in the form

$$\left\langle T\left(\Psi_\gamma^\dagger(x)\Psi_\gamma(x)\Psi_\alpha(x)\Psi_\beta^\dagger(x')\right)\right\rangle$$
$$\approx -\left\langle T\left(\Psi_\alpha(x)\Psi_\gamma(x')\right)\right\rangle\left\langle T\left(\Psi_\gamma^\dagger(x)\Psi_\beta^\dagger(x')\right)\right\rangle \qquad (3.90)$$

(here the electrons' scattering processes and the renormalization of their chemical potential are neglected).

3.3.2 Off-Diagonal Long-Range Order

We can now introduce the anomalous, nondiagonal Green's functions

$$F_{\alpha\beta}^+(x,x') = \left\langle T\left(\Psi_\alpha^\dagger(x)\Psi_\beta^\dagger(x')\right)\right\rangle, \quad F_{\alpha\beta}(x,x') = \left\langle T\left(\Psi_\alpha(x)\Psi_\beta(x')\right)\right\rangle. \quad (3.91)$$

Their presence indicates that the quantum states with N and $N \pm 1$ paired particles (Cooper pairs) are indistinguishable. The last circumstance is connected with the abovementioned macroscopic occupation of the paired states, and allows one to neglect the fluctuations in the number of pairs. It is convenient to write the propagators, which describe the superconducting state, in a matrix form

$$\widehat{G}_{\alpha\beta}(x,x') = \begin{pmatrix} G_{\alpha\beta}(x,x') & F_{\alpha\beta}(x,x') \\ -F_{\alpha\beta}^+(x,x') & \overline{G}_{\alpha\beta}(x,x') \end{pmatrix}, \qquad (3.92)$$

where the function $\overline{G}_{\alpha\beta}(x,x') = G_{\beta\alpha}(x',x)$ corresponds to the Feynman diagram with the reversed direction of arrows. The appearance of F-functions, which are non-diagonal in the Hilbert space of single-particle states, is connected with the phase coherence of the superconducting electrons [the "off-diagonal long-range order", introduced by Landau (see discussion by Ginzburg [40]) and independently by Penrose and Onsager [41].

Note that in a spatially homogeneous and stationary state the propagators (3.92) depend on the difference $(x - x')$ only. Introducing in this case the notation

$$F_{\alpha\beta}(0+) = \lim_{\mathbf{r}\to\mathbf{r}'t\to t'+0} F_{\alpha\beta}(x - x'), \qquad (3.93)$$

one can rewrite the equation (3.89) in the form

$$\left\{i\frac{\partial}{\partial t} + \frac{\nabla^2}{2m} + \epsilon_F\right\} G_{\alpha\beta}(x,x') - i\zeta F_{\alpha\gamma}(0+)F_{\gamma\beta}^+(x - x') = \widehat{1}_{\alpha\beta}\delta(x - x'). \quad (3.94)$$

The equation for $F_{\alpha\beta}^+(x - x')$ follows analogously:

$$\left\{i\frac{\partial}{\partial t} - \frac{\nabla^2}{2m} - \epsilon_F\right\} F_{\alpha\beta}^+(x,x') + i\zeta F_{\alpha\gamma}^+(0+)G_{\gamma\beta}(x - x') = 0. \qquad (3.95)$$

3.3.3   Spin-Singlet Pairing

We can now exclude the dependence on the spin variables (this is permitted in the case of interactions, which do not depend explicitly on the spins of particles). Green's functions may be presented in this case as the products of orbital and spin parts. The diagonal Green's function $G_{\alpha\beta}(x - x')$ is proportional to the unity matrix $\hat{1} \equiv \delta_{\alpha\beta}$:

$$G_{\alpha\beta}(x - x') = G(x - x')\delta_{\alpha\beta}, \tag{3.96}$$

whereas the off-diagonal functions F^+ and F are proportional to the matrix, which is antisymmetric in the spin indices

$$F^+_{\alpha\beta}(x - x') = I_{\alpha\beta}F^+(x - x'), \tag{3.97}$$

$$F_{\alpha\beta}(x - x') = -I_{\alpha\beta}F(x - x'), \tag{3.98}$$

where $I_{\alpha\beta} = i(\sigma_y)_{\alpha\beta}$ is related to the second of the Pauli matrices:

$$\hat{\sigma}_x = \begin{pmatrix} 0 & 1 \\ 1 & 0 \end{pmatrix}, \quad \hat{\sigma}_y = \begin{pmatrix} 0 & -i \\ i & 0 \end{pmatrix}, \quad \hat{\sigma}_z = \begin{pmatrix} 1 & 0 \\ 0 & -1 \end{pmatrix}. \tag{3.99}$$

This antisymmetry characterizes the singlet pairing of the electrons, assumed in the BCS-model, and we will adopt it in further analysis. [In the case of triplet pairing, the choice of $I_{\alpha\beta}$ is ambiguous (see, e.g., [42, 43]) and leads to states with different free energies]. The system of general equations for the superconductors now acquires the form

$$\left\{ i\frac{\partial}{\partial t} + \frac{\nabla^2}{2m} + \epsilon_F \right\} G(x - x') - i\zeta F(0+)F^+(x - x')$$
$$= \delta(x - x'), \tag{3.100}$$

$$\left\{ i\frac{\partial}{\partial t} - \frac{\nabla^2}{2m} - \epsilon_F \right\} F^+(x - x') + i\zeta F^+(0+)G(x - x') = 0, \tag{3.101}$$

where $F^+(0+) = F(0+)^*$.

3.3.4   Solutions in Momentum Representation

In the momentum space, (3.100), (3.101) may be rewritten as $(P = \mathbf{p}, \varepsilon)$:

$$\left\{ \varepsilon - p^2/2m + \epsilon_F \right\} G(P) - i\zeta F(0+)F^+(P) = 1, \tag{3.102}$$

$$\left\{ \varepsilon + p^2/2m - \epsilon_F \right\} F^+(P) + i\zeta F^+(0+)G(P) = 0. \tag{3.103}$$

or, counting the energy $\xi_\mathbf{p}$ from the Fermi energy level ϵ_F, $\xi_\mathbf{p} = p^2/2m - \epsilon_F \approx v_F(p - p_F)$:

$$\{\varepsilon - \xi_\mathbf{p}\} G(P) - i\zeta F(0+)F^+(P) = 1, \tag{3.104}$$

$$\{\varepsilon + \xi_\mathbf{p}\} F^+(P) + i\zeta F^+(0+)G(P) = 0. \tag{3.105}$$

The solution of (3.104), (3.105) has the form

$$G(\mathbf{p}, \epsilon) = \frac{\varepsilon + \xi_\mathbf{p}}{\varepsilon^2 - \xi_\mathbf{p} - |\Delta|^2}, \quad F^+(\mathbf{p}, \varepsilon) = -\frac{\Delta^*}{\varepsilon^2 - \xi_\mathbf{p} - |\Delta|^2}, \tag{3.106}$$

where

$$\Delta = -i\zeta F(0+), \quad \Delta^* = i\zeta F^+(0+). \tag{3.107}$$

[one should bear in mind that since the case of $T = 0$ is being considered here, $\Delta = \Delta_0 = \Delta(T = 0)$].

The rules to bypass the poles in (3.106) are defined by the Landau theorem (see, e.g., [39]), using which one can obtain

$$G(\mathbf{p}, \varepsilon) = \frac{\varepsilon + \xi_\mathbf{p}}{(\varepsilon - \varepsilon_\mathbf{p} + i\delta)(\varepsilon + \varepsilon_\mathbf{p} - i\delta)}, \tag{3.108}$$

$$F^+(\mathbf{p}, \varepsilon) = -\frac{\Delta^*}{(\varepsilon - \varepsilon_\mathbf{p} + i\delta)(\varepsilon + \varepsilon_\mathbf{p} - i\delta)}, \tag{3.109}$$

where $\xi_\mathbf{p}$ is the excitation spectrum of the superconductor

$$\varepsilon_\mathbf{p} = \sqrt{\xi_\mathbf{p} + |\Delta|^2} \tag{3.110}$$

with a gap $|\Delta|$.

3.3.5 Self-Consistency Equation

To find the value of the gap, definition (3.107) may be used. Substituting (3.109) into (3.107), one arrives (with $|\Delta| \neq 0$) at the self-consistency equation

$$1 = -\frac{\zeta}{2(2\pi)^3} \int \frac{d^3\mathbf{p}}{\sqrt{\xi_\mathbf{p}^2 + |\Delta|^2}}. \tag{3.111}$$

Integration in (3.111) over the angles and energy leads to divergence of the integral at large energies. Integrating in symmetric limits over $\xi_\mathbf{p}$, one finds

$$1 = -\frac{\zeta}{2\pi^2} m p_F \ln \frac{2\bar{\epsilon}}{|\Delta|}, \tag{3.112}$$

where $\bar{\epsilon}$ is some boundary value of $|\xi_{\mathbf{p}}|$, which depends on the model assumptions. From (3.112) it follows that

$$\Delta_0 = \Delta(T = 0) = 2\bar{\epsilon}e^{-1/\zeta_0}, \tag{3.113}$$

where

$$\zeta_0 = \frac{|\zeta| m p_F}{2\pi^2} = |\zeta| N(0). \tag{3.114}$$

In (3.114) $N(0)$ denotes the density of energy levels for the electrons on the Fermi-surface in a normal metal.

3.3.6 Isotope Effect

In the model based on the Hamiltonian (3.82), the value of Δ_0 (and, as we will see further, the critical temperature of transition, $T_c \propto \Delta_0$) may be arbitrary large if there is no restriction on the value of $\bar{\epsilon}$. In the traditional BCS model, it was assumed that only the electrons in the "Debye crust" near the Fermi surface take part in the pairing interaction, since the interaction is mediated by phonons. We will accept this assumption and put $\bar{\epsilon} = \omega_D$ in the expressions (3.112), (3.113) and further on. Since $\omega_D \propto M^{-1/2}$, where M is the lattice ion mass, it follows that $T_c \propto \Delta_0 \propto M^{-1/2}$. This leads to the difference in T_c between the same metals of different isotope composition, which is well confirmed experimentally for the usual superconductors.[6]

3.3.7 Gauge Invariance

Gauge invariance is an important property of Gor'kov's equations. Electromagnetic fields may be introduced into the system of (3.100) and (3.101) by the usual operator replacement

$$\nabla \to \nabla - ie\mathbf{A} \quad \text{or} \quad \nabla \to \nabla + ie\mathbf{A}, \tag{3.115}$$

depending on whether the space derivatives apply to the function Ψ or $\Psi^\dagger$, respectively (in the same manner the time derivation operator gains the addition of $\pm ie\varphi$). The equations for G and F^+ may be written as

---

[6]We should notice that there are exclusions from this rule even for the case of phonon-mediated pairing, e.g., in the case of PdH alloy, where the effects of the phonon anharmonicity are essential [44]. The situation with high-temperature superconductors is more complicated, and both anharmonicity and some additional effects may be significant there [45, 46].

$$\left\{ i\frac{\partial}{\partial t} + e\varphi + \frac{1}{2m}\left(\frac{\partial}{\partial \mathbf{r}} - ie\mathbf{A}\right)^2 + \epsilon_F \right\} G(x, x') + \Delta F^+(x, x') = \delta(x - x'),$$

(3.116)

$$\left\{ i\frac{\partial}{\partial t} - e\varphi - \frac{1}{2m}\left(\frac{\partial}{\partial \mathbf{r}} + ie\mathbf{A}\right)^2 - \epsilon_F \right\} F^+(x, x') + \Delta^*(x, x') G(x, x') = 0.$$

(3.117)

The functions G, F and F^+ in the presence of an external field depend on each of the variables x, x', and under the gauge transformation

$$\varphi \to \varphi - \frac{\partial \chi}{\partial t}, \qquad \mathbf{A} \to \mathbf{A} + \nabla \chi \qquad (3.118)$$

they transform according to the rules

$$G(x, x') \to G(x, x)e^{ie\{\chi(\mathbf{r}) - \chi(\mathbf{r}')\}}, \qquad F(x, x') \to F(x, x)e^{ie\{\chi(\mathbf{r}) + \chi(\mathbf{r}')\}}. \qquad (3.119)$$

which follow from the transformation rules for the field operators Ψ and $\Psi^\dagger$.

The transformation rules for $F^+(x, x')$ and, consequently, for Δ^* and Δ, are also defined by (3.119). So, the gauge invariance of (3.117) is straightforward. It must be stressed that in certain cases it becomes possible to make the value of Δ real by special choice of the gauge. In such cases this value coincides with the parameter in the excitation spectrum, which was introduced in the pioneering BCS theory. But in general, $\Delta = \Delta(x)$ is a complex variable and contains additional physical information. This circumstance is one of the important consequences of the Gor'kov theory, as we will see later in Sect. 3.4.

3.3.8 Description at Finite Temperatures

We now generalize the theory to the case of finite temperatures. To do this, it is necessary to apply the Matsubara technique [39], which introduces the imaginary time coordinate τ. The emerging equations are analogous to (3.100) and (3.101):

$$\left\{ -\frac{\partial}{\partial \tau} + \frac{\nabla^2}{2m} + \epsilon_F \right\} \mathfrak{G}(x - x') + \Delta\mathfrak{F}^+(x - x') = \delta(x - x'), \qquad (3.120)$$

$$\left\{ \frac{\partial}{\partial \tau} + \frac{\nabla^2}{2m} + \epsilon_F \right\} \mathfrak{F}^+(x - x') + \Delta^*\mathfrak{G}(x - x') = 0, \qquad (3.121)$$

where

$$\Delta = |\zeta|\mathfrak{F}(0+), \quad \Delta^* = |\zeta|\mathfrak{F}^+(0+). \tag{3.122}$$

Using discrete imaginary frequencies

$$\varepsilon = \varepsilon_n = i(2n + 1)\pi T, \quad n = 0, \pm 1, \pm 2, \ldots, \tag{3.123}$$

according to relations of the type

$$\mathfrak{F}^+(x - x') = T \sum_n e^{-\varepsilon_n \tau} \int \frac{d^3\mathbf{p}}{(2\pi)3} e^{i\mathbf{p}\mathbf{r}} \mathfrak{F}_\varepsilon^+(\mathbf{p}) \tag{3.124}$$

[and analogously for $\mathfrak{F}(x - x')$ and $\mathfrak{G}(x - x')$], one can find from (3.120) and (3.121) the system of equations

$$(\varepsilon - \xi_\mathbf{p})\mathfrak{G}_\varepsilon(\mathbf{p}) + \Delta\mathfrak{F}_\varepsilon^+(\mathbf{p}) = 1, \tag{3.125}$$

$$(\varepsilon + \xi_\mathbf{p})\mathfrak{F}_\varepsilon^+(\mathbf{p}) + \Delta^*\mathfrak{G}_\varepsilon(\mathbf{p}) = 0. \tag{3.126}$$

They have the solutions

$$\mathfrak{G}_\varepsilon(\mathbf{p}) = \frac{\varepsilon + \xi_\mathbf{p}}{\varepsilon^2 - \xi_\mathbf{p}^2 - |\Delta|^2}, \tag{3.127}$$

$$\mathfrak{F}_\varepsilon^+(\mathbf{p}) = \frac{-\Delta^*}{\varepsilon^2 - \xi_\mathbf{p}^2 - |\Delta|^2}. \tag{3.128}$$

3.3.9   Weak-Coupling Ratio $2\Delta(T = 0)/T_c$

The value $|\Delta|$ may be defined from any of the relations (3.122) and with the help of (3.128) one can obtain at $|\Delta| \neq 0$ an equation

$$1 = \frac{|\zeta|T}{(2\pi)^3} \sum_n \int \frac{d^3\mathbf{p}}{-\varepsilon_n^2 + \xi_\mathbf{p}^2 + |\Delta|^2}. \tag{3.129}$$

The summation over the frequencies may be carried out using the expression

$$\sum_{n=-\infty}^{\infty} \left[(2n + 1)^2\pi^2 + a^2\right]^{-1} = \frac{1}{2a} \tanh \frac{a}{2}. \tag{3.130}$$

As a result we obtain from (3.129) the self-consistency equation

$$1 = \zeta_0 \int_0^{\omega_D} \frac{d\xi_\mathbf{p}}{\sqrt{\xi_\mathbf{p}^2 + |\Delta|^2}} \tanh \frac{\sqrt{\xi_\mathbf{p} + |\Delta|^2}}{2T}, \tag{3.131}$$

which determines the at arbitrary temperatures. Note that (3.131) may be presented in more transparent form ($\varepsilon = \sqrt{\xi_\mathbf{p}^2 + |\Delta|^2}$):

$$1 = \zeta_0 \int_{|\Delta|}^{\omega_D} \frac{d\varepsilon(1 - 2n_\varepsilon^0)}{\sqrt{\varepsilon^2 - |\Delta|^2}}, \tag{3.132}$$

where the distribution function of Fermi excitations is given by the formula

$$n_\varepsilon^0 = \frac{1}{\exp(|\varepsilon|/T) + 1}. \tag{3.133}$$

Setting $T = 0$, one obtains from (3.132) the relation (3.111) for the gap $|\Delta_0|$. At $T = T_c$ the gap $|\Delta|$ vanishes and (3.132) reduces to the equation defining T_c:

$$1 = \zeta_0 \int_0^{\omega_D} \frac{d\varepsilon}{\varepsilon} \tanh \frac{\varepsilon}{2T_c}, \tag{3.134}$$

from which one obtains

$$T_c = 2\frac{\gamma}{\pi} \omega_D e^{-1/\zeta_0}, \tag{3.135}$$

where γ is the Euler constant: $\gamma \approx 1.78$.

Comparing the quantities (3.113) and (3.135), we find

$$\frac{2|\Delta(T = 0)|}{T_c} = \frac{2\pi}{\gamma} \approx 3.53. \tag{3.136}$$

This relation provides an empirical criterion of quantitative validity of the BCS model. It is fulfilled for most superconductors with weak electron-phonon coupling, for which

$$\zeta_0 \ll 1. \tag{3.137}$$

The BCS theory describes quite satisfactorily many phenomena occurring in superconductors at thermodynamic equilibrium, even in cases, where the inequality (3.137) is violated. In these phenomena, the BCS interaction potential does not reveal itself directly. The inelastic collisions, which were omitted in the simplified BCS picture, are also not important for this class of phenomena. We note now that the BCS-Gor'kov model may be modified to remove these shortcomings. The Migdal–Eliashberg model, which is more realistic and better applicable to the problems of nonequilibrium superconductivity, is considered in detail in Chap. 5.

3.4   Self-Consistent Pair-Field: Microscopic Justification of G–L Equations

Initially, the idea that the gap in the energy spectrum of superconductors may serve as the superconducting order parameter of the phenomenological Ginzburg–Landau theory, was contained in the foundational work of Bardeen, Cooper, and Schrieffer [47]. This idea was confirmed by Gor'kov [48], whose work has assigned the status of microscopic theory to the Ginzburg–Landau study. In the next section we follow the derivation presented by Gor'kov [48].

3.4.1   Iterated Equations

The microscopic equations of superconductivity considered in the preceding section may be written in the form

$$\left\{ \varepsilon + \frac{1}{2m} \left(\frac{\partial}{\partial \mathbf{r}} - i e \mathbf{A}(\mathbf{r}) \right)^2 + \epsilon_F \right\} \mathfrak{G}_\varepsilon(\mathbf{r}, \mathbf{r}')$$

$$+ \Delta(\mathbf{r}) \mathfrak{F}_\varepsilon^+(\mathbf{r}, \mathbf{r}') = \delta(\mathbf{r}, \mathbf{r}'), \tag{3.138}$$

$$\left\{ -\varepsilon + \frac{1}{2m} \left(\frac{\partial}{\partial \mathbf{r}} - i e \mathbf{A}(\mathbf{r}) \right)^2 + \epsilon_F \right\} \mathfrak{F}_\varepsilon^+(\mathbf{r}, \mathbf{r}')$$

$$- \Delta^*(\mathbf{r}) \mathfrak{G}_\varepsilon(\mathbf{r}, \mathbf{r}') = 0. \tag{3.139}$$

The parameter Δ^* (and analogously Δ) is connected with an anomalous Green's function by the relation

$$\Delta^*(\mathbf{r}) = |\zeta| T \sum_\varepsilon \mathfrak{F}_\varepsilon^+(\mathbf{r}, \mathbf{r}'), \tag{3.140}$$

where a summation over the frequencies $\varepsilon = i\pi T (2n + 1)$ spreads up to $|\varepsilon| < \omega_D$. Making iterations over Δ in (3.138) and (3.139), one can obtain the result, which is convenient to display diagrammatically:

$$\mathfrak{G}(\mathbf{r}, \mathbf{r}') = \quad \longrightarrow - - \longrightarrow \wedge \longleftarrow \wedge \longrightarrow + \cdots \equiv \longrightarrow \longrightarrow , \tag{3.141}$$

$$\mathfrak{F}^+(\mathbf{r}, \mathbf{r}') = \quad \longleftarrow \wedge \longrightarrow - \longleftarrow \wedge \longrightarrow \wedge \longleftarrow \wedge \longrightarrow + \cdots \equiv \longleftarrow \longrightarrow . \tag{3.142}$$

Here the right arrow corresponds to the normal state function $\mathfrak{G}^0$, the left arrow— to $\overline{\mathfrak{G}}^0$, the vertex Δ corresponds to $\Delta(r)$ or $\Delta^*(r)$, depending on the convergence or divergence of neighboring arrows. Additionally, the sign of the diagram changes, if the vertex Δ enters the diagram twice. This rule should be followed constantly;

however, we omit the sign $(-)$ on the diagram. Taking into account all the terms in the expansions of (3.141) and (3.142) allows us to present the equations (3.138) and (3.139) in the integral form:

$$\mathfrak{G}_\varepsilon(\mathbf{r}, \mathbf{r}') = \mathfrak{G}_\varepsilon^0(\mathbf{r}, \mathbf{r}') - \int \mathfrak{G}_\varepsilon^0(\mathbf{r}, \mathbf{r}_1)\Delta(\mathbf{r}_1)\mathfrak{F}_\varepsilon^+(\mathbf{r}_1, \mathbf{r}')d^3\mathbf{r}_1, \qquad (3.143)$$

$$\mathfrak{F}_\varepsilon^+(\mathbf{r}, \mathbf{r}') = \int \overline{\mathfrak{G}}_\varepsilon^0(\mathbf{r}, \mathbf{r}_1)\Delta^*(\mathbf{r}_1)\mathfrak{G}_\varepsilon(\mathbf{r}_1, \mathbf{r}')d^3\mathbf{r}_1, \qquad (3.144)$$

though only the first terms of this expansion, depicted in (3.141) and (3.142), are needed. The appropriate expressions are

$$\mathfrak{G}_\varepsilon(\mathbf{r}, \mathbf{r}') = \mathfrak{G}_\varepsilon^0(\mathbf{r}, \mathbf{r}')$$
$$- \int \mathfrak{G}_\varepsilon^0(\mathbf{r}, \mathbf{r}_1)\Delta(\mathbf{r}_1)\overline{\mathfrak{G}}_\varepsilon^0(\mathbf{r}, \mathbf{r}_2)\Delta^*(\mathbf{r}_2)\mathfrak{G}_\varepsilon^0(\mathbf{r}_2, \mathbf{r}')\,d^3\mathbf{r}_1 d^3\mathbf{r}_2, \qquad (3.145)$$

$$\mathfrak{F}_\varepsilon^+(\mathbf{r}, \mathbf{r}') = \int \overline{\mathfrak{G}}_\varepsilon^0(\mathbf{r}, \mathbf{r}_1)\Delta^*(\mathbf{r}_1)\mathfrak{G}_\varepsilon^0(\mathbf{r}_1, \mathbf{r}')d^3\mathbf{r}_1$$
$$- \int \overline{\mathfrak{G}}_\varepsilon^0(\mathbf{r}, \mathbf{r}_1)\Delta^*(\mathbf{r}_1)\mathfrak{G}_\varepsilon^0(\mathbf{r}_1, \mathbf{r}_2)\Delta(\mathbf{r}_2)\overline{\mathfrak{G}}_\varepsilon^0(\mathbf{r}_2, \mathbf{r}_3)\Delta^*(\mathbf{r}_3)$$
$$\times \mathfrak{G}_\varepsilon^0(\mathbf{r}_3, \mathbf{r}')\,d^3\mathbf{r}_1 d^3\mathbf{r}_2 d^3\mathbf{r}_3. \qquad (3.146)$$

For further calculations one must know the function $\mathfrak{G}_\varepsilon^0(\mathbf{r}, \mathbf{r}')$. We will find it first in the absence of the magnetic field, putting $\mathbf{A} = 0$ and denoting the corresponding Green's function by $\mathfrak{G}_\varepsilon^{00}(r)$.

Using the definition of $\mathfrak{G}_\varepsilon^{00}(r)$ and making straightforward calculations, one obtains

$$\mathfrak{G}_\varepsilon^{00}(r) = \int \frac{d^3\mathbf{p}}{(2\pi)^3}e^{i\mathbf{p}\cdot\mathbf{r}}\frac{1}{\varepsilon - \epsilon_\mathbf{p}}$$
$$= \int_0^\infty \int_{-1}^1 \frac{p^2\,dp\,d\cos\theta}{(2\pi)^2}e^{ipr\cos\theta}\frac{1}{\varepsilon - \epsilon_\mathbf{p}}$$
$$= -i\int_0^\infty \frac{p\,dp}{(2\pi)^2 r}\left(e^{ipr} - e^{-ipr}\right)\frac{1}{\varepsilon - \epsilon_\mathbf{p}}$$
$$= -\frac{m}{2\pi r}\exp\left\{ip_F r\frac{\varepsilon}{|\varepsilon|} - \frac{|\varepsilon|}{v_F}r\right\}. \qquad (3.147)$$

In deriving (3.147) the relations

$$pdp = md\xi_p, \quad p = p_F + \frac{\xi_p}{v_F}. \qquad (3.148)$$

were used.

3.4.2 Magnetic Field Inclusion

In presence of a magnetic field, the function $\mathfrak{G}^0$ differs from $\mathfrak{G}^{00}$ by the phase factor $\varphi(\mathbf{r}, \mathbf{r}')$

$$\mathfrak{G}^0_\varepsilon(\mathbf{r}, \mathbf{r}') = e^{-\varphi(\mathbf{r},\mathbf{r}')} \mathfrak{G}^{00}_\varepsilon(|\mathbf{r} - \mathbf{r}'|), \tag{3.149}$$

where $\varphi(\mathbf{r}, \mathbf{r}) = 0$. The function $\varphi(\mathbf{r}, \mathbf{r}')$ obeys the equation

$$\frac{\partial \varphi}{\partial \mathbf{r}} \cdot \frac{\partial \mathfrak{G}^{00}}{\partial \mathbf{r}} = e\mathbf{A} \cdot \frac{\partial \mathfrak{G}^{00}}{\partial \mathbf{r}}, \tag{3.150}$$

which may be established from (3.149), (3.138) by taking into account the quasiclassic conditions: $|\mathbf{r} - \mathbf{r}'| \gg 1/p_F$, $p_F \gg e\mathbf{A} \sim eH\lambda_L$. Because of the spatial homogeneity of $\mathfrak{G}^{00}_\varepsilon$, the relation follows from (3.150):

$$\mathbf{n} \cdot \frac{\partial}{\partial \mathbf{r}} \varphi(\mathbf{r}, \mathbf{r}') = e\mathbf{n} \cdot \mathbf{A}, \qquad \mathbf{n} = \frac{\mathbf{r} - \mathbf{r}'}{|\mathbf{r} - \mathbf{r}'|}, \tag{3.151}$$

which would be used further.

3.4.3 Slow Variation Hypothesis

As is evident from (3.147), the function $\mathfrak{G}^{00}_\varepsilon$ (and consequently $\mathfrak{G}^0_\varepsilon$) decreases over distances on the order of $\xi_0 \sim v_F/|\varepsilon|$. However, the field $\mathbf{A}$ near T_c varies over distances greatly exceeding $\xi_0 : A(\mathbf{r}) \sim H\lambda_L \sim (1 - T/T_c)^{-1/2}$. This allows us to present the function φ in the form

$$\varphi(\mathbf{r}, \mathbf{r}') = e\mathbf{A}(\mathbf{r}) \cdot (\mathbf{r} - \mathbf{r}'). \tag{3.152}$$

Now we are able to derive the equation for parameter Δ. Substituting (3.146) into (3.140) we find

$$\Delta^*(\mathbf{r}) = |V|T \sum_\varepsilon \int \overline{\mathfrak{G}}^0_\varepsilon(\mathbf{r}, \mathbf{r}_1) \Delta^*(\mathbf{r}_1) \mathfrak{G}^0_\varepsilon(\mathbf{r}_1, \mathbf{r}) \, d^3\mathbf{r}_1 - |\zeta|T \sum_\varepsilon \int \overline{\mathfrak{G}}^0_\varepsilon(\mathbf{r}, \mathbf{r}_1)$$

$$\times \Delta^*(\mathbf{r}_1) \mathfrak{G}^0_\varepsilon(\mathbf{r}_1, \mathbf{r}_2) \Delta(\mathbf{r}_2) \overline{\mathfrak{G}}^0_\varepsilon(\mathbf{r}_2, \mathbf{r}_3) \Delta^*(\mathbf{r}_3) \mathfrak{G}^0_\varepsilon(\mathbf{r}_3, \mathbf{r}) \, d^3\mathbf{r}_1 \, d^3\mathbf{r}_2 \, d^3\mathbf{r}_3. \tag{3.153}$$

Let us suppose that the pair field $\Delta^*(\mathbf{r})$ weakly varies over distances comparable with ξ_0 (this supposition, as we will see, would be confirmed). In the integrand of the first term in (3.153) one can make a series expansion over the parameter $\Delta^*(\mathbf{r})$

$$\Delta^*(\mathbf{r}') = \Delta^*(\mathbf{r}) + \frac{\partial \Delta^*(\mathbf{r})}{\partial \mathbf{r}} (\mathbf{r}' - \mathbf{r}) + \frac{1}{2} \frac{\partial^2 \Delta^*(\mathbf{r})}{\partial \mathbf{r}^2} (\mathbf{r}' - \mathbf{r})^2 + \cdots, \tag{3.154}$$

and also keep only the first item in the analogous expansion of the second term in (3.153). Substitution of (3.154) into (3.153) taking into account expressions (3.149) and (3.152), which also could be expanded in powers of the vector-potential $\mathbf{A}(\mathbf{r})$, yields the expression

$$
\Delta^*(\mathbf{r}) = |\zeta| T \sum_\varepsilon \{ \Delta^*(\mathbf{r}) \int \overline{\mathfrak{G}}_\varepsilon^{00}(r_1) \mathfrak{G}_\varepsilon^{00}(r_1) \mathrm{d}^3 \mathbf{r}_1
$$

$$
+ \frac{1}{6} \left(\frac{\partial}{\partial \mathbf{r}} + 2ie\mathbf{A}(\mathbf{r}) \right)^2 \Delta^*(\mathbf{r}) \int \overline{\mathfrak{G}}_\varepsilon^{00}(r_1) \mathfrak{G}_\varepsilon^{00}(r_1) r_1^2 \mathrm{d}^3 \mathbf{r}_1
$$

$$
- |\Delta|^2 \Delta^*(\mathbf{r}) \int \overline{\mathfrak{G}}_\varepsilon^{00}(|\mathbf{r} - \mathbf{r}_1|) \mathfrak{G}_\varepsilon^{00}(|\mathbf{r}_1 - \mathbf{r}_2|)
$$

$$
\times \overline{\mathfrak{G}}_\varepsilon^{00}(|\mathbf{r}_2 - \mathbf{r}_3|) \mathfrak{G}_\varepsilon^{00}(|\mathbf{r}_3 - \mathbf{r}|) \mathrm{d}^3 \mathbf{r}_1 \, \mathrm{d}^3 \mathbf{r}_2 \, \mathrm{d}^3 \mathbf{r}_3. \tag{3.155}
$$

By direct summation over the discrete frequencies it can be established that

$$
T \sum_\varepsilon \overline{\mathfrak{G}}_\varepsilon^{00}(r) \mathfrak{G}_\varepsilon^{00}(r) = \frac{m^2 T}{(2\pi r)^2} / \sinh \frac{2\pi T r}{v_F}, \tag{3.156}
$$

and owing to this, the first of the integrals in (3.155) would diverge, if one does not take into account "the smearing" of the coordinate r over the distances $r \lesssim \xi_0$. In the momentum space one can cut off the summation in (3.155) (at $|\varepsilon| < \omega_D$), which corresponds to this smearing. Using this circumstance, one may write

$$
T \sum_\varepsilon \int \overline{\mathfrak{G}}_\varepsilon^{00}(r') \mathfrak{G}_\varepsilon^{00}(r') \, \mathrm{d}^3 \mathbf{r}' = T \sum_\varepsilon \int \frac{\mathrm{d}^3 \mathbf{p}}{(2\pi)^3} \frac{1}{(\xi_\mathbf{p} + \varepsilon)(\xi_\mathbf{p} - \varepsilon)}
$$

$$
= \frac{m p_F}{2\pi^2} T \sum_\varepsilon \int_0^{\omega_D} \frac{\mathrm{d}\xi_\mathbf{p}}{\xi_\mathbf{p} - \varepsilon^2}
$$

$$
= \frac{m p_F}{2\pi^2} \int_{-\omega_D}^{\omega_D} \frac{\mathrm{d}\xi_\mathbf{p}}{\xi_\mathbf{p}} \tanh \frac{\xi_\mathbf{p}}{2T} = \frac{m p_F}{2\pi^2} \left(1 + \ln \frac{T_c}{T} \right). \tag{3.157}
$$

The last equality here is obtained by taking into account (3.134) for the critical temperature. The second integral in (3.155) may be evaluated, using the formula (3.156):

$$
T \sum_\varepsilon \int r_1^2 \overline{\mathfrak{G}}_\varepsilon^{00}(r_1) \mathfrak{G}_\varepsilon^{00}(r_1) \, \mathrm{d}^3 \mathbf{r}_1 = \frac{7}{8} \frac{m p_F}{2\pi^2} \frac{\zeta(3) v_F^2}{(\pi T)^2}, \tag{3.158}
$$

where $\zeta(3)$ is the Riemann zeta-function. The third integral is also not too difficult to evaluate:

$$T \sum_\varepsilon \int d^3r_1 \, d^3r_2 \, d^3r_3 \, \mathfrak{G}_\varepsilon^{00}(\mathbf{r} - \mathbf{r}_1) \mathfrak{G}_\varepsilon^{00}(\mathbf{r}_1 - \mathbf{r}_2) \mathfrak{G}_\varepsilon^{00}(\mathbf{r}_2 - \mathbf{r}_3) \mathfrak{G}_\varepsilon^{00}(\mathbf{r}_3 - \mathbf{r})$$

$$= \frac{mp_F}{2\pi^2} T \sum_\varepsilon \int d\xi_\mathbf{p} \, \frac{1}{(\xi_\mathbf{p}^2 - \varepsilon^2)^2} = \left(\frac{mp_F}{2\pi^2}\right) \frac{7\zeta(3)}{8(\pi T)^2}. \tag{3.159}$$

Gathering the results, one obtains after complex conjugation the equation for Δ:

$$\left\{ -\frac{1}{4m} \left(-i\frac{\partial}{\partial\mathbf{r}} - 2e\mathbf{A}(\mathbf{r}) \right)^2 + \left(\frac{7\zeta(3)}{6(\pi T_c)^2} \epsilon_F \right)^{-1} \right.$$

$$\left. \times \left[\frac{T - T_c}{T_c} + \frac{7\zeta(3)}{8(\pi T_c)^2} |\Delta(\mathbf{r})|^2 \right] \right\} \Delta(\mathbf{r}) = 0. \tag{3.160}$$

The BCS potential disappears from the final result, which has the form of the Ginzburg–Landau equation for the wave function (3.48) if one associates Δ with ψ.

3.4.4   Computation of Phenomenological Parameters

Microscopic derivation permits one to determine the phenomenological parameters in (3.48). First, the doubled value of the electron's charge should be noticed in (3.160): $e_* = 2e$, this is the consequence of the Cooper pairing. For this reason $m_* = 2m$ was chosen in (3.160) and as may be found in comparison with (3.37),

$$\left(\frac{\partial\alpha}{\partial T} \right)_{T_c} = \frac{6\pi^2 T_c}{7\zeta(3)\epsilon_F}. \tag{3.161}$$

The value of the coefficient β is sensitive to the normalization of the Ψ-function. In Sect. 3.2 we have adopted a normalization, with $|\Psi|^2$ corresponding to the density of pairs [see (3.53)]:

$$\mathbf{j} = 2eN_s\mathbf{v}_s. \tag{3.162}$$

The microscopic treatment is based on the initial expression for the current

$$\mathbf{j} = \lim_{\mathbf{r}\to\mathbf{r}'\tau\to-0} 2 \left\{ \frac{ie}{2m}(\nabla_\mathbf{r} - \nabla_{\mathbf{r}'}) + \frac{e^2}{m}\mathbf{A}(\mathbf{r}) \right\} T \sum_\varepsilon \mathfrak{G}_\varepsilon(\mathbf{r}, \mathbf{r}')e^{-\varepsilon\tau}. \tag{3.163}$$

Substituting here $\mathfrak{G}_\varepsilon(\mathbf{r}, \mathbf{r}')$ from (3.145), and using the quasiclassical conditions mentioned above (for details see [39]), one can find:

$$\mathbf{j} = \frac{7\zeta(3)N}{16(\pi T_c)^2} \left\{ -\frac{ie}{m} \left(\Delta^*\frac{\partial\Delta}{\partial\mathbf{r}} - \Delta\frac{\partial\Delta^*}{\partial\mathbf{r}} \right) - \frac{4e^2|\Delta|}{m}\mathbf{A}(\mathbf{r}) \right\}, \tag{3.164}$$

where N is the total density of electrons, which coincides with the normal state value. Comparing (3.164) with (3.51) one can find a relation between Ψ and Δ:

$$\Psi = \left[\frac{7\zeta(3)N}{8(\pi T_c)^2}\right]^{1/2}\Delta. \tag{3.165}$$

Now the parameter β may be obtained with the help of (3.160), (3.165), and (3.48):

$$\beta = \frac{6}{7}\frac{(\pi T_c)^2}{\zeta(3)N\epsilon_F}. \tag{3.166}$$

Another relation to be noted is

$$N_s = \frac{7\zeta(3)|\Delta|^2}{8(\pi T_c)^2}N, \tag{3.167}$$

which follows from (3.162) and (3.165) and connects the density of pairs, N_s, near T_c with the total density of electrons in a normal metal, N.

Thus, the microscopic theory not only laid the foundation for the Ginzburg–Landau theory, but also defined the phenomenological coefficients entering it. In particular, the temperature-dependent coherence length $\xi(T)$ and the penetration depth $\lambda_L(T)$ may be calculated. One can ascertain now the self-consistency of the assumptions made earlier, that at temperatures $T \approx T_c$ these lengths greatly exceed the correlation length $\xi_0 \sim v_F/T_c$. For the London superconductors these expressions are still valid for temperatures below T_c, though they fail quickly for the Pippard superconductors, if the temperature falls. It must be noted that a large class of a superconducting metals, containing nonmagnetic impurities, may be attributed to the London-type superconductors, as we will see in Chap. 4. Hence, the area of applicability of the Ginzburg–Landau equations actually is rather wide.

3.4.5 Flux Quantization

In their basic paper [29] Ginzburg and Landau acknowledged that they have no reasons to assume that e_* is different from the electron's charge e. As we discussed above, microscopic theory by Bardeen, Cooper, and Schrieffer (BCS) [47] assumes that current in superconductors is due to the Cooper pairs and Gor'kov was able to prove that $e_* = 2e$. In contrast to the mass m_*, which in GL system of equations may be changed by a simple renormalization of Ψ, the charge e_* enters the equations (3.48) and (3.51) additively and its magnitude cannot be chosen arbitrarily. Moreover, its value can be determined in experiments. In Part I of this book we demonstrated how the value of the flux in superconductors is becoming quantized (see Problem 3.5). Here we will consider the phenomenon of magnetic flux quantization quantitatively.

Fig. 3.2 A hollow
superconducting cylinder in
a magnetic field. C—the
contour of integration,
λ_L—the penetration depth

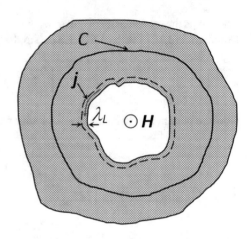

Let us imagine a massive superconductor with a cylindrical cavity placed in a
magnetic field **H**, which is parallel to the cylinder's axis. Consider a contour C (see
Fig. 3.2) that encloses the cavity and lies entirely within the depth of the superconductor. Owing to the Meissner effect, the superconducting current vanishes at distances
from the surface essentially larger than the London penetration depth λ_L. Thus, as
follows from (3.51), on the contour C one has

$$\mathbf{j} = \frac{\hbar e_*}{m_*}|\Psi|^2\nabla\theta - \frac{e_*^2}{m_*c}\Psi^2\mathbf{A} = \mathbf{0}. \tag{3.168}$$

Integrating **j** along this contour and using (3.168), we find

$$\oint \nabla\theta \cdot d\mathbf{l} = \frac{e_*}{\hbar}\oint \mathbf{A} \cdot d\mathbf{l}. \tag{3.169}$$

It must be taken into account that

$$\oint \mathbf{A}\cdot d\mathbf{l} = \int \text{curl } \mathbf{A}\cdot d\mathbf{s} = \int \mathbf{B}\cdot d\mathbf{s} = \Phi, \tag{3.170}$$

where Φ is the total magnetic flux existent in the cavity. The first of the integrals
(3.169) is the phase difference acquired while going around the contour C and it
must be a multiple to 2π:

$$\oint \nabla\theta \cdot d\mathbf{l} = 2\pi n \tag{3.171}$$

because the Ψ-function is single-valued [n is an integer in (3.171)]. As a result, one
finds from (3.169)–(3.171)

$$\Phi = \frac{hc}{e_*}n = \phi_0 n, \tag{3.172}$$

or, in other words, the magnetic flux in the cavity of a superconductor is quantized and may change only in portions $\phi_0 = hc/e_*$. This phenomenon was predicted first by London [7] and later confirmed experimentally by Deaver and Fairbank [49] and also by Doll and Näbauer [50], who found the value of e_* in (3.172) to be equal to twice the electronic charge: $e_* = 2e$. Had the doubling of the carrier's charge been known in 1950, the analysis of the Ginzburg–Landau equation for the "quantum mechanical function" ψ of superconducting electrons might significantly accelerate the subsequent development of the microscopic theory of superconductivity: the idea of pairing would become evident much before the BCS approach.

3.4.6 Failure of "Quantum-Mechanical Generalization" for Time-Dependent Problems

In the vicinity of a critical temperature T_c, the solutions of the Ginzburg–Landau equations are fully equivalent to the solutions of the BCS equations. At the same time, the Ginzburg–Landau technique is considerably simpler. As we have seen, the "wave function of superconducting electrons" $\Psi(\mathbf{r})$ is closely connected with the field $\Delta(\mathbf{r})$, which characterizes the Cooper pair condensate. If one ignores the nonlinear term in (3.128), then the equation for $\Psi(\mathbf{r})$ formally coincides with the Schrödinger equation for the particle with the charge $2e$ and the mass $2m$. The simplicity and transparency of such an analogy has lead to attempts of using the Schrödinger-type equation to describe the dynamic properties of superconductors in nonstationary fields: $\mathbf{A} = \mathbf{A}(\mathbf{r}, t)$. At first glance, the "natural" extension of (3.128) for the nonstationary case may be obtained by the "quantum-mechanical" generalization:

$$i\frac{\partial \Psi(\mathbf{r}, t)}{\partial t} = \left\{ -\frac{1}{4m} (\nabla - 2ie\mathbf{A}(\mathbf{r}, t))^2 \right.$$
$$\left. -\frac{6(\pi T_c)^2}{7\zeta(3)\epsilon_F} \left[\frac{-T_c}{T_c} + \frac{1}{N}|\Psi|^2 \right] \right\} \Psi(\mathbf{r}, t). \qquad (3.173)$$

Such a generalization, however, may immediately lead to a contradiction (mentioned by Eliashberg [51]). Indeed, the "continuity equation"

$$\frac{\partial |\Psi|^2}{\partial t} \propto \mathrm{div} \left\{ |\Psi|^2 \left(\mathbf{A} - \frac{1}{2e} \nabla \theta \right) \right\} \qquad (3.174)$$

follows in the usual manner from (3.173). However, such an equation is inconsistent with the action of nonstationary fields on superconductors. Indeed, based on Gauss' theorem, one can derive from (3.174) that at an action of the time-dependent field $\mathbf{A}(t)$, $|\Psi|^2$ inside of any finite volume of superconductor can change only because of the flow of the currents $\mathbf{j}$ through the surrounding surface, which is obviously incorrect physically. To break this conservation of particles (i.e., the Cooper pairs), the

Schrödinger-type equation should possess a non-Hermitian Hamiltonian. Indeed, as we will see in the next chapters, the correct equation has the structure of Schrödinger-type equation with a conjugated imaginary Hamiltonian, which breaks the particle conservation: in nonstationary fields the Cooper pairs will be allowed to convert into the single-particle excitations and vice versa.

References

1. H. Kamerlingh-Onnes, Further experiments with liquid helium. G. On the electrical resistance of pure metals, etc., VI. On the sudden change in the rate at which the resistance of mercury disappears. Commun. Phys. Lab. Univ. Leiden 124c (1911)
2. W. Meissner, R. Ochsenfeld, Ein neuer Effekt bei Einritt der Supraleitfahigkeit. Naturwissenschaften **21**(44), 787–788 (1933)
3. F. London, H. London, The electromagnetic equations of the superconductor. Proc. R. Soc. Lond. **A149**, 71–88 (1935)
4. Y. Aharonov, D. Bohm, Significance of electromagnetic potentials in the quantum theory. Phys. Rev. **115**(3), 485–491 (1959)
5. W. Ehrenberg, R.E. Siday, The refractive index in electron optics and the principles of dynamics. Proc. Phys. Soc. B **62**, 8–21 (1949)
6. R. Feynman, *The Feynman Lectures on Physics*, vol. 2 (Addison-Wesley Publishing Company Inc, Reading, 1964), pp. 15–8
7. F. London, Macroscopical interpretation of superconductivity. Proc. R. Soc. Lond. **A152**, 24–34 (1935)
8. P.A.M. Dirac, *The Principles of Quantum Mechanics*, 4th ed. (Clarendon Press, Oxford, 1958), pp. 253–275
9. C. Kittel, *Quantum Theory of Solids* (Wiley, New York, 1963), pp. 286–288
10. J.D. Bjorken, S.D. Drell, *Relativistic Quantum Mechanics* (McGraw-Hill, New York, 1964), pp. 1–44
11. C. Kittel, *Introduction to Solid State Physics*, 4th ed. (Wiley, New York, 1971), pp. 727–729
12. P.G. De Gennes, *Superconductivity of Metals and Alloys* (W.A. Benjamin, New York, 1966), pp. 137–170
13. D. Lurié, S. Cremer, Zitterbewegung of quasiparticles in superconductors. Physica **50**, 224–240 (1970)
14. C.J. Pethick, H. Smith, Relaxation and collective motion in superconductors: a two-fluid description. Ann. Phys. **119**(1), 133–169 (1979)
15. O. Klein, Die Reflection von Elektronen an einen Potentialssprung nach der relativistischen Dynamic von Dirac. Z. Phys. **53**, 157–165 (1929)
16. A.F. Andreev, The thermal conductivity of the intermediate state in superconductors. Sov. Phys. JETP **19**(5), 1228–1231 (1964) [Zh. Eksp. i Teor. Fiz. **46**(5) 1823–1828 (1964)]
17. I.P. Krylov and Yu.V. Sharvin, Radio-frequency size effect in a layer of normal metal bounded by its superconducting phase. Sov. Phys. JETP **37**(3), 481–486 (1973) [Zh. Eksp. i Teor. Fiz. **64**(3) 946–957 (1973)]
18. P.W. Anderson, When the electron falls apart. Phys. Today 42–47 (1997)
19. G.E. Blonder, M. Tinkham, T.M. Klapwijk, Transition from metallic to tunneling regimes in superconducting microconstrictions: excess current, charge imbalance, and supercurrent conversion. Phys. Rev. B **25**(7), 4515–4532 (1982)
20. S. Sinha, K.-W. Ng, Zero bias conductance peak enhancement in $Bi_2Sr_2CaCu_2O_8/Pb$ tunneling junctions. Phys. Rev. Lett. **80**(6), 1296–1299 (1998)
21. Y. Tanaka, S. Kashiwaya, Theory of tunneling spectroscopy of d-wave superconductors. Phys. Rev. Lett. **74**(17), 3451–3454 (1995)

22. S. Kashiwaya, Y. Tanaka, M. Koyanagi, K. Kajimura, Theory for tunneling spectroscopy of anisotropic superconductors. Phys. Rev. B **53**(5), 2667–2676 (1996)

23. T.P. Devereaux, P. Fulde, Multiple Andreev scattering in superconductor normal-metal superconductor junctions as a test for anisotropic electron pairing. Phys. Rev. B **47**(21), 14638–14641 (1993)

24. C.-R. Hu, Midgap surface states as a novel signature for $d_{x_a^2 - x_b^2}$-wave superconductivity. Phys. Rev. Lett. **72**(10), 1526–1529 (1994)

25. J.H. Xu, J.H. Miller Jr., C.S. Ting, Conductance anomalies in a normal-metal − d-wave superconductor junction. Phys. Rev. B **53**(6), 3604–3612 (1996)

26. J.M. Hergenrother, M.T. Tuominen, M. Tinkham, Charge transport by Andreev reflection through a mesoscopic superconducting island. Phys. Rev. Lett. **72**(11), 1742–1745 (1994)

27. B.J. van Wees, P. de Vries, P. Magnée, T.M. Klapwijk, Excess conductance of superconductor-semiconductor interfaces due to phase conjugation between electrons and holes. Phys. Rev. Lett. **69**(3), 510–513 (1992)

28. F.W.J. Hekking, Y.V. Nazarov, Interference of two electrons entering a superconductor. Phys. Rev. Lett. **71**(10), 1625–1628 (1993)

29. V.L. Ginzburg, L.D. Landau, To the theory of superconductivity. Zh. Eksp. i Teor. Fiz. **20**(12), 1064–1082 (1950). In Russian

30. L.D. Landau, E.M. Lifshitz, *Statistical Physics*, 2nd ed. (Addison-Wesley, Reading, 1970), pp. 424–454

31. E.M. Lifshitz, L.P. Pitaevskii, *Statistical Physics*, Part 2 (Addison-Wesley, Reading, 1980), pp. 450–468

32. A.A. Abrikosov, *Fundamentals of the Theory of Metals* (North-Holland, Amsterdam, 1988), p. 389

33. L. Cooper, Bound electron pairs in a degenerate Fermi gas. Phys. Rev. **104**(4), 1189–1190 (1956)

34. L.I. Schiff, *Quantum Mechanics*, 3rd edn. (McGraw-Hill, New York, 1968), p. 247

35. S.T. Belyayev, Application of the methods of quantum field theory to a system of Bosons. Sov. Phys. JETP **7**(2), 289–299 (1958) [Zh. Eksp. i Teor. Fiz. **34**(2), 417–432 (1958)]

36. P.W. Anderson, *Basic Notions of Condensed Matter Physics* (Benjamin/Cummings, London, 1984), pp. 229–248

37. L.P. Gor'kov, On the energy spectrum of superconductors. Sov. Phys. JETP **7**(3), 505–508 (1964) [Zh. Eksp. i Teor. Fiz. **34**(3) 735–739 (1958)]

38. J.R. Schrieffer, *Theory of Superconductivity* (W.A. Benjamin, New York, 1964), pp. 1–282

39. A.A. Abrikosov, L.P. Gor'kov, I.E. Dzyaloshinskii, *Quantum Field Theoretical Methods in Statistical Physics*, 2nd ed. (Pergamon, Oxford, 1965), pp. 42–439

40. V.L. Ginzburg, Superconductivity and Superfluidity (What is done and what is not done). Sov. Phys. Uspekhi **40**(4), 407–432 (1997) [Usp. Fiz. Nauk **167**(4), 429–454 (1997)]

41. O. Penrose, L. Onsager, Bose-Einstein condensation and liquid helium. Phys. Rev. **104**(3), 576–584 (1956)

42. V.P. Mineev, Superfluid ^3He: introduction to the subject. Sov. Phys. Uspekhi **26**(2), 160–175 (1963) [Usp. Fiz. Nauk **139**(2), 303–332 (1983)]

43. G.E. Volovik, L.P. Gor'kov, Superconducting classes in heavy-fermion systems. Sov. Phys. JETP **61**(4), 843–854 (1985) [Zh. Eksp. i Teor. Fiz. **88**(4), 1412–1429 (1985)]

44. E.G. Maximov, in *High-Temperature Superconductivity* ed. by V.L. Ginzburg, D.A, Kitzhnitz (Consultants Beureau, New York, 1982), pp. 1–364

45. V.Z. Kresin, S.A. Wolf, Microscopic model for the isotope effect in high- T_c oxides. Phys. Rev. B **49**(5), 3652–3654 (1994)

46. A. Bill, V.Z. Kresin, S.A. Wolf, Isotope effect in high- T_c materials: role of non-adiabaticity and magnetic impurities. Z. Phys. B **104**(4), 759–763 (1997)

47. J. Bardeen, L. Cooper, J. Schrieffer, Theory of superconductivity. Phys. Rev. **108**(5), 1175–1204 (1957)

48. L.P. Gor'kov, Microscopic derivation of the Ginzburg-Landau equations in the theory of superconductivity, Sov. Phys. JETP **9**(6), 1364–1367 (1959) [Zh. Eksp. i Teor. Fiz. **36** (6), 1918–1923 (1959)]

49. B.S. Deaver, W.M. Fairbank Jr., Experimental evidence for quantized flux in superconducting cylinders. Phys. Rev. Lett. **7**(2), 43–46 (1961)
50. R. Doll, M. Näbauer, Experimental proof of magnetic flux quantization in superconducting ring. Phys. Rev. Lett. **7**(2), 51–52 (1961)
51. G.M. Eliashberg, Theory of nonequilibrium states and nonlinear electrodynamics of superconductors. Thesis Sov. Doct, Degree, Chernogolovka - Moscow, 1971

Chapter 4
Superconductors with Impurities

In this Chapter, we will continue our field-theoretical description of superconductors, taking into consideration ordinary and magnetic impurities. The inclusion of ordinary impurities will simplify the dynamic description of superconductivity in many ways. First of all, the isotropic Fermi-liquid model of metals will become valid. Second, superconductors will become "Londonized", i.e., their dynamics will become local. The inclusion of magnetic impurities will deliver a surprising result: gapless superconductivity, which will shine additional light onto the essence of Bose-condensate of Cooper pairs. For this type of superconductor, it will become possible to derive the time-dependent Ginzburg–Landau equations. This is the simplest case of validity for these equations, which we exclusively used in Part I. It contains almost all the essential features of more general TDGL equations which will be derived in Chap. 7.

4.1 Scattering on Ordinary Impurities

4.1.1 Magnetic and Nonmagnetic Impurities

The interaction Hamiltonian of electrons with impurity atoms may be written as

$$\widehat{H}_{int} = \sum_{\alpha} \int \Psi^{+}(x) \widehat{\mathfrak{B}} V(\mathbf{r} - \mathbf{r}_\alpha) \Psi(x) \, d^3\mathbf{r}, \tag{4.1}$$

where α indicates the impurity atoms, and the potential $\widehat{\mathfrak{B}}$ is

$$\widehat{\mathfrak{B}}(\mathbf{r}) = u_1(\mathbf{r}) + u_2(\mathbf{r})(\mathbf{S} \cdot \widehat{\sigma}). \tag{4.2}$$

In expression (4.2) the potentials $u_1(\mathbf{r})$ and $u_2(\mathbf{r})$ stand for the exchange interactions of electrons with nonmagnetic and magnetic impurities, respectively; $\widehat{\sigma}$ is the spin matrix of the electron; $\mathbf{S}$ is the magnetic moment of the impurity atom.

© Springer Nature Switzerland AG 2020
A. Gulian, *Shortcut to Superconductivity*,
https://doi.org/10.1007/978-3-030-23486-7_4

4.1.2 Diagram Expansion and Spatial Averaging for Normal Metals

A diagram technique for the scattering of electrons on impurity atoms may be constructed in the usual manner–by the series expansion of the S-matrix. We use here the approach developed by Abrikosov and Gor'kov [1–3]. It is convenient to consider the properties of this diagram expansion on the example of normal metal, using in (4.2) $\widehat{\mathfrak{B}} = u_1$. Using an $\times$ (cross) to mark the interaction vertex of electrons with impurities, we obtain the diagram series

$$\underset{0}{\xrightarrow{\hspace{0.8cm}}} = \underset{0}{\overset{\delta(\mathbf{p}\text{-}\mathbf{p}')}{\xrightarrow{\hspace{0.8cm}}}} + \underset{1}{\overset{\mathbf{p}\quad\mathbf{p}'}{\xrightarrow{\hspace{0.3cm}}\times\xrightarrow{\hspace{0.3cm}}}} + \underset{2}{\overset{\mathbf{p}\quad\mathbf{p}''\quad\mathbf{p}'}{\xrightarrow{\hspace{0.2cm}}\times\xrightarrow{\hspace{0.2cm}}\times\xrightarrow{\hspace{0.2cm}}}} + \cdots = \xrightarrow{\hspace{0.8cm}} + \xrightarrow{\hspace{0.2cm}}\times\xrightarrow{\hspace{0.2cm}}\xrightarrow{\hspace{0.2cm}} \tag{4.3}$$

or in analytic form[1]

$$G(\mathbf{p}, \mathbf{p}') = \delta(\mathbf{p} - \mathbf{p}')G^0(\mathbf{p})$$
$$+ \sum_\alpha G^0(\mathbf{p}) \int u(\mathbf{p} - \mathbf{p}'')e^{i(\mathbf{p}-\mathbf{p}'')\cdot\mathbf{r}_\alpha} G(\mathbf{p}'', \mathbf{p}') \frac{d^3\mathbf{p}''}{(2\pi)^3} \tag{4.4}$$

[the combination $u(\mathbf{q})e^{i\mathbf{q}\cdot\mathbf{r}_\alpha}\delta(\varepsilon - \varepsilon')$ with summing over α corresponds to the interaction vertex, where $\mathbf{q}$ is the momentum transferred, and $u(\mathbf{q})$ is the Fourier-component of the potential u_1]. Equation (4.4) should be averaged over the impurity coordinates, assuming their chaotic spatial distribution. The averaged values will be denoted by bars above the symbols. Because the averaging procedure is applied to a large volume with many impurity atoms,

$$\overline{G(\mathbf{p}, \mathbf{p}')} = G(\mathbf{p})\delta(\mathbf{p} - \mathbf{p}'). \tag{4.5}$$

After the averaging, diagram 2 in the series (4.3) becomes proportional to the potential $u(\mathbf{p}'' - \mathbf{p}')u(\mathbf{p} - \mathbf{p}'')e^{i(\mathbf{p}-\mathbf{p}'')\cdot\mathbf{r}_\alpha+i(\mathbf{p}''-\mathbf{p}')\cdot\mathbf{r}_\beta}$, and the averaging yields an expression analogous to the one from diagram 1 in all cases, except when $\mathbf{r}_\alpha = \mathbf{r}_\beta$ and $\mathbf{p} = \mathbf{p}'$. As a result, the averaging of diagram 2 gives

$$\overline{G^{(2)}(\mathbf{p}, \mathbf{p}')} = G^0(\mathbf{p}) \overline{\sum_\alpha e^{i(\mathbf{p}-\mathbf{p}')\cdot\mathbf{r}_\alpha}} \int \frac{d^3\mathbf{p}''}{(2\pi)^3} u(\mathbf{p} - \mathbf{p}'')u(\mathbf{p}'' - \mathbf{p}')G^0(\mathbf{p}'')G^0(\mathbf{p}')$$
$$= n \int |u(\mathbf{p} - \mathbf{p}')|^2 G^0(\mathbf{p}') \frac{d^3\mathbf{p}'}{(2\pi)^3} \left[G^0(\mathbf{p})\right]^2 \delta(\mathbf{p} - \mathbf{p}'), \tag{4.6}$$

where $n = N_i/V_0$ is the density of impurity atoms. Using the explicit expression for $G^0(\mathbf{p}, \varepsilon) = \left[\varepsilon - \xi_\mathbf{p} + i\delta\,\text{sign}(\xi_\mathbf{p})\right]^{-1}$, one can find from (4.6)

---

[1]Because the impurity field is a static one, we omit in this Section the variable ε in propagators $G(\mathbf{p}, \mathbf{p}', \varepsilon)$, $G^0(\mathbf{p}, \varepsilon)$ etc., showing this variable explicitly only when its presence is essential.

$$\overline{G^{(2)}(\mathbf{p}, \varepsilon)} = \left[G^0(\mathbf{p}, \varepsilon)\right]^2 \frac{i \, \text{sign}(\varepsilon)}{2\tau}, \tag{4.7}$$

where

$$\frac{1}{\tau} = \frac{nmp_F}{(2\pi)^2} \int |u(\theta)|^2 d\Omega \tag{4.8}$$

is the electron elastic scattering time. Thus the main contribution arises from the diagrams containing crosses, which correspond to the same atoms. It is convenient to link these crosses by broken lines. The diagrams with three crosses provide nothing new. The fourth order of the perturbation theory generally is represented by the diagram

$$= \sum_{\alpha\beta\gamma\delta} \int u(\mathbf{p} - \mathbf{p}_1)e^{i(\mathbf{p}-\mathbf{p}_1)\cdot\mathbf{r}_\alpha} G^0(\mathbf{p}_1)$$

$$\times u(\mathbf{p}_1 - \mathbf{p}_2)e^{i(\mathbf{p}_1-\mathbf{p}_2)\cdot\mathbf{r}_\beta} G^0(\mathbf{p}_2)u(\mathbf{p}_2 - \mathbf{p}_3)e^{i(\mathbf{p}_2-\mathbf{p}_3)\cdot\mathbf{r}_\gamma} G^0(\mathbf{p}_3)$$

$$\times e^{i(\mathbf{p}_3-\mathbf{p}')\cdot\mathbf{r}_\delta} u(\mathbf{p}_3 - \mathbf{p}')\frac{d^3\mathbf{p}_1 \, d^3\mathbf{p}_2 \, d^3\mathbf{p}_3}{(2\pi)^9}. \tag{4.9}$$

A comparison of contributions from the diagrams

$$\tag{4.10}$$

shows that the diagrams with intersected broken lines contain a small parameter $1/(\epsilon_F\tau) \sim 1/(p_Fl)$, where l is the electron's mean free path. Indeed, for the first of the diagrams in (4.10) we have

$$G_1^{(4)} \sim \sum_{\alpha,\gamma} \int u(\mathbf{p} - \mathbf{p}_1)e^{i(\mathbf{p}-\mathbf{p}_1)\cdot\mathbf{r}_\alpha} G^0(\mathbf{p}_1)u(\mathbf{p}_1 - \mathbf{p}_2)e^{i(\mathbf{p}_1-\mathbf{p}_2)\cdot\mathbf{r}_\alpha}$$

$$\times G^0(\mathbf{p}_2)u(\mathbf{p}_2 - \mathbf{p}_3)e^{i(\mathbf{p}_2-\mathbf{p}_3)\cdot\mathbf{r}_\gamma} G^0(\mathbf{p}_3)e^{i(\mathbf{p}_3-\mathbf{p}')\cdot\mathbf{r}_\gamma} u(\mathbf{p}_3 - \mathbf{p}')$$

$$\times \frac{d^3\mathbf{p}_1 d^3\mathbf{p}_2 d^3\mathbf{p}_3}{(2\pi)^9}. \tag{4.11}$$

After the averaging over the impurity positions, this transforms to

$$\overline{G_1^{(4)}} = \sum_{\alpha,\gamma} \int u(\mathbf{p} - \mathbf{p}_1)\overline{e^{i(\mathbf{p}-\mathbf{p}_2)\cdot\mathbf{r}_\alpha} e^{i(\mathbf{p}_2-\mathbf{p}')\cdot\mathbf{r}_\gamma}} G^0(\mathbf{p}_1)G^0(\mathbf{p}_2)$$

$$\times G^0(\mathbf{p}_3)u(\mathbf{p}_1 - \mathbf{p}_2)u(\mathbf{p}_2 - \mathbf{p}_3)u(\mathbf{p}_3 - \mathbf{p}')\frac{d^3\mathbf{p}_1 \, d^3\mathbf{p}_2 \, d^3\mathbf{p}_3}{(2\pi)^9}$$

$$
= \frac{(2\pi)^6}{V_0^2} \sum_{\alpha,\gamma} \int |u(\mathbf{p} - \mathbf{p}_1)|^2 G^0(\mathbf{p}_1) G^0(\mathbf{p}) G^0(\mathbf{p}_3) |u(\mathbf{p}' - \mathbf{p}_3)|^2 \frac{d^3\mathbf{p}_1 \, d^3\mathbf{p}_3}{(2\pi)^6}
$$

$$
\times \, \delta(\mathbf{p} - \mathbf{p}') \sim G^0(p) \frac{1}{\tau^2} \delta(\mathbf{p} - \mathbf{p}'). \tag{4.12}
$$

At the same time, the third of the diagrams in (4.10) yields

$$
G_3^{(4)} \sim \sum_{\alpha,\gamma} \int u(\mathbf{p} - \mathbf{p}_1) e^{i(\mathbf{p} - \mathbf{p}_1)\cdot\mathbf{r}_\alpha} G^0(\mathbf{p}_1) u(\mathbf{p}_1 - \mathbf{p}_2) e^{i(\mathbf{p}_1 - \mathbf{p}_2)\cdot\mathbf{r}_\gamma}
$$

$$
\times \, G^0(\mathbf{p}_2) u(\mathbf{p}_2 - \mathbf{p}_3) e^{i(\mathbf{p}_2 - \mathbf{p}_3)\cdot\mathbf{r}_\alpha}
$$

$$
\times G^0(\mathbf{p}_3) e^{i(\mathbf{p}_3 - \mathbf{p}')\cdot\mathbf{r}_\gamma} u(\mathbf{p}_3 - \mathbf{p}') \frac{d\mathbf{p}_1 \, d\mathbf{p}_2 \, d\mathbf{p}_3}{(2\pi)^9}, \tag{4.13}
$$

or, after averaging over the impurities,

$$
\overline{G_3^{(4)}} \sim \frac{(2\pi)^6}{V_0^2} \sum_{\alpha,\gamma} \int \delta(\mathbf{p} - \mathbf{p}_1 + \mathbf{p}_2 - \mathbf{p}_3) \delta(\mathbf{p}_1 - \mathbf{p}_2 + \mathbf{p}_3 - \mathbf{p}')
$$

$$
\times \, G^0(\mathbf{p}_1) G^0(\mathbf{p}_2) G^0(\mathbf{p}_3) u(\mathbf{p} - \mathbf{p}_1) u(\mathbf{p}_1 - \mathbf{p}_2) u(\mathbf{p}_2 - \mathbf{p}_3) u(\mathbf{p}_3 - \mathbf{p}')
$$

$$
\times \, \frac{d^3\mathbf{p}_1 d^3\mathbf{p}_2 d^3\mathbf{p}_3}{(2\pi)^9} = \frac{(2\pi)^6}{V_0^2} \sum_{\alpha,\gamma} \int |u(\mathbf{p}_1 - \mathbf{p}_2)|^2 |u(\mathbf{p} - \mathbf{p}_1)|^2 G^0(\mathbf{p}_1)
$$

$$
\times \, G^0(\mathbf{p}_2) G^0(\mathbf{p} - \mathbf{p}_1 + \mathbf{p}_2) \frac{d^3\mathbf{p}_1 \, d^3\mathbf{p}_2}{(2\pi)^9} \delta(\mathbf{p} - \mathbf{p}'). \tag{4.14}
$$

Restrictions that follow from the angle integration in (4.14) require that one of the G-functions be of the order $G \sim 1/\epsilon_F$. Meanwhile in expression (4.13) the same function is of the order $G \sim \tau$ in the region important for integration. This circumstance confirms the statement on diagrams with intersections. The situation is analogous to the case of the second and third diagrams (4.10).

4.1.3   Born's Approximation

Apart from the diagrams considered (4.10) there is another one of the fourth order:

$$
\tag{4.15}
$$

The contribution of such diagrams is essential in the case of non-Born scattering. We consider the opposite situation

$$
p_F^3 \int u(\mathbf{r}) d^3\mathbf{r} \ll \epsilon_F, \tag{4.16}
$$

where one can omit contributions such as Exp. $(4.15)^2$.

Let us sum now the selected diagrams. We have

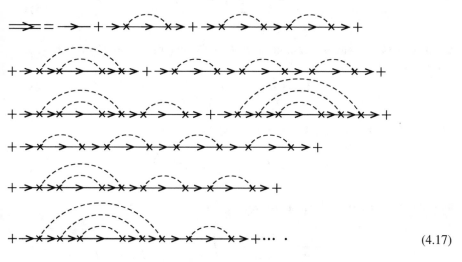

$$(4.17)$$

It is not difficult to see that the result of graphical summation of series (4.17) may be depicted as

$$(4.18)$$

or in analytic form

$$G(\mathbf{p}) = G^0(\mathbf{p}) + nG^0(\mathbf{p}) \int |u(\mathbf{p} - \mathbf{p}')|^2 G(\mathbf{p}') \frac{d^3\mathbf{p}'}{(2\pi)^3} G(\mathbf{p}). \qquad (4.19)$$

The solution of Dyson's equation (4.19), as usual, may be presented in the form

$$G(\mathbf{p}, \varepsilon) = \frac{1}{\left[G^0(\mathbf{p}, \varepsilon)\right]^{-1} - \Sigma^{\text{imp}}}, \qquad (4.20)$$

where, since it follows from (4.19) and (4.10), the self-energy part Σ^{imp} is defined by the equation

$$\Sigma^{\text{imp}} = n \int \frac{d^3\mathbf{p}'}{(2\pi)^3} |u(\mathbf{p} - \mathbf{p}')|^2 \frac{1}{\left[G^0(\mathbf{p}, \varepsilon)\right]^{-1} - \Sigma^{\text{imp}}}. \qquad (4.21)$$

---

[2]This assumption is satisfactory for our immediate purposes. However, when we discuss electron scattering by magnetic impurities later, keeping this assumption results in a failure to predict the Kondo effect [4] and its very interesting consequences for transport phenomena in metals, especially in the case of thermoelectricity [5–10].

Assuming Σ^{imp} is purely imaginary, $\Sigma^{\text{imp}} = -i\beta$, we find in analogy with (4.7):

$$\beta = \frac{\text{sign}(\beta)}{2\tau}. \tag{4.22}$$

Comparing (4.22) with the limiting case $G \to G^0$, one finds $\beta = \text{sign}(\varepsilon)/2\tau$ and, consequently,

$$G(\mathbf{p}, \varepsilon) = \frac{1}{\varepsilon - \xi_{\mathbf{p}} + \frac{i}{2\tau}\,\text{sign}(\varepsilon)}. \tag{4.23}$$

Moving now from (4.23) to the coordinate representation:

$$G(\mathbf{r}, t) = \int e^{i\mathbf{p}\mathbf{r} - i\varepsilon t} G(\mathbf{p}, \varepsilon) \frac{d\varepsilon\, d^3\mathbf{p}}{(2\pi)^4} \tag{4.24}$$

and taking into account formula (3.148), we rewrite (4.24) in the form

$$G(\mathbf{r}, t) = \frac{m}{i(2\pi)^2 r} \int_{-\infty}^{\infty} \frac{d\varepsilon}{2\pi} e^{-i\varepsilon t} \int_{-\infty}^{\infty} \frac{d\xi_{\mathbf{p}}}{\varepsilon - \xi_{\mathbf{p}} + i\,\text{sign}(\varepsilon)/2\tau}$$
$$\times \left[\exp\left\{ i\left(p_F + \frac{\xi_P}{v_F} \right) r \right\} - \exp\left\{ -i\left(p_F + \frac{\xi_P}{v_F} \right) r \right\} \right]. \tag{4.25}$$

Closing the integration contour over $\xi_{\mathbf{p}}$ in the upper and lower half-planes for the first and second integrals in (4.25), respectively, and noting that the first integral is nonzero at $\varepsilon > 0$ and the second at $\varepsilon < 0$ only, we find

$$G(\mathbf{r}, t) = -\frac{m}{2\pi r} e^{-r/2l} \left\{ \int_0^{\infty} \frac{d\varepsilon}{2\pi} e^{-i\varepsilon t} e^{ip_F r + i\varepsilon r/v_F} + \int_{-\infty}^0 \frac{d\varepsilon}{2\pi} e^{-i\varepsilon t} e^{-ip_F r - i\varepsilon r/v_F} \right\}$$
$$= e^{-r/2l} \left\{ \int_{-\infty}^{\infty} \frac{d\varepsilon}{2\pi} e^{-i\varepsilon t} \frac{m}{2\pi r} \left[e^{i(p_F + \varepsilon/v_F)r\,\text{sign}(\varepsilon)} \right] \right\} = e^{-r/2l} G^0(\mathbf{r}, t), \tag{4.26}$$

where $l = v_F \tau$ is the electrons mean free path.

4.1.4   Equations for a Superconducting State

For superconductors we have the system of equations

$$\tag{4.27}$$

$$\text{(diagram)} = \text{(diagram)} + \text{(diagram)} + \text{(diagram)} +$$

$$+ \text{(diagram)} + \text{(diagram)} . \tag{4.28}$$

in analogy with (4.18) (here we set $\widehat{\mathfrak{B}} = u_1$). Let us make the series expansion in (4.27) and (4.28) in powers of the Bose-field Δ, for example, in an equation for G-function:

$$\text{(diagram)} = \text{(diagram)} + \text{(diagram)} + \cdots + \text{(diagram)} +$$

$$+ \text{(diagram)} + \cdots . \tag{4.29}$$

Separating the free line $\longrightarrow$ in this diagrams, we obtain a remaining series $\{1+\Delta[\text{(diagram)}+\cdots+\text{(diagram)}+\cdots]\}$ plus a class of diagrams with even numbers of Δ, which enter the sum after the sign of the cross (we do not distinguish here between Δ and Δ^*): $\text{(diagram)} +\cdots$. Two options are possible when the external broken line is separated from such a diagram: there is either an even or an odd number of vertices Δ in the inner and in the outer regions of this broken line. Summation of these two classes of diagrams yields

$$\text{(diagram)} \quad \text{and} \quad \text{(diagram)} \tag{4.30}$$

Thus, we get the equation for the G-function

$$G_0^{-1} G = 1 - \Delta F^+ + \Sigma_1^{\text{imp}} G - \Sigma_2^{\text{imp}} F^+ \tag{4.31}$$

(we have used here the rule concerning the diagram signs mentioned in Sect. 3.4).

In the same manner one can obtain for F^+ the expansion

$$\text{(diagram)} = \text{(diagram)} + \text{(diagram)} + \cdots +$$

$$+ \text{(diagram)} + \cdots + \text{(diagram)} + \cdots \tag{4.32}$$

Again, the left free line may be separated, which is followed by the vertex Δ or by a cross. Summing the first class of diagrams, one obtains ΔG. If a vertex (diagram) follows an arrow $\longrightarrow$, one can separate the first external broken line: (diagram) (in the inner part of the diagram series there are broken lines and vertices Δ). If there is an even number of Δ in the inner part, then summation of the diagrams gives the function Σ_1^{imp}, and the function F^+ obviously emerges in the outer part. If the broken line embraces the odd numbers of Δ, then this class of diagrams yields the function $\Sigma_2^{+\,\text{imp}}$, and the function G emerges in the outer part.

Thus

$$\left(\overline{G^0}\right)^{-1} F^+ = \Delta^* G + \Sigma_1^{imp} F^+ + \left(\Sigma_2^+\right)^{imp} G, \tag{4.33}$$

where $\overline{G^0} = (\varepsilon + \xi_{\mathbf{p}})^{-1}$. Taking into account the definition of functions Σ_1^{imp} and Σ_2^{+imp}:

$$\Sigma_1^{imp} = n \int |u(\mathbf{p} - \mathbf{p}')|^2 G(\mathbf{p}') \frac{d^3 \mathbf{p}'}{(2\pi)^3}, \tag{4.34}$$

$$\Sigma_2^{+imp} = n \int |u(\mathbf{p} - \mathbf{p}')|^2 F^+(\mathbf{p}') \frac{d^3 \mathbf{p}'}{(2\pi)^3}, \tag{4.35}$$

and analogously for other two elements of the $\widehat{\Sigma}$-matrix (3.92), we obtain the system of equations for superconductors with impurities:

$$\begin{pmatrix} \xi - \varepsilon - \Sigma_1^{imp} & -(\Delta + \Sigma_2^{imp}) \\ \Delta^* + \Sigma_2^{+imp} & \xi + \varepsilon - \Sigma_1^{imp} \end{pmatrix} \begin{pmatrix} G & F \\ -F^+ & G \end{pmatrix}_p = \widehat{1}. \tag{4.36}$$

The self-energy matrix here is:

$$\widehat{\Sigma}^{imp} = \begin{pmatrix} \Sigma_1 & \Sigma_2 \\ -\Sigma_2^+ & \Sigma_1 \end{pmatrix}^{imp} = \frac{1}{2\pi\tau} \frac{2\pi^2}{mp_F} \int \frac{d^3 \mathbf{p}}{(2\pi)^3} \widehat{G}(\mathbf{p}, \varepsilon), \tag{4.37}$$

as follows from (4.34), (4.35), and (4.8).

4.1.5 Anderson's Theorem

The solution of (4.36) and (4.37) may be found in the same manner as done above for the normal state. It gives the same formal result: the appearance of exponential factors $e^{-|\mathbf{r}-\mathbf{r}'|/2l}$ in the Green's functions. However, the gap in the energy spectrum of the superconductor is subject to the self-consistency (3.122), which includes the superconducting propagator at $\mathbf{r} = \mathbf{r}'$. So evidently nonmagnetic impurities do not influence the thermodynamics of a superconductor. This result is called "the Anderson theorem" [11].

4.1.6 "Londonization" by Elastic Scattering

Another important consequence follows from the comparison of (3.147) and (4.26). At $l < v_F / T_c$, the electron correlation radius in superconductors becomes less than ξ_0. We have mentioned this circumstance in Chap. 3 as the "Londonization" of superconductors by the elastic scattering on impurities. This aspect of the influence of impurities is important for superconductors, making their electrodynamics local.

4.2 Magnetic Impurities

When the paramagnetic part of the potential $\widehat{\mathfrak{B}}(\mathbf{r})$ (4.2) is "switched on", the interaction becomes explicitly dependent on the electrons' spins. Consequently, the spin variables should be preserved in the intermediate calculations of Sect. 4.1. Using the Hamiltonian

$$\widehat{H} = -\int \left\{ -\left(\Psi^\dagger(\mathbf{r})\frac{\nabla^2}{2m}\Psi(\mathbf{r})\right) + \frac{\zeta}{2}\left(\Psi^\dagger(\mathbf{r})\left(\Psi^\dagger(\mathbf{r})\Psi(\mathbf{r})\right)\Psi(\mathbf{r})\right) \right.$$
$$\left. + \left(\Psi^\dagger(\mathbf{r})\widehat{\mathfrak{B}}\Psi(\mathbf{r})\right) \right\} d^3\mathbf{r}, \tag{4.38}$$

the following equations of motion for the Heisenberg operators $\Psi(x)$ and $\Psi^\dagger(x)$ can be obtained:

$$\left\{ i\frac{\partial}{\partial t} - \frac{\nabla^2}{2m} \right\}\Psi_\alpha^\dagger(x) + \zeta\Psi_\alpha^\dagger(x)\left(\Psi^\dagger(x)\Psi(x)\right) + \Psi_\beta^\dagger\sigma_{\beta\alpha} = 0 \tag{4.39}$$

$$\left\{ i\frac{\partial}{\partial t} + \frac{\nabla^2}{2m} \right\}\Psi_\alpha(x) - \zeta\left(\Psi^\dagger(x)\Psi(x)\right)\Psi_\alpha(x) - \Psi_\beta\sigma_{\alpha\beta} = 0 \tag{4.40}$$

[as earlier, $x = (\mathbf{r}, t)$; we have included the summation over impurities and the exchange potential in $\sigma_{\alpha\beta}$]. Starting from the definition of Green's functions, taking the derivative of (3.88) and account of (4.38) and using Gor'kov's decoupling (3.90), one obtains the equation

$$\left\{ i\frac{\partial}{\partial t} + \frac{\nabla^2}{2m} + \epsilon_F \right\}G_{\alpha\beta}(x, x') - \sigma_{\alpha\gamma}G_{\gamma\beta}(x, x')$$
$$- i\zeta F_{\alpha\gamma}(x, x)F_{\gamma\beta}^+(x, x') = \delta_{\alpha\beta}\delta(x - x'), \tag{4.41}$$

where the function $F_{\alpha\beta}^+(x, x') = \langle T\Psi_\alpha^\dagger(x)\Psi_\beta^\dagger(x')\rangle$ obeys the equation

$$\left\{ i\frac{\partial}{\partial t} - \frac{\nabla^2}{2m} - \epsilon_F \right\}F_{\alpha\beta}^+(x, x') + \sigma_{\alpha\gamma}^\dagger F_{\gamma\beta}^+(x, x') + i\zeta F_{\alpha\delta}^+(x, x)G_{\delta\beta}(x, x') = 0. \tag{4.42}$$

Now, making the series expansion for the functions G and F^+ on the basis of (4.39) and (4.40), one can see that σ enters into the diagrams $\longrightarrow \times \longrightarrow$, whereas σ^{tr} enters into the diagrams $\longleftarrow \times \longleftarrow$.

4.2.1   Averaging over Spin Directions

Let us return now to the definition of functions

$$\Sigma_1^{imp} = \text{<diagram>} , \quad \Sigma_2^{imp} = \text{<diagram>} , \quad \Sigma^{+\,imp}_{\ 2} = \text{<diagram>} \tag{4.43}$$

(see Sect. 4.1). Besides averaging over the spatial distribution of impurities, one must also average over the spin directions, assuming their random orientation. In the absence of impurities it follows that $\overline{G_{\alpha\beta}} = \overline{G}\delta_{\alpha\beta}$. If the dashed line is spin dependent, then

$$\overline{G_{\alpha\beta}} = \overline{\sigma_{\alpha\gamma}G_{\gamma\delta}\sigma_{\delta\beta}} = \overline{G}\sigma_{\alpha\gamma}\delta_{\gamma\delta}\sigma_{\delta\beta} = \overline{G}\delta_{\alpha\beta}. \tag{4.44}$$

Thus, the averaging of diagrams for G-functions adds the term $u_2^2 S(S+1)/3$ to the potential u_1^2. The situation with functions $\overline{F}$ and $\overline{F^+}$ is different. The corresponding diagrams contain an additional factor Δ and a line $\longleftarrow$, for example,

$$\text{<diagram>} \tag{4.45}$$

As a result, the part of the diagram represented by a dashed line and containing $\sigma^{tr}\widehat{\Delta}\sigma$ will have a sign opposite to $\widehat{\Delta}$. Indeed, $\widehat{\Delta}_{\alpha\beta} = \Delta I_{\alpha\beta}$ (see 3.97 and 3.98). But $I^{-1}\sigma^{tr}I = -\sigma$, and this causes a chain of relations: $\sigma^{tr}I = -I\sigma$; $\sigma^{tr}I\sigma = -I\sigma^2 = -I$.

4.2.2   Spin-Flip Time τ_S

The "magnetic" part in the averaged diagrams for F and F^+ has a sign opposite to the part produced by the usual impurities. Accounting for this, the value τ^{-1} in the diagonal components of $\widehat{\Sigma}$ in (4.37) is replaced by

$$\tau_1^{-1} = \frac{nmp_F}{(2\pi)^2} \int \left\{ |u_1(\theta)|^2 + |u_2(\theta)|^2 \frac{1}{3} S(S+1) \right\} d\Omega, \tag{4.46}$$

and in nondiagonal components by

$$\tau_2^{-1} = \frac{nmp_F}{(2\pi)^2} \int \left\{ |u_1(\theta)|^2 - |u_2(\theta)|^2 \frac{1}{3} S(S+1) \right\} d\Omega. \tag{4.47}$$

The difference in values (4.46) and (4.47) is due exclusively to the magnetic part of the interaction and defines the reciprocal spin-flip time τ_S^{-1}:

$$\tau_1^{-1} - \tau_2^{-1} = 2\tau_S^{-1}. \tag{4.48}$$

4.2.3 Reduction of Transition Temperature

Let us now consider the influence of paramagnetic impurities on the thermodynamic properties of superconductors. The initial equations (prior to the impurity averaging) in the representation of the imaginary discrete frequencies $\varepsilon = \varepsilon_n = i\pi T(2n+1)$ have the form

$$\left\{ \varepsilon + \frac{\nabla^2}{2m} + \epsilon_F \right\} \widehat{\mathfrak{G}}_\varepsilon(\mathbf{r}, \mathbf{r}') - \widehat{\mathfrak{B}}\mathfrak{G}_\varepsilon(\mathbf{r}, \mathbf{r}') + \widehat{\Delta}(\mathbf{r})\widehat{\mathfrak{F}}_\varepsilon^+(\mathbf{r}, \mathbf{r}') = \widehat{1} \cdot \delta(\mathbf{r} - \mathbf{r}'),$$

$$\tag{4.49}$$

$$\left\{ \varepsilon - \frac{\nabla^2}{2m} - \epsilon_F \right\} \widehat{\mathfrak{F}}_\varepsilon^+(\mathbf{r}, \mathbf{r}') + \widehat{\mathfrak{B}}^{tr}\widehat{\mathfrak{F}}_\varepsilon^+(\mathbf{r}, \mathbf{r}') + \widehat{\Delta}^*(\mathbf{r})\widehat{\mathfrak{G}}_\varepsilon(\mathbf{r}, \mathbf{r}') = 0, \tag{4.50}$$

$$\widehat{\Delta}^*(\mathbf{r}) = |\zeta|T \sum_\varepsilon \widehat{\mathfrak{F}}_\varepsilon^+(\mathbf{r}, \mathbf{r}'). \tag{4.51}$$

As before, we use the potential $\widehat{\mathfrak{B}}$ (4.2). It will be shown now that the critical temperature remains unchanged if $u_2 = 0$ and diminishes if $u_2 \neq 0$.

Since the temperatures close to the critical one are of interest, in the expansion of $\mathcal{F}^+$ in powers of the field Δ it is sufficient to retain only the lowest-order diagram:

$$\Longleftrightarrow \approx \overleftarrow{\Delta} \tag{4.52}$$

Substituting the corresponding analytic expression into the self-consistency (4.51), we find

$$\Delta_{\alpha\beta}^*(\mathbf{r}) = |\zeta|T \sum_\varepsilon \int \overline{\mathfrak{G}}_{\alpha\gamma}(\mathbf{r}, \mathbf{r}')\Delta_{\gamma\delta}^*(\mathbf{r}')\mathfrak{G}_{\delta\beta}(\mathbf{r}', \mathbf{r})d^3\mathbf{r}'. \tag{4.53}$$

The equation for Δ^* must have a nonzero solution at the critical temperature. Averaging (4.53) over the impurity positions and taking into account that $\Delta(\mathbf{r})$ is a smooth function and $\mathfrak{G}(\mathbf{r})$ is rapidly oscillating one, one may write

$$\overline{\Delta_{\alpha\beta}^*(\mathbf{r})} = |\zeta|T \sum_\varepsilon \int \overline{\Delta_{\gamma\delta}^*(\mathbf{r}')\mathfrak{G}_{\alpha\gamma}(\mathbf{r}, \mathbf{r}')\mathfrak{G}_{\delta\beta}(\mathbf{r}', \mathbf{r})}d^3\mathbf{r}'. \tag{4.54}$$

The averaging procedure in (4.54) produces the broken lines connecting the crosses not only on the same propagation line, but also on different lines (recall that the potential $\widehat{\mathfrak{B}}^{tr}$ corresponds to crosses on the G-function's line). In the first case, we have $\sigma_{\alpha\gamma}\sigma_{\gamma\beta} = \delta_{\alpha\beta}$, and a factor $|u_1(\mathbf{q})|^2 + \frac{1}{3}S(S+1)|u_2(\mathbf{q})|^2$ arises for the diagram. This leads to the substitution $l \to l_1 = v_F\tau_1$ in expression (4.26):

$$\overline{\mathfrak{G}}_\varepsilon(\mathbf{r} - \mathbf{r}') = \mathfrak{G}_\varepsilon^0(\mathbf{r} - \mathbf{r}') \exp\left(-\frac{|\mathbf{r} - \mathbf{r}'|}{2l_1}\right). \tag{4.55}$$

Correspondingly, the Fourier component of (4.55) has the form [compare (4.23)]:

$$\widetilde{\mathfrak{G}}_\varepsilon(\mathbf{p}) = \frac{1}{\varepsilon\eta_1 - \xi_\mathbf{p}}, \qquad \eta_1 = 1 + \frac{1}{2\tau_1|\varepsilon|}. \tag{4.56}$$

In the second case, one must calculate in (4.54) a "ladder" diagram of "dressed" functions:

$$(\ \ ) + \xi - \xi + \xi = \xi + \xi = \xi + \ldots \equiv (\ \ ) = (\ \ )\{1 + \xi - \xi\} \tag{4.57}$$

It is expedient to introduce the functions $K_{\alpha\beta}(\mathbf{p}_1, \mathbf{p}_2)$ by the relation

$$\overline{\mathfrak{G}_{\alpha\gamma}(\mathbf{r}, \mathbf{r}'')I_{\gamma\delta}\mathfrak{G}_{\delta\beta}(\mathbf{r}'', \mathbf{r})} = \int K_{\alpha\beta}(\mathbf{p}_1, \mathbf{p}_2)e^{i\mathbf{p}_1\cdot(\mathbf{r}-\mathbf{r}'')-i\mathbf{p}_2\cdot(\mathbf{r}''-\mathbf{r}')}\frac{d^3\mathbf{p}_1 d^3\mathbf{p}_2}{(2\pi)^6}. \tag{4.58}$$

Then (4.57) can be presented in the form

$$K_{\alpha\beta}(\mathbf{p}_1, \mathbf{p}_2) = \widetilde{\mathfrak{G}}_\varepsilon(\mathbf{p}_1)\widetilde{\mathfrak{G}}_\varepsilon(\mathbf{p}_2)\left\{I_{\alpha\beta} + n\int \frac{d^3\mathbf{p}}{(2\pi)^3}\left[|u_1|^2 \right.\right.$$
$$\left.\left. +\frac{1}{3}S(S + 1)|u_2|^2\sigma_{\alpha\delta}\sigma_{\lambda\beta}^{\text{tr}}K_{\delta\lambda}(\mathbf{p}_1', \mathbf{p}_2')\right]\right\}, \tag{4.59}$$

where $\mathbf{p}_2'$ is defined from the momentum conservation law: $\mathbf{p}_1 + \mathbf{p}_2 = \mathbf{p}_1' + \mathbf{p}_2'$. The spin part of $K_{\alpha\beta}$ can be separated further: $K_{\alpha\beta} = I_{\alpha\beta}K$. After that, a combination of the type $\sigma_{\alpha\delta}\sigma_{\lambda\beta}I_{\delta\lambda}$ appears on the right hand side of (4.59), which, as noted earlier, is equal to $(-I_{\alpha\beta})$. So one can write (4.59) in the form

$$K(\mathbf{p}_1, \mathbf{p}_2) = \widetilde{\mathfrak{G}}(\mathbf{p}_1)\widetilde{\mathfrak{G}}(\mathbf{p}_2)\{1 + L_\varepsilon\}, \tag{4.60}$$

where

$$L_\varepsilon = n\int \left[|u_1|^2 - \frac{1}{3}S(S + 1)|u_2|^2\right]K(\mathbf{p}_1', \mathbf{p}_2')\frac{d^3\mathbf{p}'}{(2\pi)^3}. \tag{4.61}$$

Multiplying (4.60) by $n\left[|u_1|^2 - \frac{1}{3}S(S + 1)|u_2|^2\right]/(2\pi)^3$ and integrating over $d^3\mathbf{p}_1$, we obtain

$$L_\varepsilon = (1 + L_\varepsilon)n\int \left[|u_1|^2 - \frac{1}{3}S(S + 1)|u_2|^2\right]\widetilde{\mathfrak{G}}(\mathbf{p}_1)\widetilde{\mathfrak{G}}(\mathbf{p}_2)\frac{d\mathbf{p}_1}{(2\pi)^3}. \tag{4.62}$$

Keeping in mind the self-consistency (4.53), we put $\mathbf{p}_1 = \mathbf{p}_2$ in (4.62) and obtain

$$L_\varepsilon = \frac{1}{2\tau_2 \varepsilon \eta_S}, \qquad \eta_S = 1 + \frac{1}{2\tau_S |\varepsilon|}, \tag{4.63}$$

taking into account (4.48) and (4.56).

We return now to (4.54) and move the factor $\overline{\Delta}^*$ out from under the integral operator (as in Sect. 4.4, in what follows we will discard the bar above the symbol Δ). Using the expressions (4.58)–(4.63) we find in this way

$$\Delta^*(\mathbf{r}) = \Delta^*(\mathbf{r}) |\zeta| T \sum_\varepsilon \int K(\mathbf{p} - \mathbf{p}_1) \frac{d^3 \mathbf{p}_1}{(2\pi)^3}$$

$$= \Delta^*(\mathbf{r}) |\zeta| T \sum_\varepsilon \int \left(1 + \frac{1}{2\varepsilon \tau_2 \eta_S}\right) \widetilde{\mathfrak{G}}(\mathbf{p}_1) \widetilde{\widetilde{\mathfrak{G}}}(\mathbf{p}_1) \frac{d^3 \mathbf{p}_1}{(2\pi)^3}, \tag{4.64}$$

from which the equation for the critical temperature T_c follows:

$$1 = |\zeta| T_c \sum_\varepsilon \int \frac{(\eta_1/\eta_S)}{\xi_\mathbf{p}^2 + (\eta_1 |\varepsilon|)^2} \frac{d^3 \mathbf{p}}{(2\pi)^3}. \tag{4.65}$$

One can see from this expression that T_c will not change in the absence of magnetic impurities. Indeed, at $u_2 = 0$, $\tau_2 = \tau$, $\eta_S = 1$, (4.65) transforms to

$$1 = |\zeta| \frac{m p_F}{(2\pi)^2} T_c \sum_\varepsilon \int \frac{\eta_1 d\xi_\mathbf{p}}{|\varepsilon|^2 \eta_1^2 + \xi_\mathbf{p}^2}. \tag{4.66}$$

Making in (4.66) the replacement $(\xi_\mathbf{p} \eta_1) \to \xi_\mathbf{p}$, we arrive at (4.37) determining T_c, which corresponds to a "pure" sample, $\eta_1 = 1$. This verifies the unshifted value of T_c.

The situation is different for $u_2 \neq 0$ when

$$\eta_S = 1 + \frac{1}{2\tau_1 |\varepsilon|} - \frac{1}{2\tau_2 |\varepsilon|} = 1 + \frac{1}{\tau_S |\varepsilon|} > 1. \tag{4.67}$$

Restricting ourselves to a crude approximation, we take η_S as a constant.[3] In this approximation one arrives at the equation

$$1 = \frac{|\zeta|}{\eta_S} T_c \sum_\varepsilon \int \frac{\eta_1 d^3 \mathbf{p}}{\xi_\mathbf{p}^2 + (\eta_1 |\varepsilon|)^2}, \tag{4.68}$$

which coincides with (4.66), but with a smaller interaction constant and hence [see (3.135)] with smaller T_c.

---

[3] The exact calculations [2] lead to the following expression for the critical temperature T_c : $\ln(T_{co}/T_c) = \psi(1/2 + \rho/2) - \psi(1/2)$ where ψ is the logarithmic derivative of the Γ-function, $\rho = 1/(\pi \tau_s T_c)$, T_{co} is the critical temperature in absence of impurities.

4.2.4   Energy-Gap Suppression

In the presence of impurities, the single-particle spectrum of the system is not a well-defined quantity, because $\mathbf{p}$ is a bad quantum number. So, we will try to determine the value of a gap on the base of reasonings, that are principally different from those used in Sect. 3.3. Namely, we return to the expression for Green's function of a "pure" superconductor at temperature $T = 0$. We present (3.108) in the form ($\mathcal{P}$ indicates the principal value)

$$G = \mathcal{P}\frac{\varepsilon + \xi_{\mathbf{p}}}{\varepsilon^2 - \varepsilon_{\mathbf{p}}^2} - i\pi\delta\left[\varepsilon - \sqrt{\Delta^2 + \xi_{\mathbf{p}}^2}\,\mathrm{sign}(\xi_{\mathbf{p}})\right]. \tag{4.69}$$

The imaginary part of Green's function in (4.69) is determined by the δ-function. The first excited state in the system (at $\xi_{\mathbf{p}} \to 0$) may be found as the minimal positive value of ε at which Green's function acquires an imaginary part. This conclusion, as was shown by Migdal and Galitzkiy [12], remains valid in the general case.

Solving the system (4.36) and (4.37) with the help of (4.46) and (4.47), one can find for $G_{\varepsilon}(\mathbf{p})$ and $F_{\varepsilon}(\mathbf{p})$ (in a real order parameter gauge, $\Delta^* = \Delta$) the expressions:

$$G_{\varepsilon}(\mathbf{p}) = \frac{\bar{\varepsilon} + \xi_{\mathbf{p}}}{\bar{\varepsilon}^2 - \xi^2 - \bar{\Delta}^2}, \qquad F_{\varepsilon}(\mathbf{p}) = \frac{\bar{\Delta}}{\bar{\varepsilon}^2 - \xi^2 - \bar{\Delta}^2}. \tag{4.70}$$

Here

$$\bar{\varepsilon} = \varepsilon + \frac{i}{2\tau_1}\frac{u}{\sqrt{1 - u^2}}, \qquad \bar{\Delta} = \Delta + \frac{i}{2\tau_2}\frac{u}{\sqrt{1 - u^2}}, \qquad u = \frac{\bar{\varepsilon}}{\bar{\Delta}}. \tag{4.71}$$

The equation for the function $u(\varepsilon)$ follows from (4.71):

$$\frac{\varepsilon}{\Delta} = u\left(1 - \frac{1}{\tau_S\Delta\sqrt{1 - u^2}}\right). \tag{4.72}$$

At $\tau_S\Delta > 1$ and for sufficiently small values of ε, both functions u and ε are real. The right-hand side of (4.72) has the maximum at

$$u_0 = \left[1 - (1/\tau_S\Delta)^{2/3}\right]^{1/2} \tag{4.73}$$

at the corresponding value

$$\varepsilon_0 = \Delta\left[1 - (1/\tau_S\Delta)^{2/3}\right]^{3/2}. \tag{4.74}$$

For larger ε, the solutions u are complex and the quantity $G(\varepsilon)$ acquires an imaginary part. So the quantity ε_0 (4.74) determines the value of the gap in superconductors with paramagnetic impurities.

4.2.5 Gapless Superconductivity

As follows from (4.74), the value of the gap vanishes at

$$\tau_S \Delta = 1, \tag{4.75}$$

which is possible for $\Delta \neq 0$. This means that for superconductors with paramagnetic impurities, the order parameter Δ does not coincide with the value of the gap.

Let us now determine, at what concentration of impurities Δ vanishes. The self-consistency equation for the order parameter may be written in the form

$$F(0) = \int F_\varepsilon(\mathbf{p}) \frac{\mathrm{d}\varepsilon \mathrm{d}^3 \mathbf{p}}{(2\pi)^4}. \tag{4.76}$$

Integrating (4.76) with the help of (4.71)–(4.73) and (3.107) (for details see [2]) we arrive at the expression (Δ_0 is the gap in a "pure" superconductor):

$$\ln \frac{\Delta}{\Delta_0} = \begin{cases} -\pi/(4\tau_S \Delta_0) & \text{at } \tau_S \Delta \geq 1, \\ -\ln\left[\left(1 + \sqrt{1 - (\tau_S \Delta)^2}\right)/(\tau_S \Delta)\right] \\ -[1/(2\tau_S \Delta)]\arcsin \tau_S \Delta + \sqrt{1 - (\tau_S \Delta)^2}/2 & \text{at } \tau_S \Delta < 1. \end{cases} \tag{4.77}$$

Setting $\Delta \to 0$ in (4.77), we have

$$\ln \frac{\Delta}{\Delta_0} = \ln \frac{\tau_S \Delta}{1 + \sqrt{1 - (\tau_S \Delta)^2}}. \tag{4.78}$$

It follows from (4.78) that the critical concentration of impurities at which the superconducting order parameter vanishes is determined by the relation

$$\left.\frac{1}{\tau_S}\right|_{\Delta=0} = \frac{\Delta_0}{2}. \tag{4.79}$$

At the same time, as follows from (4.77) and (4.75), the gap vanishes when

$$\left.\frac{1}{\tau_S}\right|_{\varepsilon_0=0} = \Delta_0 \exp\left(-\frac{\pi}{4}\right). \tag{4.80}$$

Because $e^{-\pi/4} < 1/2$, one can conclude that superconducting correlations remain in the superconductor while the gap has disappeared. Hence, there is a certain interval of paramagnetic impurity concentration in which "gapless superconductivity" can be realized. In this gapless superconductors, the quantum correlations in the self-consistent are strong enough to maintain the superfluid nature of the condensate motion (or, in other words, to maintain discussed above "off-diagonal long-range order") despite the absence of the gap in single-electron excitation spectrum.

If the impurity concentration is increased, the gap singularity in a single-particle density of states smears out simultaneously with vanishing of the gap, as may be seen from (4.70) and (4.72) (detailed calculations may be found in [13] and the corresponding figures—in [14]). This property permits us to derive the nonstationary equations of the Ginzburg–Landau type for alloys with magnetic impurities.

4.3 Nonstationary Ginzburg–Landau Equations

As in a stationary case (see Sect. 3.4), the self-consistency equation

$$\Delta_\omega^*(\mathbf{k}) = |\zeta| T \sum_n \int \frac{d^3\mathbf{p}}{(2\pi)^3} \mathfrak{F}_{\varepsilon\varepsilon-\omega}^+(\mathbf{p}_1, \mathbf{p} - \mathbf{k}), \qquad \omega = 2n\pi T i \qquad (4.81)$$

may serve as a starting point for derivation of nonstationary equation for the order parameter $\Delta(\mathbf{r}, t)$. The idea of calculations in a nonstationary situation is to present $\mathfrak{F}_{\varepsilon\varepsilon-\omega}^+$ as a series expansion in powers of Δ_ω and Δ_ω^* and in powers of electromagnetic field potentials, considering all these Bose fields as classical. As a result, an equation would follow from (4.81):

$$\Delta_\omega^* = B^{(1)}(\omega)\Delta_\omega^* + \sum_{\omega'+\omega''+\omega'''=\omega} B^{(3)}(\omega', \omega'', \omega''')\Delta_{\omega'}^* \Delta_{\omega''} \Delta_{\omega'''}^* + \dots \qquad (4.82)$$

where the coefficients $B^{(i)}$ represent the response of the system to the action of the classical field Δ.

4.3.1 Causality Principle and Nonlinear Problems

In nonequilibrium conditions, an equation of the kind (4.81) may be obtained in a real-time representation using the Keldysh technique [15]. The same result may be obtained by the Gor'kov–Eliashberg technique [16], which is a generalization of the usual procedure of analytical continuation to the nonlinear case. The underlying principle at the base of this technique asserts that the response of system (4.82) in a real-time representation must contain the values of the field Δ in the moments preceding the current time. This demand can be satisfied if the coefficients $B^{(i)}(\omega', \omega'', \dots, \omega^i)$, which are determined in the Matsubara technique on the imaginary axis, are analytically continued onto the upper half-plane for all the frequencies $\omega', \omega'', \dots, \omega^i$. One can verify this assertion in a manner analogous to the case of linear response (e.g., in the case of derivation of the Kramers–Krönig relations). In the next section we trace the calculations of Gor'kov and Eliashberg [16].

4.3.2 Equations on Imaginary Axis

Allowing for the time-dependence of the fields, the Gor'kov equations can be represented in the form $[\varepsilon = (2n + 1)\pi T i, \omega = 2m\pi T i]$

$$
\begin{pmatrix} \frac{1}{2m}\left(-i\frac{\partial}{\partial \mathbf{r}}\right)^2 + H_1 - \epsilon_F - \varepsilon & -\Delta_\omega(\mathbf{r}) \\ \Delta_\omega^*(\mathbf{r}) & \frac{1}{2m}\left(-i\frac{\partial}{\partial \mathbf{r}}\right)^2 + \overline{H}_1 - \epsilon_F + \varepsilon \end{pmatrix} \begin{pmatrix} \mathfrak{G} & \mathfrak{F} \\ -\mathfrak{F}+ & \mathfrak{G} \end{pmatrix} = \widehat{1} \cdot (2\pi)^4 \delta(\mathbf{r} - \mathbf{r}')\delta(\omega),
$$

(4.83)

where the function $\Delta_\omega^*(\mathbf{r})$ is defined by equation (4.81). In expression (4.83) the values of H_1 and $\overline{H}_1$ are given by the relations:

$$
H_1 = -\frac{e}{c}(\mathbf{v} \cdot \mathbf{A}_\omega(\mathbf{r})) + e\varphi_\omega(\mathbf{r}), \quad \overline{H}_1 = \frac{e}{c}(\mathbf{v} \cdot \mathbf{A}_\omega(\mathbf{r})) + e\varphi_\omega(\mathbf{r}).
$$

(4.84)

Equation (4.83) and the expressions for the current and the density of particles

$$
\mathbf{j}_\omega(\mathbf{k}) = -\frac{2e}{m}T\sum_\varepsilon \int \frac{d^3\mathbf{p}}{(2\pi)^3}\mathbf{p}\mathfrak{G}_{\varepsilon\varepsilon-\omega}(\mathbf{p}_+, \mathbf{p}_-) - \frac{e^2}{mc^2}(NA)_{\omega,\mathbf{k}},
$$

(4.85)

$$
N_\omega(\mathbf{k}) = -2T\sum_\varepsilon \int \frac{d^3\mathbf{p}}{(2\pi)^3}\mathfrak{G}_{\varepsilon\varepsilon-\omega}(\mathbf{p}_+, \mathbf{p}_-), \quad \mathbf{p}_\pm = \mathbf{p}\pm\frac{\mathbf{k}}{2}
$$

(4.86)

are basic for further calculations.[4] Using (4.83), one can establish the diagram expansions for the functions $\mathfrak{G}$ and $\mathfrak{F}+$:

(4.87)

(4.88)

or in analytic form

$$
\mathfrak{G}_{\varepsilon\varepsilon-\omega}(\mathbf{p}, \mathbf{p}') =
$$
$$
(2\pi)^4 \frac{\delta(\mathbf{p} - \mathbf{p}')\delta(\omega)}{\xi_\mathbf{p} - \varepsilon} + \frac{e}{mc}\frac{\mathbf{p}' \cdot \mathbf{A}_\omega(\mathbf{p} - \mathbf{p}')}{\xi_\mathbf{p} - \varepsilon}\frac{1}{\xi_{\mathbf{p}'} - \varepsilon + \omega} + \dots
$$

(4.89)

$$
\mathfrak{F}_{\varepsilon\varepsilon-\omega}^+(\mathbf{p}, \mathbf{p}') = \frac{1}{\xi_\mathbf{p} + \varepsilon}\Delta_{\omega-\omega'}^*(\mathbf{p} - \mathbf{p}')\frac{1}{\xi_{\mathbf{p}'} - \varepsilon} - \dots
$$

(4.90)

---

[4] Notice the difference in signs between (4.83) and its static analogue (3.120), (3.121). In both cases we have retained the notation of the original works [17, 18] to maintain the connection between these equations and many other original investigations. The difference in propagators' signs is unimportant here.

[the summation in (4.89) and (4.90) includes all the intermediate energies and integration over all the intermediate momenta]. Note that to derive the nonstationary Ginzburg–Landau equations (as in the static case), one may keep only the first few terms of the decomposition (4.88), with subsequent substitution into (4.81) and (4.85). However, it is expedient to consider the problem from a more general point of view, retaining all terms in (4.89) and (4.90).

4.3.3 Analytical Continuation Procedure

In expressions (4.81), (4.85) and (4.86), it is necessary to carry out analytical continuation over ω from the upper half-plane onto the real axis. Note, that the analytical structure of diagrams is insensitive to the directions of arrows and to the presence of vertices Δ_{ω_i} and $H_{1\omega_i}$. Let us consider a general term of the series:

$$T \sum_{\varepsilon} \mathfrak{G}_\varepsilon \mathfrak{G}_{\varepsilon-\omega_1} \mathfrak{G}_{\varepsilon-\omega_1-\omega_2} \cdots \mathfrak{G}_{\varepsilon-\omega}, \tag{4.91}$$

where the summation is also assumed over the internal frequencies, subject to the condition $\sum \omega_i = \omega$. The procedure of analytical continuation of (4.89) over all ω_i onto the real axis should not depend on the order in which the continuation over each of the frequencies proceeds. Then the problem of analytical continuation of the whole structure will be solved.

Let us transform the sum in (4.89) into the contour integral

$$T \sum_{n} \mathfrak{G}_{\varepsilon_n} \mathfrak{G}_{\varepsilon_n-\omega_1} \cdots \mathfrak{G}_{\varepsilon_n-\omega} = \oint_C \frac{dz}{4\pi i} \tanh \frac{z}{2T} \mathfrak{G}_z \mathfrak{G}_{z-\omega_1} \cdots \mathfrak{G}_{z-\omega}, \tag{4.92}$$

where the contour C encloses all the poles of the hyperbolic tangent and does not contain the poles of the $\mathfrak{G}$-functions (Fig. 4.1). Consider a diagram of the Nth power of the field and make the cuts between the singularities of the integrand in (4.92) produced by the functions $\mathfrak{G}_z$, $\mathfrak{G}_{z-\omega_1}$ and so on. Transform the integration contour C into a new one C', which goes along the banks of the cut (Fig. 4.1) and along the arcs of large circles. The contributions from the latter disappear [owing to the factor $(1/z)^{N+1}$] in all the diagrams, except the zero-order one for the $\mathfrak{G}$-function (which does not depend explicitly on the time variable). On horizontal parts of the integration we have $z = \omega_i + \varepsilon$ where ε is a real variable and ω_i is a fixed imaginary frequency. Shifting the integration variable and taking into account that $\tanh(x + i\pi n) = \tanh x$, we can write

Fig. 4.1 Transformation of the integration contour: dashed lines indicate cuts in the z-plane (in absence of the cut, e.g., at Im z = 0, the total contribution over lines a and b equals to zero)

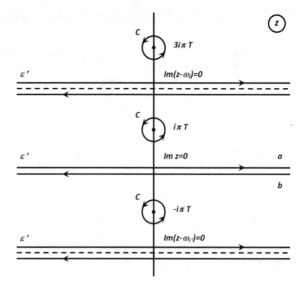

$$\oint \frac{dz}{4\pi i} \tanh \frac{z}{2T} \mathfrak{G}_z \mathfrak{G}_{z-\omega_1} \ldots \mathfrak{G}_{z-\omega} =$$

$$\int_{-\infty}^{\infty} \frac{d\varepsilon}{4\pi i} \tanh \frac{\varepsilon}{2T} \left\{ \left(G_\varepsilon^R - G_\varepsilon^A \right) \mathfrak{G}_{\varepsilon-\omega_1} \ldots \mathfrak{G}_{\varepsilon-\omega} \right.$$

$$+ \mathfrak{G}_{\varepsilon+\omega_1} \left(G_\varepsilon^R - G_\varepsilon^A \right) \mathfrak{G}_{\varepsilon-\omega_2} \ldots \mathfrak{G}_{\varepsilon-\omega+\omega_1}$$

$$+ \mathfrak{G}_{\varepsilon+\omega_1+\omega_2} \mathfrak{G}_{\varepsilon+\omega_2} \left(G_\varepsilon^R - G_\varepsilon^A \right) \mathfrak{G}_{\varepsilon-\omega_3} \ldots \mathfrak{G}_{\varepsilon-\omega+\omega_1+\omega_2}$$

$$+ \mathfrak{G}_{\varepsilon+\omega} \mathfrak{G}_{\varepsilon+\omega-\omega_1} \ldots \left(G_\varepsilon^R - G_\varepsilon^A \right) \right\}, \tag{4.93}$$

where the functions G^R and G^A have the well known analytical properties:

$$G_\varepsilon^R = \left[\xi_\mathbf{p} \overset{(-)}{\underset{(+)}{-}} (\varepsilon + i\delta) \right]^{-1}, \quad G_\varepsilon^A = \left[\xi_\mathbf{p} \overset{(-)}{\underset{(+)}{-}} (\varepsilon - i\delta) \right]^{-1}, \quad \delta \to +0. \tag{4.94}$$

Such representation allows us to determine analytical properties for all factors to the left and to the right of G^R and G^A, if all the frequencies ω_i belong to the same half-plane. In particular, if all the frequencies ω_i belong to the upper half-plane (which corresponds to the general causality principle mentioned earlier), then all the functions in (4.93) to the left of the factor $\left(G_\varepsilon^R - G_\varepsilon^A \right)$ would be retarded, and those to the right would be advanced. Indicating these functions by the letters R and A, we can now move to the real frequencies $\omega_i \to \omega_i + i\delta, \delta > 0$.

4.3.4 Anomalous Propagators and Dyson Equations

We have formulated the procedure of analytical continuation, which is independent of the ordering in ω_i. The final result may be written after once again shifting the integration variable in the integrands:

$$
\begin{aligned}
T \sum_\varepsilon \mathcal{G}_\varepsilon \mathcal{G}_{\varepsilon-\omega_1} \cdots \mathcal{G}_{\varepsilon-\omega} &= \int_{-\infty}^{\infty} 4\pi i \left\{ -\tanh\frac{\varepsilon}{2T} G_\varepsilon^A G_{\varepsilon-\omega_1}^A \cdots G_{\varepsilon-\omega}^A \right. \\
&+ G_\varepsilon^R G_{\varepsilon-\omega_1}^R \cdots G_{\varepsilon-\omega}^R \tanh\frac{\varepsilon-\omega}{2T} + G_\varepsilon^R \left(\tanh\frac{\varepsilon}{2T} - \tanh\frac{\varepsilon-\omega_1}{2T}\right) G_{\varepsilon-\omega_1}^A \cdots G_{\varepsilon-\omega}^A \\
&+ G_\varepsilon^R G_{\varepsilon-\omega_1}^R \left(\tanh\frac{\varepsilon-\omega_1}{2T} - \tanh\frac{\varepsilon-\omega_1-\omega_2}{2T}\right) G_{\varepsilon-\omega_1-\omega_2}^A \cdots G_{\varepsilon-\omega}^A \\
&+ \left. G_\varepsilon^R G_{\varepsilon-\omega_1}^R \cdots G_{\varepsilon-\omega+\omega_1}^R \left(\tanh\frac{\varepsilon-\omega+\omega_1}{2T} - \tanh\frac{\varepsilon-\omega}{2T}\right) G_{\varepsilon-\omega}^A \right\}.
\end{aligned}
\tag{4.95}
$$

This expression may be rewritten as

$$
T \sum_\varepsilon \mathcal{G}_{\varepsilon\varepsilon-\omega} \to \int_{-\infty}^{\infty} \frac{d\varepsilon}{4\pi i} G_{\varepsilon\varepsilon-\omega},
\tag{4.96}
$$

where

$$
G_{\varepsilon\varepsilon-\omega} = G_{\varepsilon\varepsilon-\omega}^R \tanh\frac{\varepsilon-\omega}{2T} - \tanh\frac{\varepsilon}{2T} G_{\varepsilon\varepsilon-\omega}^A + G_{\varepsilon\varepsilon-\omega}^{(a)}.
\tag{4.97}
$$

The regular functions $G^{R(A)}$ in (4.95) and (4.97) are determined from diagram expansions in which all the functions are retarded (or advanced). On the contrary, the analytic structure of the "anomalous" function $G^{(a)}$ is much more complicated. Taking into account the directions of arrows in the diagrams, one can find for $G^{(a)}$ the graphical expression

$$
G^{(a)} = \qquad\qquad\qquad\qquad\qquad ,
\tag{4.98}
$$

where all the field vertices are multiplied by $\{\tanh\left[(\varepsilon - \omega_1)/(2T)\right] - \tanh\left[(\varepsilon - \omega_1 - \omega_2)/(2T)\right]\}$. The retarded propagator corresponds to the line lying to the left of the vertex, and the advanced propagator corresponds to the line lying to the right. In analytic form we have

$$
\begin{aligned}
G_{\varepsilon\varepsilon-\omega}^{(a)}(p_1, p-k) &= \iint \frac{d\omega_1 d\omega_2}{(2\pi)^2} \iint \frac{d^3 k_1}{(2\pi)^3} \frac{d^3 k_2}{(2\pi)^3} \left(\tanh\frac{\varepsilon-\omega_1}{2T} - \tanh\frac{\varepsilon-\omega_1-\omega_2}{2T}\right) \\
&\times \left\{ -G^R \Delta F^{+A} - F^R \Delta^* G^A - G^R H_1 G^A + F^R \overline{H}_1 F^{+A} \right\},
\end{aligned}
\tag{4.99}
$$

where

$$\{G^R \Delta F^{+A}\} \equiv G^R_{\varepsilon\varepsilon-\omega_1}(\mathbf{p}_1, \mathbf{p} - \mathbf{k}_1)\Delta_{\omega_2}(\mathbf{k}_2) F^{+A}_{\varepsilon-\omega_1-\omega_2,\varepsilon-\omega}(\mathbf{p} - \mathbf{k}_1 - \mathbf{k}_2, \mathbf{p} - \mathbf{k}).$$

$$(4.100)$$

The expressions for nondiagonal Green's functions may be found analogously. For $F^{+(a)}_{\varepsilon\varepsilon-\omega}$ we obtain

$$F^{+(a)}_{\varepsilon\varepsilon-\omega} = \iint \frac{d\omega_1 d\omega_2}{(2\pi)^3} \iint \frac{d^3\mathbf{k}_1}{(2\pi)^3} \frac{d^3\mathbf{k}_2}{(2\pi)^3} \left(\tanh \frac{\varepsilon - \omega_1}{2T} - \tanh \frac{\varepsilon - \omega_1 - \omega_2}{2T} \right)$$

$$\times \left\{ -\overline{G}^R \overline{H}_1 F^{+A} - F^{+R} H_1 G^A + \overline{G}^R \Delta^* G^A - F^{+R} \Delta F^{+A} \right\}.$$

$$(4.101)$$

The Dyson equations may also be found for anomalous functions. A graphic representation is useful for this purpose. Let us present (4.98) in the form

$$\mathfrak{G}^{(a)} = \quad$$ $$.$$

$$(4.102)$$

Here the upper lines correspond to the retarded propagator, and the lower ones to the advanced propagators; the right vertices are multiplied by ($\tanh \frac{\varepsilon}{2T} - \tanh \frac{\varepsilon-\omega}{2T}$), where ε, $\varepsilon - \omega$ are the frequencies corresponding to adjacent lines. Specifying these diagrams, say, in the following way

$$\mathfrak{G}^{(a)} = \quad$$ $$+ \cdots,$$

$$(4.103)$$

and detaching the upper free-line $(\xi_\mathbf{p} - \varepsilon - i\delta)^{-1}$ (shown by a dashed line), one obtains the following Dyson-type equation

$$(\xi_\mathbf{p} - \varepsilon)G^{(a)}_{\varepsilon\varepsilon-\omega}(\mathbf{p}, \mathbf{p} - \mathbf{k}) = - \int \frac{d^4 k_1}{(2\pi)^4} \left\{ \left[H_1(k_1)G^A(p - k_1, p - k) \right. \right.$$

$$\left. + \Delta(k_1)F^{+A}(p - k_1, p - k) \right] \left(\tanh \frac{\varepsilon}{2T} - \tanh \frac{\varepsilon - \omega_1}{2T} \right)$$

$$+ H_1(k_1)G^{(a)}(p - k_1, p - k) + \Delta(k_1)F^{+(a)}(p - k_1, p - k) \right\}. \quad (4.104)$$

Using the Dyson equations (which the functions $G^{R(A)}$ and $F^{+R(A)}$ obey) and definitions like (4.97), one can exclude from consideration the anomalous functions obtaining the closed equation for G-function:

$$(\xi - \varepsilon)G_{\varepsilon\varepsilon-\omega} = - \int \frac{d^4 k_1}{(2\pi)^4} \left[H_1(k_1)G_{\varepsilon-\omega_1\varepsilon-\omega} + \Delta(k_1)F^+_{\varepsilon-\omega_1\varepsilon-\omega} \right], \quad (4.105)$$

or in a concise notation

$$(\xi - \varepsilon)G_{\varepsilon\varepsilon-\omega} = -\left\{H_1 G + \Delta F^+\right\}_{\varepsilon\varepsilon-\omega}. \tag{4.106}$$

This equation coincides in form with the equations for retarded and advanced propagators. Being homogeneous, these equations to some extent are deficient without certain additional conditions. We will consider one called a "normalization condition" in Chap. 5.

4.3.5 Regular Terms

Returning now to the problem of derivation of the Ginzburg–Landau type dynamic equations, we substitute an expression

$$F^+_{\varepsilon\varepsilon-\omega} = F^{+R}_{\varepsilon\varepsilon-\omega} \tanh \frac{\varepsilon - \omega}{2T} - \tanh \frac{\varepsilon}{2T} F^{+A}_{\varepsilon\varepsilon-\omega} + F^{+(a)}_{\varepsilon\varepsilon-\omega} \tag{4.107}$$

into the self-consistency equation, which now acquires the form

$$\Delta^*_\omega(\mathbf{k}) = |\zeta| \int_{-\omega_D}^{\omega_D} \frac{d\varepsilon}{4\pi i} \frac{d^3\mathbf{p}}{(2\pi)^3} F^+_{\varepsilon\varepsilon-\omega}(\mathbf{p}, \mathbf{p} - \mathbf{k}). \tag{4.108}$$

Because the functions F^{+R} and F^{+A} are analytical in the upper and lower half-planes correspondingly, respectively, one can move again to the summation over $\varepsilon_n = i\pi T(2n+1)$ in the first two ("regular") terms. As a result we obtain

$$\Delta^*_\omega(\mathbf{k}) = |\zeta| T \left[\sum_{n \geq 0} \int \frac{d^3\mathbf{p}}{(2\pi)^3} \mathfrak{F}^+_{\varepsilon_n + \omega \varepsilon_n}(\mathbf{p}, \mathbf{p} - \mathbf{k}) + \sum_{n < 0} \int \frac{d^3\mathbf{p}}{(2\pi)^3} \mathfrak{F}^+_{\varepsilon_n \varepsilon_n - \omega}(\mathbf{p}, \mathbf{p} - \mathbf{k}) \right]$$

$$+ |\zeta| \int_{-\omega_D}^{\omega_D} \frac{d\varepsilon}{4\pi i} \frac{d^3\mathbf{p}}{(2\pi)^3} F^{+(a)}_{\varepsilon\varepsilon-\omega}(\mathbf{p}, \mathbf{p} - \mathbf{k}) \tag{4.109}$$

(ω is real now!). Further manipulations of the regular terms in (4.109) are similar to those considered in Sect. 3.4 for the static case. As follows from that discussion, it is enough to consider only the diagrams

$$\tag{4.110}$$

Unlike the static case, the field vertices in (4.110) are time-dependent (e.g., $\Delta = \Delta_\omega$), so the diagrams explicitly depend on time. We will discuss the most simple and important case of alloys with paramagnetic impurities, when the impurity concentration is sufficiently high ($\Delta \tau_S \ll 1$), so $\varepsilon \sim \omega \sim \Delta^2 \tau_S \ll \Delta$. In this case the time dependence may be kept only in the first diagram on the right-hand side of (4.110), inserting in the others $\omega = 0$ and returning to the static case. Simultaneously, only the first (linear) term may be kept in this selected diagram in its expansion over ω. Substituting these expressions into (4.109), one finds for the regular contribution from the first diagram (4.110):

$$\Delta_\omega^{*(1)}(\mathbf{r}) = |\zeta| T \left\{ \sum_{n \geq 0} \int d^3 \mathbf{r}_1 \left[\overline{\mathfrak{G}}_{\varepsilon_n + \omega}(\mathbf{r} - \mathbf{r}_1) \Delta_\omega^*(\mathbf{r}_1) \mathfrak{G}_{\varepsilon_n}(\mathbf{r} - \mathbf{r}_1) \right] \right.$$
$$\left. + \sum_{n < 0} \int d^3 \mathbf{r}_1 \left[\mathfrak{G}_{\varepsilon_n}(\mathbf{r} - \mathbf{r}_1) \Delta_\omega^*(\mathbf{r}_1) \overline{\mathfrak{G}}_{\varepsilon_n - \omega}(\mathbf{r} - \mathbf{r}_1) \right] \right\}, \quad (4.111)$$

where the $\mathfrak{G}$-functions are defined according to (3.147). The series over n arising in (4.111) may be summed, yielding

$$\Delta_\omega^{*(1)}(\mathbf{r}) = |\zeta| T \int d^3 \mathbf{r}_1 \Delta^*(\mathbf{r}_1) \exp\left\{ i\omega \frac{|\mathbf{r} - \mathbf{r}_1|}{v_F} \right\} \frac{m^2}{(2\pi|\mathbf{r} - \mathbf{r}_1|)^2 \sinh \frac{2\pi T |\mathbf{r} - \mathbf{r}_1|}{v_F}}.$$
$$(4.112)$$

As is clear from (4.112), the expansion of the exponent would occur in powers of the factor ω / T_c. In addition to the terms obtained in Sect. 3.4, we will obtain the term

$$\Delta_\omega^{*(1)}(\mathbf{r}) = i\omega \frac{|\zeta| T}{v_F} \Delta_\omega^*(\mathbf{r}) \int d^3 \mathbf{r}_1 \frac{m^2}{(2\pi)^2 |\mathbf{r} - \mathbf{r}_1| \sinh \frac{2\pi T |\mathbf{r} - \mathbf{r}_1|}{v_F}}, \quad (4.113)$$

which is integrable in analytic form.

It should be noted that the scalar potential φ escapes from the regular terms' contributions, as one may verify calculating the second and third diagrams in (4.110). (We will not present here these straightforward but sufficiently tedious calculations.) Note also, that the imaginary unit i in (4.113) causes (after the Fourier-transformation) the dynamic equation to be of the diffusion type (thus the difficulty described at the end of Chap. 3 is avoided). In writing down this equation, the presence of impurities makes it necessary to take into account a renormalization of regular terms. This procedure also renormalizes the coefficients[5] of the static (3.160). As a result, the equation for the order parameter acquires the form

---

[5] We omit here the details of the calculations, and trace only the principal issues of derivation of time-dependent Ginzburg–Landau equations (one can find certain details in [16]). The more general case will be considered in detail in Chap. 7.

$$\frac{\partial \Delta^*}{\partial t} + \frac{\tau_S}{3}\{[-\pi^2(T_c^2 - T^2) + \frac{|\Delta|^2}{2}]\Delta^* - \frac{v_F^2 \tau_1}{\tau_S}(\nabla + \frac{2ie}{c}A^2)\Delta^*\} - 2ie\Delta^*\Phi = 0,$$

$$(4.114)$$

$$\Phi = -\frac{i}{e\Delta^*\tau_S} \frac{2\pi^2}{|\zeta|mp_F} \int \frac{d^3\mathbf{k}}{(2\pi)^3} \frac{d\omega}{2\pi} e^{ikx - i\omega t} \int_{-\infty}^{\infty} \frac{d\varepsilon}{4\pi i} \int \frac{d^3\mathbf{p}}{(2\pi)^3} F^{(a)}(p, p-k) = 0.$$

$$(4.115)$$

4.3.6 TDGL Equations for Gapless Superconductors

We must account now for the contribution to (4.114) from the anomalous part $F^{(a)}$ in (4.115). In analogy to (4.102), the equation for $F_{\varepsilon\varepsilon-\omega}^{(a)}$ can be written in a form:

$$(4.116)$$

The expression

$$F_{\varepsilon\varepsilon-\omega}^{(a)} = \int \frac{d\varepsilon_1 \, d\omega_1}{(2\pi)^2}\left(\tanh\frac{\varepsilon_1}{2T} - \tanh\frac{\varepsilon_1 - \omega_1}{2T}\right) F_{\varepsilon\varepsilon_1}^R \Delta_{\omega_1} F_{\varepsilon_1 - \omega_1 \, \varepsilon - \omega}^A \qquad (4.117)$$

corresponds to the last diagram in (4.116). For pure superconductors at $\Delta \ll T$, $\omega_1 \ll T$ and $\varepsilon, \varepsilon_1 \ll T$, one may write in (4.117):

$$\tanh\frac{\varepsilon_1}{2T} - \tanh\frac{\varepsilon_1 - \omega_1}{2T} \approx \frac{\omega_1}{2T}\cosh^{-2}\frac{\varepsilon_1}{2T} \approx \frac{\omega_1}{2T} \qquad (4.118)$$

and consequently:

$$\Phi(t) \sim F^{(a)}(\mathbf{r}_1\mathbf{r}; t_1 t) \sim \int d^3\mathbf{r}_1 \, dt_1 \, F^R(\mathbf{r}_1\mathbf{r}; t_1 t)\Delta(t_1, \mathbf{r}_1)F^A(\mathbf{r}_1\mathbf{r}; t_1 t). \quad (4.119)$$

Further transformation of (4.119) seems to be impossible, because the functions $F^{R,A}$ oscillate in time with frequency $|\Delta|$.

In the presence of paramagnetic impurities, the situation differs qualitatively. In this case Green's functions decay exponentially for times $t_1 - t \sim \tau_S$, the kernels of the integral equations for Δ become local in time and that makes it possible to use a technique, analogous to the static case. Without further calculations we note only that at sufficiently high concentration of paramagnetic impurities, the result has the form $\Phi = -\varphi$. Then the equation for Δ may be written as[6]

---

[6]Similar equation for superconductors was originally derived on less rigorous basis by Schmid [19].

$$\left(\frac{\partial}{\partial t} + 2i\varphi\right)\Delta^* + \frac{\tau_S}{3}\left\{\left[-\pi^2\left(T_c^2 - T^2\right) + \frac{|\Delta|}{2}\right]\Delta^* - \frac{v_F\tau_1}{\tau_S}\left(\nabla + \frac{2ie}{c}\mathbf{A}\right)^2\Delta^*\right\} = 0.$$

(4.120)

The expression for the current in the gapless superconductors has a form characteristic for a two-fluid model:

$$\mathbf{j} = \mathbf{j}_s + \mathbf{j}_n, \qquad \mathbf{j}_s = \frac{2\sigma\tau_S}{c}|\Delta|^2\mathbf{Q}, \qquad \mathbf{j}_n = \sigma\mathbf{E},$$

(4.121)

where $\mathbf{E} = -ie\dot{\mathbf{A}} - \nabla\varphi$ is the electric field strength and $\mathbf{Q} = 2m\mathbf{v}_s$ is the superfluid's momentum, which is related to the superfluid velocity (3.54). In the case of superconductors with a finite gap, some additional terms arise in the current that correspond to the interference of normal and superfluid motions (see Chap. 7).

Thus the dynamic generalization of the Ginzburg–Landau equation for the order parameter has the form of a diffusion-type equation. Clearly, there is an essential difference between (4.120) and the diffusion equation (or the equation for the heat transfer), because in the case of superconductivity (4.120) is connected with (4.121) and with the Maxwell equations that comprise a strongly non-linear set of equations. The solutions of these equations (as we have seen in Chap. 2) can be periodic in space and time, revealing the remarkable properties of nonequilibrium superconductors.

References

1. A.A. Abrikosov, L.P. Gor'kov, On the theory of superconducting alloys. Sov. Phys. JETP **8**(6), 1090–1108 (1958) [*Zh. Eksp. i Teor. Fiz.* **35**[6(12)], 1558–1571 (1958)]
2. A.A. Abrikosov, L.P. Gor'kov, Contribution to the theory of superconducting alloys with paramagnetic impurities. Sov. Phys. JETP **12**(6), 1243–1253 (1961) [*Zh. Eksp. i Teor. Fiz.* **39**[6(12)], 1782–1796 (1960)]
3. A.A. Abrikosov, L.P. Gor'kov, I.E. Dzyaloshinskiy, *Quantum Field Theoretical Methods in Statistical Physics*, 2nd Edn. (Pergamon Press, Oxford, 1965), pp. 323–338
4. J. Kondo, Resistance minimum in dilute magnetic alloys. Progr. Theor. Phys. **32**(1), 37–49 (1964)
5. J. Kondo, Giant thermo-electric power of dilute magnetic alloys. Progr. Theor. Phys. **34**(3), 372–382 (1965)
6. A.A. Abrikosov, Electron scattering on magnetic impurities in metals and anomalous resistivity effects. Physics **2**(1), 5–20 (1965)
7. K. Fischer, Self-consistent treatment of the Kondo effect. Phys. Rev. **158**(3), 613–622 (1967)
8. D.K.C. Mac Donald, *Thermoelectricity: An Introduction to the Principles*, (John Wiley and Sons, New York, 1962), pp.1–133
9. R.D. Barnard, *Thermoelectricity in metals and alloys* (Taylor and Francis, LTD., London, 1972), pp. 1–259
10. J. Kondo, *Solid State Physics*, vol. 23, ed. by F. Seitz, D. Turnbull, H. Ehrenreich (Academic, New York, 1969), pp. 184–280
11. P.G. De Gennes, *Superconductivity of Metals and Alloys* (W.A. Benjamin Inc, New York, 1966), pp. 157–159
12. A.B. Migdal, V.M. Galitzkiy, Application of quantum field theory methods to the many body problem. *Sov. Phys. JETP* **7**(1), 96–104 (1958) [*Zh. Eksp. i Teor. Fiz.* **34**(1), 139–150 (1958)]

13. A.V. Svidzinskii, *Spatially-Inhomogeneous Problems in the Theory of Superconductivity* (Nauka, Moscow, 1982), pp. 78–83, in Russian

14. A.A. Abrikosov, *Fundamentals of the theory of metals* (North-Holland, Amsterdam, 1988), pp. 502–532

15. A.I. Larkin, Yu.N. Ovchinnikov, Nonlinear effects during the motion of vortices in superconductors. *Sov. Phys. JETP* **46**(1), 155–162 (1977) [*Zh. Eksp. i Teor. Fiz.* **73**[1(7)], 299–312 (1977)]

16. L.P. Gor'kov, G.M. Eliashberg, Generalization of the Ginzburg–Landau equations for nonstationary problems in the case of alloys with paramagnetic impurities. *Sov. Phys. JETP* **27**(2), 328–334 (1968) [*Zh. Eksp. i Teor. Fiz.* **54**(2), 612–626 (1968)]

17. L.P. Gor'kov, On the energy spectrum of superconductors. *Sov. Phys. JETP* **7**(3), 505–508 (1964) [*Zh. Eksp. i Teor. Fiz.* **34**(3) 735–739 (1958)]

18. G.M. Eliashberg, Inelastic electron collisions and nonequilibrium stationary states in superconductors. *Sov. Phys. JETP* **34**(3), 668–676 (1972) [*Zh. Eksp. i Teor. Fiz.* **61**[3(9)], 1254–1272 (1971)]

19. A. Schmid, A time-dependent Ginzburg-Landau equation and its application to the problem of resistivity in the mixed state. Phys. Kond. Materie **5**, 302–317 (1966)

Chapter 5
General Equations for Nonequilibrium States

The traditional theory of superconductivity relies on electron-phonon interaction. Explicitly, this was introduced into the theory of superconductivity by Eliashberg on the basis of Fröhlich's Hamiltonian and Migdal's theorem . We will use Eliashberg's model in the weak-coupling limit to derive kinetic equations, where both electrons, pairs and phonons are out of thermal equilibrium. Two methods are equally applicable for this task: the method of analytical continuation and Keldysh's technique. We will demonstrate both methods in this Chapter. The final kinetic equations will be expressed via the energy-integrated Green's functions, which in the case of equilibrium problems are called Eilenberger functions. This corresponds to a quasiclassical approximation in nonequilibrium superconductivity of superconductors that have finite gaps. The interaction of Cooper pairs with the electrons and phonons plays an important role in the action of external fields, determining both the behavior of the order parameter and the nonequilibrium effects in the electron-phonon system.

5.1 Migdal–Eliashberg Phonon Model

5.1.1 Fröhlich's Hamiltonian

The interaction of electrons with phonons in metals will be considered in this book within the isotropic model [1]. The oscillations of the ionic lattice produce lattice polarization. The interaction energy of electrons with the lattice is

$$- e \int \int n(\mathbf{r}) K(\mathbf{r} - \mathbf{r}') \operatorname{div} \mathbf{P}(\mathbf{r}) \, \mathrm{d}^3 \mathbf{r} \, \mathrm{d}^3 \mathbf{r}' \tag{5.1}$$

where $n(\mathbf{r})$ is the density of electrons at the point $\mathbf{r}$, $\mathbf{P}(\mathbf{r})$ is the polarization vector, and $K(\mathbf{r} - \mathbf{r}')$ is the interaction, having a Coulomb dependence at small distances and vanishing, owing to screening effects, at distances exceeding the lattice parameters

© Springer Nature Switzerland AG 2020
A. Gulian, *Shortcut to Superconductivity*,
https://doi.org/10.1007/978-3-030-23486-7_5

of a crystalline cell. Denoting these by a single distance parameter (a), the function $K(\mathbf{r} - \mathbf{r}')$ may be approximated as $K(\mathbf{r} - \mathbf{r}') \approx a^2 \delta(\mathbf{r} - \mathbf{r}')$. As to the polarization vector, it is proportional to the displacement $\mathbf{q}(\mathbf{r})$ of crystalline ions [1]

$$\mathbf{P}(\mathbf{r}) = C\mathbf{q}(\mathbf{r}) \equiv Ze\frac{N}{V_0}\mathbf{q}(\mathbf{r}), \tag{5.2}$$

where N/V_0 is the number of ions in a unite volume; and Ze is the ionic charge. We expand the displacement vector $\mathbf{q}(\mathbf{r}, t)$ in plane-waves

$$\mathbf{q}(\mathbf{r}, t) = \frac{1}{\sqrt{V}} \sum_{\mathbf{k}} \frac{\mathbf{k}}{|\mathbf{k}|} \left\{ q_{\mathbf{k}} e^{i(\mathbf{k}\cdot\mathbf{r} - \omega_0(\mathbf{k})t)} + q_{\mathbf{k}}^\dagger e^{-i(\mathbf{k}\cdot\mathbf{r} - \omega_0(\mathbf{k})t)} \right\} \tag{5.3}$$

and introduce the operators $b_{\mathbf{k}}$, $b_{\mathbf{k}}^\dagger$, connected with $q_{\mathbf{k}}$, $q_{\mathbf{k}}^\dagger$ by relations

$$q_{\mathbf{k}} = \frac{b_{\mathbf{k}}}{\sqrt{2\rho\omega_0(\mathbf{k})}}, \qquad q_{\mathbf{k}}^\dagger = \frac{b_{\mathbf{k}}^\dagger}{\sqrt{2\rho\omega_0(\mathbf{k})}}, \tag{5.4}$$

where ρ is the mass density of medium, and $\omega(\mathbf{k})$ is the phonons dispersion law.

Taking into account that $\rho\dot{q}(\mathbf{r}, t)$ is the momentum density of the medium, and also the quantum-mechanical commutation rule

$$\rho\left[\dot{q}_i(\mathbf{r}, t), q_k(\mathbf{r}', t)\right]_- = -i\delta(\mathbf{r} - \mathbf{r}')\delta_{ik}, \tag{5.5}$$

one can verify that the quantities $b_{\mathbf{k}}$ and $b_{\mathbf{k}}^\dagger$ (5.4) are Bose operators. Because the kinetic energy operator is

$$W_{\text{kin}} = \frac{\rho}{2} \int [\dot{q}_i(\mathbf{r}, t)]^2 \, d^3\mathbf{r} \tag{5.6}$$

and the mean kinetic energy of oscillations is equal to the mean potential energy, we have

$$\overline{H} = 2\overline{W}_{\text{kin}} = \sum_{\mathbf{k}} \omega_0(\mathbf{k}) \left(N_{\mathbf{k}} + \frac{1}{2} \right), \tag{5.7}$$

where $N_{\mathbf{k}} = \langle b_{\mathbf{k}}^\dagger b_{\mathbf{k}} \rangle$. The operator of a free phonon field is defined by the relation

$$\varphi(x) = \frac{1}{\sqrt{V_0}} \sum_{\mathbf{k}} \sqrt{\frac{\omega(\mathbf{k})}{2}} \left\{ b_{\mathbf{k}} e^{i(\mathbf{k}\cdot\mathbf{r} - \omega_0(\mathbf{k})t)} + b_{\mathbf{k}}^\dagger e^{-i(\mathbf{k}\cdot\mathbf{r} - \omega_0(\mathbf{k})t)} \right\}. \tag{5.8}$$

---

[1] Because the interaction energy is proportional to div$\mathbf{P} \propto$ div$\mathbf{q}$, one may conclude that (in the isotropic model only!) the electrons interact with longitudinal phonons only.

Note that φ is a real quantity [in the Debye model the summation in (5.8) is restricted by the condition $|\mathbf{k}| < k_D$]. The Hamiltonian of the electron-phonon interaction may then be written as

$$H_{e-ph} = g \int \Psi_\alpha^\dagger(\mathbf{r})\Psi_\alpha(\mathbf{r})\varphi(\mathbf{r})d^3\mathbf{r}, \qquad (5.9)$$

where the interaction constant g is defined by

$$g = \frac{ea^2 C}{u_0\sqrt{\rho}}, \qquad (5.10)$$

and $u_0 = \omega_0(\mathbf{k})/k$ is the sound velocity. The Hamiltonian (5.9) in the theory of metals is usually called "the Fröhlich Hamiltonian".

5.1.2 Migdal Diagram Expansion

The interaction of electrons with phonons in normal metals was considered in the diagram approach by Migdal [2], who used the Fröhlich Hamiltonian (5.9). In this approach, the Dyson equation for Green's function for electrons has the form

$$(5.11)$$

As shown by Migdal, even in the case of strong electron-phonon interaction the vertex remains "bare"[2]

$$\Gamma = \Gamma_0\left(1 + O\sqrt{\frac{m}{M}}\right), \qquad (5.12)$$

where M is the ionic mass in the crystalline lattice. If one starts from the Green function G_0 for noninteracting electrons and uses for the free phonon field Green's function

$$\mathfrak{D}_0(x_1 - x_2) = -i\langle T\left(\varphi(x_1)\varphi(x_2)\right)\rangle, \qquad (5.13)$$

where $\varphi(x)$ is defined by (5.8), then based on (5.11) and a corresponding equation for the $\mathfrak{D}$-function

$$(5.14)$$

---

[2]This point was critically reconsidered by Alexandrov and Ranninger [3]. They have developed an approach (the so-called bipolaron theory of superconductivity), based on violation of (5.12), which was accepted and developed further by other investigators. We will not consider this possibility, see references in [4].

one obtains the renormalized expressions for electron and phonon spectra and also for the damping of electron and phonon excitations. These renormalizations become important when the dimensionless interaction parameter

$$\lambda = \frac{mp_F}{2\pi^2} g^2 \tag{5.15}$$

is of the order of unity. The experiment shows that the renormalizations indeed occur in an electron system, whereas they are almost unobservable in a phonon system (at $\epsilon_F \gg \omega_D$). Some doubts were expressed in this connection concerning the adequacy of the Fröhlich Hamiltonian for this problem. As was shown further [5], the renormalization of the phonon spectrum in the above calculation scheme would correspond to the double counting of interaction between electrons and phonons. Unlike the electron system, the phonons in the "adiabatic approximation" are well-defined. The same is valid for the electron system if the parameter λ (5.15) is small.

We will consider further only metals with a weak electron-phonon interaction, assuming

$$\lambda \ll 1 \tag{5.16}$$

and neglecting the renormalization effects. Applicability of the results to metals with strong electron-phonon coupling should be analyzed separately. The effects of renormalization are not very essential for the kinetics and can be taken into account in the initial equilibrium state.

5.1.3 Eliashberg Equations in Weak-Coupling Limit

All the conclusions concerning the vertex renormalization (Γ) remain valid in the superconducting state, because only the excitations with large energies are essential for renormalization processes. The superconducting scale of energies is much less than these high energies. In the bare vertex approximation (Γ_0) we have the following system for electrons in superconductors

$$\Longrightarrow = \longrightarrow + \longrightarrow \!\!\!\bullet\!\!\!\!\overset{\displaystyle\frown}{}\!\!\!\bullet\!\!\! \Longrightarrow + \longrightarrow \!\!\!\bullet\!\!\!\!\overset{\displaystyle\frown}{}\!\!\!\bullet\!\!\! \longleftarrow , \tag{5.17}$$

$$\Longleftarrow = \longleftarrow \!\!\!\bullet\!\!\!\!\overset{\displaystyle\frown}{}\!\!\!\bullet\!\!\! \Longleftarrow + \longleftarrow \!\!\!\bullet\!\!\!\!\overset{\displaystyle\frown}{}\!\!\!\bullet\!\!\! \Longrightarrow . \tag{5.18}$$

For the phonon Green function $\sim\!\!\sim\!\!\sim$ in (5.17) and (5.18), the equation may also be written:

$$\approx\!\!\approx = \sim\!\!\sim + \sim\!\!\sim\!\!\bigcirc\!\!\sim\!\!\sim + \sim\!\!\sim\!\!\bigcirc\!\!\sim\!\!\sim . \tag{5.19}$$

Equations (5.17)–(5.19) were first formulated and solved by Eliashberg [6].

5.1.4 Comparison with BCS-Gor'kov Model

One may note a similarity between (5.17) and (5.18) and (3.100), (3.101), or (3.125) and (3.126), which can be made more transparent, if (5.17) and (5.18) are written in the momentum representation [$\varepsilon = \varepsilon_n = 2(n+1)\pi T i$]:

$$(\varepsilon_n - \xi - \Sigma_{1\varepsilon})\mathfrak{G}_\varepsilon(\mathbf{p}) + \Sigma_{2\varepsilon}\mathfrak{F}_\varepsilon^+(\mathbf{p}) = 1, \quad (-\varepsilon_n - \xi - \overline{\Sigma}_{1\varepsilon})\mathfrak{F}_\varepsilon^+(\mathbf{p}) + \Sigma_{2\varepsilon}^+\mathfrak{G}_\varepsilon(\mathbf{p}) = 0, \tag{5.20}$$

where

$$\Sigma_1(\varepsilon_n, \mathbf{p}) = T\sum_{n'}\int \frac{d^3\mathbf{p}'}{(2\pi)^3}\mathfrak{G}(\varepsilon_{n'}, \mathbf{p}')\mathfrak{D}(\varepsilon_n - \varepsilon_{n'}; \mathbf{p} - \mathbf{p}'), \tag{5.21}$$

$$\Sigma_2^+(\varepsilon_n, \mathbf{p}) = T\sum_{n'}\int \frac{d^3\mathbf{p}'}{(2\pi)^3}\mathfrak{F}^+(\varepsilon_{n'}, \mathbf{p}')\mathfrak{D}(\varepsilon_n - \varepsilon_{n'}; \mathbf{p} - \mathbf{p}'). \tag{5.22}$$

The interaction matrix element is incorporated into the definition of the $\mathfrak{D}$-function; hence the bare phonon Green function has the form

$$\mathfrak{D}_0(\omega_n, \mathbf{q}) = g^2\frac{2\omega_\mathbf{q}^2}{\omega_\mathbf{q}^2 - \omega_n^2} \quad \omega_\mathbf{q} = u|\mathbf{q}|, \quad \omega_n = 2\pi n T i. \tag{5.23}$$

The above-mentioned similarity becomes more complete if one neglects the renormalization of the electron spectrum, letting

$$\xi + \Sigma_1 \approx \xi + \overline{\Sigma}_1 \approx \xi. \tag{5.24}$$

After that, the self-consistency equation (3.122) acquires the form

$$\Delta = \Sigma_2 = T\sum_{n'}\int \frac{d^3\mathbf{p}'}{(2\pi)^3}\mathfrak{F}(\varepsilon_{n'}, \mathbf{p}')\mathfrak{D}(\varepsilon_n - \varepsilon_{n'}; \mathbf{p} - \mathbf{p}'). \tag{5.25}$$

Thus, it is clear that all the equilibrium results of the BCS-Gor'kov theory are contained in the phonon model of superconductors. At the same time, the latter model is much richer and may serve as a basis for the study of electron and phonon kinetics in real superconductors. Besides, in the Migdal–Eliashberg model, the critical parameters of a superconductor are expressed in terms of the parameters of a normal metal. In particular, the critical temperature in the weak coupling phonon model (5.17), (5.18) is given by the relation, analogous to (3.135), where ζ_0 is replaced by the parameter λ (5.15). The same replacement occurs in the expression (3.113) for the gap at zero temperature, and in addition $\bar{\epsilon}$ is replaced by ω_D.[3]

5.2 Equations for Nonequilibrium Propagators

5.2.1 Phonon Heat-Bath: Applicability

We continue theoretical study of nonequilibrium superconductivity with the simplest case, where the phonons play the role of a heat-bath for the electron system. In what cases is this phonon heat-bath model applicable? We examine this question in the particular case of a thin film with thickness d. Let us assume $d \sim \xi_0 \sim v_F / T_c$. Because the wavelength of the phonon is $\lambda_{\rm ph} \sim u / T$, where u is the velocity of sound, then at $T \sim T_c$ we have $\lambda_{\rm ph} \ll d$ so that the "geometric-acoustical" approximation could be used to describe the phonon's propagation. (Note that this approximation becomes invalid at $T \to 0$.) If the "acoustical densities" ρu of the film and $\rho' u'$ of its environment coincide, then phonons in the superconductor lose their energy at each collision with the specimen's walls [8]. (Evidently, if $\rho u = \rho' u'$, the phonons leave the film without reflection at the boundary.) However, as was shown in [2], the lifetime of thermal phonons, owing to their interaction with the conduction electrons in the metal, is $\tau_{\rm ph-e} \sim v_F / (uT)$ and consequently the scattering length of the phonon is $L \sim v_F / T$, which has an order of ξ_0. Thus (if $d < \xi_0$) the nonequilibrium phonons, emitted during the relaxation processes by electrons have enough time to leave the film without producing an influence back on the electron system. It must be stressed, that the phonon heat-bath model can be used in various situations. In each case an analysis of its applicability is required. For example, at $T \ll |\Delta|$ and for weak external pumping, the number of excess electron excitations is small and the electrons shift the phonons from equilibrium only slightly, even in thick films. In the case of massive superconductor placed in an external electromagnetic field, the picture is spatially inhomogeneous. There diffusion plays the main role in the relaxation processes in single-electron systems. The phonons remain in equilibrium if their scattering length exceeds the diffusion length of electron excitations.

5.2.2 Expansion over External Field Power

We pass now to a formal description of superconductor electrodynamics on the basis of Eliashberg equations in the framework of the phonon heat-bath model. In a static case $[\mathbf{A} = \mathbf{A}(\mathbf{r})]$ the initial equations in the spatial representation have the form

---

[3]These expressions for T_c and $\Delta(0)$, as well as their ratio change significantly in the strong coupling limit (see, e.g., [7]).

$$\left(\begin{matrix} \frac{1}{2m}\left(\widehat{\mathbf{p}} - \frac{e}{c}\mathbf{A}\right)^2_{\mathbf{r}} - \epsilon_F - \varepsilon - \Sigma_1 & -\Sigma_2 \\ \Sigma_2^+ & \frac{1}{2m}\left(\widehat{\mathbf{p}} + \frac{e}{c}\mathbf{A}\right)^2_{\mathbf{r}} + \epsilon_F + \varepsilon - \overline{\Sigma}_1 \end{matrix} \right) \left(\begin{matrix} \mathfrak{G} & \mathfrak{F} \\ -\mathfrak{F}^+ & \overline{\mathfrak{G}} \end{matrix} \right) = \widehat{1} \cdot \delta(\mathbf{r} - \mathbf{r}'), \quad (5.26)$$

and the self-energy parts are defined by relations

$$\widehat{\Sigma}_{\varepsilon,\mathbf{r},\mathbf{r}'} = \left(\begin{matrix} \mathfrak{t}_1 & \mathfrak{t}_2 \\ -\mathfrak{t}_2^+ & \overline{\mathfrak{t}}_1 \end{matrix} \right)_{\varepsilon',\mathbf{rr}'} = T \sum_{\varepsilon'} \mathfrak{D}_{\varepsilon-\varepsilon'} \left(\begin{matrix} \mathfrak{G} & \mathfrak{F} \\ -\mathfrak{F}^+ & \overline{\mathfrak{G}} \end{matrix} \right)_{\varepsilon',\mathbf{rr}'}, \quad (5.27)$$

where $\varepsilon = \varepsilon_m \equiv (2m+1)\pi T i$ and the matrix product is understood as a convolution over the internal variables, e.g.:

$$\Sigma \mathfrak{G} = \int d^3\mathbf{r}_1 \Sigma(\mathbf{r}, \mathbf{r}_1)\mathfrak{G}(\mathbf{r}, \mathbf{r}_1). \quad (5.28)$$

The phonon propagator in (5.27) is taken as an equilibrium one (5.23):

$$\mathfrak{D}(\omega_n, \mathbf{q}) \rightarrow \mathfrak{D}_0(\omega_n, \mathbf{q}). \quad (5.29)$$

5.2.3 Analytical Continuation: Causal Propagators

Using the technique of analytical continuation, introduced in Sect. 4.3, we will first obtain the expressions for the functions $\Sigma_\varepsilon^{R(A)}$, analytical in the upper (and, correspondingly, in the lower) half-plane. For this purpose we represent $\widehat{\Sigma}_\varepsilon$ (5.27) in the form

$$\widehat{\Sigma}_\varepsilon = T \sum_{\varepsilon'} \mathfrak{D}_{\varepsilon-\varepsilon'} \widehat{\mathfrak{G}}_{\varepsilon'} = T \sum_{\omega} \mathfrak{D}_\omega \widehat{\mathfrak{G}}_{\varepsilon-\omega} = \oint_C \frac{dz}{4\pi i} \coth \frac{z}{2T} \mathfrak{D}_z \widehat{\mathfrak{G}}_{\varepsilon-z}, \quad (5.30)$$

where the contour C encloses the poles of a hyperbolic cotangent and does not include the singularities of the function $\mathfrak{D}_z \widehat{\mathfrak{G}}_{\varepsilon-z}$ in the z-plane. Making cuts in this plane along the lines $\mathrm{Im}\, z = 0$ and $\mathrm{Im}(\varepsilon - z) = 0$, one can transform the integration contour C into another one, going along the arcs of large circles and along the banks of these cuts (Fig. 4.1). Taking into account the fact that the integrals along the arcs of large circles vanish when the radii tend to infinity, we obtain, using Fig. 4.1, the result

$$\widehat{\Sigma}_\varepsilon^{R(A)}(\mathbf{r}, \mathbf{r}_1) = \int_{-\infty}^{\infty} \frac{d\varepsilon'}{4\pi i} \left\{ \coth \frac{\varepsilon' - \varepsilon}{2T} (D^R - D^A)_{\varepsilon'-\varepsilon} G_{\varepsilon'}^{R(A)} \right.$$
$$\left. + D_{\varepsilon'-\varepsilon}^{A(R)} \tanh \frac{\varepsilon'}{2T} \left(\widehat{G}_{\varepsilon'}^R - \widehat{G}_{\varepsilon'}^A\right) \right\}_{\mathbf{rr}'}. \quad (5.31)$$

The hyperbolic tangent appears in (5.31) owing to the shift of the integration variable by the imaginary frequency ε. Because the analytical properties of propagators entering (5.31) are now definite, the variable ε' may be considered as real .

5.2.4 Phonon Heat-Bath: Consequences

Let us study the phonon propagators in detail. As follows from (5.19),

$$\mathfrak{D}(q) = \mathfrak{D}_0(q) + \mathfrak{D}_0(q)\Pi(q)\mathfrak{D}(q), \tag{5.32}$$

where $\Pi(q)$ is the polarization operator. We can rewrite (5.32) in the form

$$\mathfrak{D}_\omega(\mathbf{q}) = \frac{1}{\left[\mathfrak{D}_\omega^0(\mathbf{q})\right]^{-1} - \Pi_\omega(\mathbf{q})}. \tag{5.33}$$

The real part of the polarization operator Re $\Pi(q)$ is connected with the renormalization of the sound velocity. It is governed by the total mass of electrons; the range of temperature smearing of the Fermi-step gives a correction $\propto T/\epsilon_F$. Corrections, connected with the superconducting transition, have the same smallness, $\propto |\Delta|/\epsilon_F$. This is also true for renormalization caused by an electromagnetic field. As noted in Sect. 5.1, these renormalizations could be assumed as being already made. However, the imaginary part Im $\Pi_\omega(\mathbf{q})$ is wholly defined by the vicinity of Fermi surface and thus is very sensitive to the distribution of electron excitations. To realize the assumption concerning the phonon equilibrium, it would be necessary in deriving the dynamic equations to take into consideration in an explicit form a sink for the relaxation of phonons that is stronger than the source producing the deviation of phonons from equilibrium, which is caused by processes in the electron system. However, one can use the following artificial method: maintain the equilibrium distribution of phonons by keeping the initial presentation of discrete phonon frequencies and completely neglecting collisions of phonons with electrons in the equations for the phonon propagator.

Such an approach was proposed by Eliashberg [9] and we outline it here.

We will generalize the discussion, assuming the external field in (5.26) and in Green's functions there to depend on $\mathbf{r}, t$: $\mathbf{A} = \mathbf{A}(\mathbf{r}, t)$. Consequently, on the right side of (5.26) the additional factor $\delta(\tau - \tau')$ appears and Green's functions will acquire dependence on τ and τ'.

The modified system (5.26) can be expanded in a series over the external field. The diagrams consist of transit lines with different directions of arrows, containing field vertices ⋛— and phonon insertions ——〰〰〰——. As in Sect. 4.3, the directions of the arrows are not important for the procedure of analytical continuation.

Consider a diagram of Nth power in the external field for Green's function of electrons. Two types of diagrams may arise, depending on whether the diagram

Fig. 5.1 Diagrams for electron propagator without (**a**) and with (**b**) the phonon insertion

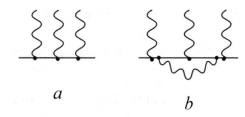

a

b

contains the phonon insertions (Fig. 5.1). Any diagram will depend on the frequencies of its extreme lines ε and $\varepsilon - \omega$ and of the field vertices ω_l.

The analytical structure of $\mathfrak{G}^{(N)}$ as the function of the variable ε at fixed ω_l should be found. Owing to the causality principle considered in Chap. 4, the necessary analytical continuation must be made over all ω_l from the upper half-plane, so in all the expressions of the type $\omega_l = 2l\pi T i$ we will make $\omega_l > 0$. First, we will consider the diagrams without Σ-insertions (as in Fig. 5.1a). The analytical properties of such diagrams are described by a simple composition:

$$\mathfrak{G}^{(N)}_{\varepsilon\varepsilon-\omega} \propto \mathfrak{G}_\varepsilon \mathfrak{G}_{\varepsilon-\omega_1} \mathfrak{G}_{\varepsilon-\omega_1-\omega_2} \cdots \mathfrak{G}_{\varepsilon-\omega}, \tag{5.34}$$

and their singularities (the poles) lay on the lines $\mathrm{Im}\,\varepsilon = 0$, $\mathrm{Im}\,\varepsilon = \omega_1$, $\mathrm{Im}\,\varepsilon = \omega_1 + \omega_2$,..., $\mathrm{Im}\,\varepsilon = \omega$, which are parallel to the abscissa. We will ascertain that these lines are singular for the arbitrary type of diagram $\mathrm{Im}\,G^{(N)}$. For this purpose it is sufficient to verify that the function $\Sigma^{(N)}_{\varepsilon\varepsilon-\omega}$, as the function of its external argument ε, has the same analytical structure as $\mathrm{Im}\,G^{(N)}_{\varepsilon\varepsilon-\omega}$, if the same set of field vertices is included. In other words, the singularities of Σ are determined by the singularities of its internal electron Green function. To see this, we again transform the sum over frequencies in (5.27) to the contour integral over the singularities of the hyperbolic tangent. Shifting the integration contour along the banks of the cuts, one obtains (in the same manner as in Section 4.3) the expression:

$$\Sigma^{(N)}_{\varepsilon\varepsilon-\omega} = \int_{-\infty}^{\infty} \frac{dz}{4\pi i} \left\{ \coth \frac{z}{2T} \left(D^R - D^A \right)_z \mathfrak{G}^{(N)}_{z+\varepsilon,z+\varepsilon-\omega} \right.$$
$$\left. + \tanh \frac{z}{2T} \left[\mathfrak{D}_{z-\varepsilon} \delta_1(\mathfrak{G}^{(N)}) + \ldots + \mathfrak{D}_{z-\varepsilon+\omega} \delta_{N+1}(\mathfrak{G}^{(N)}) \right] \right\}. \tag{5.35}$$

In (5.35) $\delta_l(\mathfrak{G}^{(N)})$ is the jump of the function $\mathfrak{G}^{(N)}$ at the bank l. Because in (5.35) z is real, one can see that $\Sigma_{\varepsilon\varepsilon-\omega}$ contains ε in the same combinations with ω_l as $\mathfrak{G}_{\varepsilon\varepsilon-\omega}$, and this proves the above statement concerning the analytic structure of an arbitrary type diagram.

5.2.5 Analytical Continuation: Anomalous Functions

Now we can carry out the analytical continuation of (5.35). Continuing analytically onto the real axis from the upper bank of the uppermost cut, we obtain the function $[\Sigma^{(N)}]^R$, and continuing from the lower bank of the lowermost cut, we get $[\Sigma^{(N)}]^A$. (In these cases all the functions have definite signs of imaginary parts, hence the subsequent continuation over each ω_l does not depend on the value of $\omega_l > 0$.) Thus, making $\mathrm{Im}\,\varepsilon > \omega$ ($\mathrm{Im}\,\varepsilon < 0$), shifting the integration variable to restore the initial notation of the arguments, summing over N and denoting

$$
G_{\varepsilon\varepsilon-\omega} = \sum_{N=0}^{\infty}\sum_{l=1}^{N+1} \delta_l(\mathfrak{G}^{(N)}) \tanh \frac{\varepsilon_l}{2T}, \tag{5.36}
$$

where $\varepsilon_l = \varepsilon - \omega_1 - \omega_2 - \ldots - \omega_{l-1}$, we arrive at

$$
\widehat{\Sigma}_{\varepsilon\varepsilon-\omega}^{R(A)} = \int_{-\infty}^{\infty} \frac{d\varepsilon'}{4\pi i} \left\{ \coth \frac{\varepsilon' - \varepsilon}{2T} \left(D^R - D^A\right)_{\varepsilon'-\varepsilon} \widehat{G}_{\varepsilon'\varepsilon'-\omega}^{R(A)} + D_{\varepsilon'-\varepsilon}^{A(R)} \widehat{G}_{\varepsilon'\varepsilon'-\omega} \right\}. \tag{5.37}
$$

The static limit of (5.37) [at $\widehat{G}_{\varepsilon\varepsilon-\omega} = 2\pi\delta(\omega)\widehat{G}_\varepsilon$, where $\widehat{G}_\varepsilon = (\widehat{G}^R - \widehat{G}^A)_\varepsilon \tanh (\varepsilon/2T)$] coincides with (5.31). Because all the self-energy functions and propagators composing $\widehat{G}^{R(A)}$, are retarded (or advanced), the equation that determines $\widehat{G}^{R(A)}$ has a form (see also Sect. 4.3):

$$
\left\{ \begin{pmatrix} \frac{1}{2m}\left(\widehat{\mathbf{p}} - \frac{e}{c}\mathbf{A}\right)_{\omega,\mathbf{r}}^2 - \epsilon_F - \varepsilon + e\varphi - \Sigma_1 & -\Sigma_2 \\ \Sigma_2^+ & \frac{1}{2m}\left(\widehat{\mathbf{p}} + \frac{e}{c}\mathbf{A}\right)_{\omega,\mathbf{r}}^2 + \epsilon_F + \varepsilon + e\varphi - \overline{\Sigma}_1 \end{pmatrix} \begin{pmatrix} G & F \\ -F^+ & G \end{pmatrix} \right\}^{R,A}_{\varepsilon,\varepsilon-\omega}
$$
$$
= \widehat{1} \cdot 2\pi\delta(\omega)(\mathbf{r} - \mathbf{r}'), \tag{5.38}
$$

where, as earlier,

$$
(\Sigma G)_{\varepsilon\varepsilon-\omega,\mathbf{rr}'} \equiv \int \frac{d\omega_1}{2\pi} \int d^3\mathbf{r}_1 \,\Sigma_{\varepsilon\varepsilon-\omega_1}(\mathbf{r},\mathbf{r}_1) G_{\varepsilon-\omega_1,\varepsilon-\omega}(\mathbf{r}_1,\mathbf{r}'). \tag{5.39}
$$

As for the functions $\widehat{G}_{\varepsilon\varepsilon-\omega}$ defined by the relation (5.36), one can obtain the equation for them in a manner used earlier in the Gor'kov's model. For this purpose the terms corresponding to the upper bank of the uppermost cut and the lower bank of the lowermost cut must be separated:

$$
\widehat{G}_{\varepsilon\varepsilon-\omega} = \widehat{G}_{\varepsilon\varepsilon-\omega}^R \tanh \frac{\varepsilon - \omega}{2T} - \tanh \frac{\varepsilon}{2T} \widehat{G}^A + \widehat{G}_{\varepsilon\varepsilon-\omega}^{(a)}. \tag{5.40}
$$

For $\widehat{G}_{\varepsilon\varepsilon-\omega}^{(a)}$ the diagrams are analogous to (4.98), although the vertices Δ and Δ^* are replaced now by the functions $\Sigma_2^{(a)}$ and $\Sigma_2^{+(a)}$, and the field vertices ⎯⎯⎯

have additional terms $\Sigma_1^{(a)}$ or $\overline{\Sigma}_1^{(a)}$. Because the functions $\Sigma^{(a)}$ contain the factors $[\tanh(\varepsilon/2T) - \tanh((\varepsilon - \omega)/2T)]$, it is convenient to introduce a function $h_{\varepsilon\varepsilon-\omega} = \{\tanh(\varepsilon/2T) - \tanh[(\varepsilon - \omega)/2T]\}H_1(\omega)$. In doing so, $(-h + \Sigma_1^{(a)})$ will correspond

to the vertices $\gtrless$, and $(-\overline{h} + \overline{\Sigma}_1^{(a)})$ to $\lessgtr$. Taking into account that the products, such as FF^+ or $\Sigma_2^+ F$, change the sign of the diagram, one can write:

$$G_{\varepsilon\varepsilon-\omega}^{(a)} = \left\{ G^R(-h + \Sigma_1^{(a)})G^A - G^R \Sigma_2^{(a)} F^{+A} - F^R \Sigma_2^{+(a)}\overline{G}^A - F^R(-\overline{h} + \overline{\Sigma}_1^{(a)})F^+ \right\}_{\varepsilon\omega-\varepsilon},$$
(5.41)

where the notation

$$\{ABC\}_{\varepsilon\varepsilon-\omega} = \int\int \frac{d\omega_1 d\omega_2}{(2\pi)^2} A_{\varepsilon\varepsilon-\omega_1} B_{\varepsilon-\omega_1\varepsilon-\omega_2} C_{\varepsilon-\omega_2\varepsilon-\omega}$$
(5.42)

is used. In the same manner one has

$$F_{\varepsilon\varepsilon-\omega}^{(a)} = \left\{ G^R\left(-h + \Sigma_1^{(a)}\right)F^A + G^R \Sigma_2^{(a)}\overline{G}^A + F^R\left(-\overline{h} + \overline{\Sigma}_1^{(a)}\right)\overline{G}^A - F^R \Sigma_2^{+(a)}F^A \right\}_{\varepsilon\varepsilon-\omega},$$
(5.43)

$$F_{\varepsilon\varepsilon-\omega}^{+(a)} = \left\{ \overline{G}^R\left(-\overline{h} + \overline{\Sigma}_1^{(a)}\right)F^{+A} + F^{+R}\left(-h + \Sigma_1^{(a)}\right)G^A \right.$$
$$\left. + \overline{G}^R \Sigma_2^{+(a)}G^A - F^{+R}\Sigma_2^{(a)}F^{+A} \right\}_{\varepsilon\varepsilon-\omega},$$
(5.44)

$$\overline{G}_{\varepsilon\varepsilon-\omega}^{(a)} = \left\{ \overline{G}^R\left(-\overline{h} + \overline{\Sigma}_1^{(a)}\right)\overline{G}^A - F^{+R}\left(-h + \Sigma_1^{(a)}\right)F^A \right.$$
$$\left. + \overline{G}^R \Sigma_2^{+(a)}F^A - F^{+R}\Sigma_2^{(a)}G^A \right\}_{\varepsilon\varepsilon-\omega}.$$
(5.45)

The elements of $\widehat{\Sigma}^{(a)}$ are found from the definition of $\widehat{\Sigma}$:

$$\widehat{\Sigma}_{\varepsilon\varepsilon-\omega} = \sum_{N=0}^{\infty}\sum_{l=1}^{N+1} \delta_l\left(\widehat{\Sigma}^{(N)}\right)\tanh\frac{\varepsilon_l}{2T},$$
(5.46)

from which, in analogy with (5.40),

$$\widehat{\Sigma}_{\varepsilon\varepsilon-\omega} = \widehat{\Sigma}_{\varepsilon\varepsilon-\omega}^R \tanh\frac{\varepsilon - \omega}{2T} - \tanh\frac{\varepsilon}{2T}\widehat{\Sigma}_{\varepsilon\varepsilon-\omega}^A + \widehat{\Sigma}_{\varepsilon\varepsilon-\omega}^{(a)}.$$
(5.47)

5.2.6 Complete Set of Equations

We will find now the explicit form of the dependence between Σ and G. We will use representation (5.35) for $\Sigma_{\varepsilon\varepsilon-\omega}^{(N)}$ and calculate directly the sum (5.46). Taking into account that the phonon propagator has poles at $\text{Im}(\varepsilon - \omega_i) = 0$, writing the expression for $\delta_i \left(\Sigma_{\varepsilon\varepsilon-\omega}^{(N)} \right)$ and shifting the integration variable (subject to $\tanh[(z + \omega_i)/(2T)] = \tanh[z/(2T)]$); multiplying the result by $\tanh(\varepsilon_l/2T)$ and also taking into account the identity

$$\coth(x - x') \tanh x = -\tanh x \tanh x' + \coth(x - x') \tanh x' + 1 \qquad (5.48)$$

and summing first over i and then over all the orders of the perturbation theory, we find the expression

$$\widehat{\Sigma}_{\varepsilon\varepsilon-\omega} = \int_{-\infty}^{\infty} \frac{d\varepsilon'}{4\pi i} \left\{ \coth \frac{\varepsilon' - \varepsilon}{2T} \widehat{G}_{\varepsilon'\varepsilon'-\omega} - \left(\widehat{G}^R - \widehat{G}^A \right)_{\varepsilon'\varepsilon'-\omega} \right\} \left(D^R - D^A \right)_{\varepsilon'-\varepsilon}. \qquad (5.49)$$

To complete the set of equations, it is necessary to establish the equation for the $\widehat{G}$-function, defined by the relation (5.36). The method to be used here was described in Chap. 4. Starting from the diagram expansion for $\mathfrak{G}$-function and separating there the line corresponding to the bare propagator of electrons, we find from the $11-$element of $\widehat{G}$-matrix:

$$\left[\frac{1}{2m} \left(\widehat{\mathbf{p}} - \frac{e}{c} \mathbf{A} \right)^2 + e\varphi - \epsilon_F - \varepsilon \right] G_{\varepsilon\varepsilon-\omega}^{(a)}$$

$$= \left\{ \left(-h + \Sigma_1^{(a)} \right) G^A - \Sigma_2^{(a)} F^{+A} \right\}_{\varepsilon\varepsilon-\omega} + \left\{ \Sigma_1^R G^{(a)} - \Sigma_2^R F^{+(a)} \right\}_{\varepsilon\varepsilon-\omega}. \qquad (5.50)$$

Using the definitions (5.46), (5.40), and also (5.38) for the causal Green's functions, one can obtain on the basis of (5.50) the equation

$$\begin{pmatrix} \frac{1}{2m} \left(\widehat{\mathbf{p}} - \frac{e}{c} \mathbf{A} \right)^2 + e\varphi - \epsilon_F - \varepsilon - \Sigma_1^R & -\Sigma_2^R \\ \Sigma_2^{+R} & \frac{1}{2m} \left(\widehat{\mathbf{p}} + \frac{e}{c} \mathbf{A} \right)^2 + e\varphi - \epsilon_F + \varepsilon - \overline{\Sigma}_1^R \end{pmatrix} \begin{pmatrix} G & F \\ -F^+ & \overline{G} \end{pmatrix}$$

$$= \begin{pmatrix} \Sigma_1 & \Sigma_2 \\ -\Sigma_2^+ & \overline{\Sigma}_1 \end{pmatrix} \begin{pmatrix} G^A & F^A \\ -F^{+A} & \overline{G}^A \end{pmatrix}, \qquad (5.51)$$

or in the integral form:

$$\widehat{G} = \widehat{G}^R \, \widehat{\Sigma} \, \widehat{G}^A. \qquad (5.52)$$

Thus the closed system of (5.37), (5.38), (5.49) and (5.51) is derived for the functions $\widehat{G}, \widehat{G}^R, \widehat{G}^A$ and $\widehat{\Sigma}, \widehat{\Sigma}^R, \widehat{\Sigma}^A$, which describes the behavior of nonequilibrium superconductors in the phonon heat-bath model. The temperature enters these equations explicitly only in equations for Σ and $\Sigma^{R(A)}$ as the characteristic of the phonon heat-bath.

5.2.7 Keldysh Technique Approach

Note that one can obtain the same results by a completely different method, developed by Keldysh [10] to describe nonequilibrium states . In that case the electron Green's function is defined in the following way (we use here the notations of Volkov and Kogan [11]):

$$G^{ik}_{\mu\nu}(1, 2) = -i \langle T \Psi_\mu(1i) \Psi_\nu^\dagger(2k) \rangle. \tag{5.53}$$

Here $1 = (\mathbf{r}_1, t_1)$; and μ and ν are the Nambu indices of the field operators

$$\Psi_1(1i) \equiv \Psi_\uparrow(1i), \qquad \Psi_2(1i) \equiv \Psi_\downarrow^\dagger(1i). \tag{5.54}$$

The Keldysh indices i, k are the signs minus or plus, according to the position of the time coordinate of the Ψ-operators on each of two time-axes $(-\infty, \infty$ or $\infty, -\infty)$ [10]. The time on the second axis (the index $+$) is greater than any time on the first axis (the index $-$). For functions G, G^R and G^A, which are defined as in the case of a normal metal (see, e.g., [12]),

$$\begin{pmatrix} 0 & G^A \\ G^R & G \end{pmatrix} = \widehat{U}^{-1} \widehat{G} \widehat{U}, \ \widehat{G} = (G)^{ik}, \ \widehat{U} = \frac{\widehat{1} + i\widehat{\sigma}_y}{2}, \tag{5.55}$$

one can obtain the equations coinciding with (5.51) and (5.38). This coincidence of the results obtained by the Gor'kov–Eliashberg and the Keldysh techniques, will be demonstrated further on, when the phonon kinetics in superconductors are considered.

5.3 Quasiclassical Approximation

The equations obtained in the preceding Section may be simplified further when the phenomena occurring in superconductors involve electrons localized in the momentum space near the vicinity of the Fermi surface. (In other words, when microscopic processes of interest may be considered as macroscopic on the atomic scales of space and time.) Such a situation is typical for most of the phenomena occurring in nonequilibrium superconductors. In this case one can use a generalization of the method introduced by Eilenberger [13] for equilibrium superconductors.

5.3.1 Eilenberger Propagators

The essence of this approach may be elucidated in terms of the electron's wave-function of the superconductor. The wave-function of an electron with a momentum in the vicinity of the Fermi-surface oscillates rapidly in space and time. Under the influence of external quasiclassical perturbation, the wave-function's amplitude becomes weakly modulated. The information of interest is contained in the "enveloping curve" of the modulated signal. This allows us to ignore the "carrying" frequency and to use only the "enveloping curve".[4] In the Green's functions technique, this procedure is equivalent to the integration over the values of $|\mathbf{p}|$ or $\xi = v_F(p - p_F)$.

Consider one of the equations for the self-energy functions, for example, for Σ^R (5.37). In the momentum representation we have

$$\widehat{\Sigma}^R(P, P - K) = \int_{-\infty}^{\infty} \frac{d\varepsilon'}{4\pi i} \int \frac{d^3\mathbf{p}'}{(2\pi)^3} \left\{ \coth \frac{\varepsilon' - \varepsilon}{2T} (D^R - D^A)(P' - P)\widehat{G}^R(P', P' - K) \right.$$
$$\left. + D^A(P' - P)\widehat{G}(P', P' - K) \right\}. \tag{5.56}$$

Here $P = \{\varepsilon, \mathbf{p}\}$, $K = \{\omega, \mathbf{k}\}$. If the external momentum in (5.56) is close to the Fermi surface: $p \sim p_F$, ε and $\varepsilon' < \omega_D$, then the main contribution to integral (5.56) is provided by the region $|\mathbf{p} - \mathbf{p}'| \ll p_F$ (the integration over $\mathbf{p}'$ in the regions remote from the Fermi surface renormalizes the chemical potential, which is insensitive to details of the electron distribution in the vicinity of p_F). The D-function now depends only on the angle θ between $\mathbf{p}$ and $\mathbf{p}'$: $|\mathbf{p} - \mathbf{p}'|^2 \approx 2p_F^2(1 - \cos\theta)$. Using the chain of equalities

$$\frac{d^3\mathbf{p}}{(2\pi)^3} = \frac{p^2 dp d\Omega_\mathbf{p}}{(2\pi)^3} = \frac{d\Omega_\mathbf{p}}{2} \frac{p dp^2}{(2\pi)^3} \approx \frac{mp_F}{2\pi^2} d\xi \frac{d\Omega_\mathbf{p}}{4\pi}, \tag{5.57}$$

it is easy to establish that $\widehat{\Sigma}^R$ is expressed by Green's functions, integrated over the energy variable :

$$\widehat{g}_{\varepsilon\varepsilon-\omega}^{(R,A)}(\mathbf{p}, \mathbf{k}) = \int_{-\infty}^{\infty} d\xi \, \widehat{G}^{(R,A)}(P, P - K), \quad \widehat{g}^{(R,A)} = \begin{pmatrix} g & f \\ -f^+ & g \end{pmatrix}^{(R,A)}. \tag{5.58}$$

Similar conclusions follow for other self-energy functions, so that one has:

$$\widehat{\Sigma}_{\varepsilon\varepsilon-\omega}^{R(A)}(\mathbf{p}, \mathbf{k}) = \int_{-\infty}^{\infty} \frac{d\varepsilon'}{4\pi i} \int \frac{d\Omega_{\mathbf{p}'}}{4\pi} \left\{ \coth \frac{\varepsilon' - \varepsilon}{2T} (D^R - D^A)_{\varepsilon'-\varepsilon}(\theta) \, \widehat{g}_{\varepsilon'\varepsilon'-\omega}^{R(A)}(\mathbf{p}', \mathbf{k}) \right.$$
$$\left. + D_{\varepsilon'-\varepsilon}^{A(R)}(\theta) \, \widehat{g}_{\varepsilon'\varepsilon-\omega}(\mathbf{p}', \mathbf{k}) \right\}, \tag{5.59}$$

---

[4]Analogous procedure is applied in passing from the Bogolyubov-De Gennes equations to the Andreev equations (see Sect. 3.1).

$$\widehat{\Sigma}_{\varepsilon\varepsilon-\omega}(\mathbf{p}, \mathbf{k}) = \int_{-\infty}^{\infty} \frac{d\varepsilon'}{4\pi i} \int \frac{d\Omega_{\mathbf{p}'}}{4\pi} \left\{ \coth \frac{\varepsilon' - \varepsilon}{2T} \widehat{g}_{\varepsilon'\varepsilon'-\omega}(\mathbf{p}', \mathbf{k}) \right.$$
$$\left. - (\widehat{g}^R - \widehat{g}^A)_{\varepsilon'\varepsilon'-\omega}(\mathbf{p}', \mathbf{k}) \right\} (D^R - D^A)_{\varepsilon'-\varepsilon}(\theta). \quad (5.60)$$

5.3.2 Eliashberg Kinetic Equations

Now let us transform (5.51). Ignoring the quadratic terms in $\mathbf{A}$, and moving to the quantities H_1 and $\overline{H}_1$, one finds by multiplying (5.52) by $[\overline{G}^R]^{-1}$ from the left and by $[\overline{G}^A]^{-1}$ from the right:

$$\begin{pmatrix} \xi - \varepsilon + H_1 & 0 \\ 0 & \xi + \varepsilon + \overline{H}_1 \end{pmatrix} \widehat{G} = \widehat{\Sigma}^R \widehat{G} + \widehat{\Sigma} \widehat{G}^A, \quad (5.61)$$

$$\widehat{G} \begin{pmatrix} \xi - \mathbf{vk} - \varepsilon + \omega + H_1 & 0 \\ 0 & \xi - \mathbf{vk} + \varepsilon - \omega + \overline{H}_1 \end{pmatrix} = \widehat{G}\widehat{\Sigma}^A + \widehat{G}^R \widehat{\Sigma}, \quad (5.62)$$

where $v = p/m$. Subtracting (5.61) from (5.62) and integrating the result over ξ, we find the equation for the $\widehat{g}$-function (5.58):

$$\begin{pmatrix} (\omega - \mathbf{vk})g & (2\varepsilon - \omega - \mathbf{vk})f \\ (2\varepsilon - \omega + \mathbf{vk})f^+ & -(\omega + \mathbf{vk})\overline{g} \end{pmatrix} = \begin{pmatrix} H_1 & 0 \\ 0 & \overline{H}_1 \end{pmatrix} \widehat{g} - \widehat{g} \begin{pmatrix} H_1 & 0 \\ 0 & \overline{H}_1 \end{pmatrix}$$
$$+ \widehat{g}\widehat{\Sigma}^A - \widehat{\Sigma}^R \widehat{g} + \widehat{g}^R \widehat{\Sigma} - \widehat{\Sigma}\widehat{g}^A. \quad (5.63)$$

The set of arguments of the $\widehat{g}$-functions entering this equation is analogous to that of electron distribution function. By this reason (5.63) may be called the generalized kinetic equation. The quantity

$$\widehat{I} = \widehat{g}\widehat{\Sigma}^A - \widehat{\Sigma}^R \widehat{g} + \widehat{g}^R \widehat{\Sigma} - \widehat{\Sigma}\widehat{g}^A \quad (5.64)$$

is the collision integral (at present, between electrons and phonons). The equation for the function $\widehat{g}^{R(A)}$ [which may be obtained from (5.38) by the procedure used above] is similar to (5.63), although the quantity $\widehat{I}$ must be replaced by $\widehat{I}^{R(A)}$:

$$\widehat{I}^{R(A)} = \widehat{g}^{R(A)} \widehat{\Sigma}^{R(A)} - \widehat{\Sigma}^{R(A)}\widehat{g}^{R(A)}. \quad (5.65)$$

In the equilibrium case, when the field $\mathbf{A}$ is absent:

$$(\widehat{g}^R - \widehat{g}^A)_{\varepsilon} = 2\pi i \begin{pmatrix} \varepsilon & \Delta \\ -\Delta^* & -\varepsilon \end{pmatrix} \frac{\text{sign}\varepsilon}{\sqrt{\varepsilon^2 - |\Delta|^2}} \theta(\varepsilon^2 - |\Delta|^2), \quad (5.66)$$

i.e., $(\widehat{g}^R - \widehat{g}^A)_{\varepsilon}$ is proportional to the density of single-particle excitation's states (for this reason $\widehat{g}^R$, $\widehat{g}^A$ are called "spectral functions").

As to the functions $\widehat{g}_{\varepsilon\varepsilon-\omega}$, from (5.58) in the equilibrium case, the relation follows

$$\widehat{g}_\varepsilon = (\widehat{g}^R - \widehat{g}^A)_\varepsilon \tanh\frac{\varepsilon}{2T} = (\widehat{g}^R - \widehat{g}^A)_\varepsilon(1 - 2n_\varepsilon^F)\mathrm{sign}\varepsilon, \qquad (5.67)$$

where n_ε^F is the distribution function of the electron-like ($\varepsilon > 0$) and hole-like ($\varepsilon < 0$) Fermi excitations. In the nonequilibrium case, as we will see, n_ε in general does not necessarily coincide with $n_{-\varepsilon}$. However, the correct generalization of (5.67) cannot be achieved by the trivial replacement $1 - 2n_\varepsilon^F \to 1 - n_\varepsilon - n_{-\varepsilon}$, as might be thought. Such a replacement would retain g_ε as an odd function in ε, whereas in general case g_ε can have an even in ε part also. The necessary generalization can be obtained with the help of the normalization condition for g-functions, as was shown for the equilibrium case by Eilenberger [13] and for the nonequilibrium case by Larkin and Ovchinnikov [14].

5.3.3   Normalization Condition

In a real-time representation [see (5.53)], this condition has the form[5]

$$\breve{g} * \breve{g} = \mathrm{const} \cdot \breve{1}, \qquad (5.68)$$

where

$$\breve{g} = \begin{pmatrix} \widehat{g}^R & \widehat{g} \\ 0 & \widehat{g}^A \end{pmatrix}, \quad \breve{1} = \begin{pmatrix} \widehat{1} & 0 \\ 0 & \widehat{1} \end{pmatrix}, \qquad (5.69)$$

and the symbol $*$ is the convolution in time according to

$$A * B = \int A(t_1, t_2) B(t_3, t_2) dt_3. \qquad (5.70)$$

The normalization condition (5.68) is satisfied identically by the following substitution

$$\widehat{g} = \widehat{g}^R * \widehat{a} - \widehat{a} * \widehat{g}^A, \qquad (5.71)$$

where $\widehat{a}$ is an arbitrary 2×2−matrix function of ε, which may be represented as the sum of the Pauli matrices, the diagonal matrices only participating in this decomposition:

$$\widehat{a} = f_1\widehat{1} + f_2\widehat{\sigma}_z. \qquad (5.72)$$

---

[5]To avoid breaking the presentation, we will prove this statement in Chap. 7. It is worth to notice, that in the theory of superconductivity the normalization condition is proved only on the "physical level" of rigor.

The functions f_1 and f_2 are linked with the distribution function of electron-hole excitations n_ε. Before showing this relation, we consider some general properties of f_1 and f_2 and establish their gauge transformation laws.

5.3.4 Gauge Transformation Rules

The basic gauge transformation law for a field operator $\Psi(\mathbf{r}, t)$ under the transformation of scalar potential $\varphi \to \varphi - \dot{\chi}/2$ is

$$\Psi(\mathbf{r}, t) \to \exp[i\chi(\mathbf{r}, t)/2]\Psi(\mathbf{r}, t) \tag{5.73}$$

(where χ is an arbitrary function). From (5.73) and (5.53) it follows that the propagator $\widehat{g}$ (as well as the spectral functions $\widehat{g}^R$ and $\widehat{g}^A$) is transformed according to

$$\widehat{g} \to \exp\left[i\widehat{\sigma}_z\chi/2\right] * \widehat{g} * \exp\left[-i\widehat{\sigma}_z\chi/2\right]. \tag{5.74}$$

In the quasiclassical limit, when the propagators are fast varying functions of a difference variable $(t_1 - t_2)$ and slow varying functions of a sum $t = (t_1 + t_2)/2$, the expression (5.70) may be presented in the form

$$A * B = AB + \frac{i}{2}\{A_{,\varepsilon}\dot{B} - \dot{A}B_{,\varepsilon}\} - \frac{1}{8}\{A_{,\varepsilon\varepsilon}\ddot{B} - 2\dot{A}_{,\varepsilon}\dot{B}_{,\varepsilon} + \ddot{A}B_{,\varepsilon\varepsilon}\} + \dots . \tag{5.75}$$

In (5.75) the following notations are used: $A_{,\varepsilon} \equiv \partial A/\partial\varepsilon$, $\dot{A} \equiv \partial A/\partial t$, and the frequency ε corresponds to the Fourier-transform over the difference argument $(t_1 - t_2)$. From (5.74) and (5.75) a transformation law follows for diagonal components of propagators[6]:

$$\widehat{g}^{\text{diag}} \to \widehat{g}^{\text{diag}} + \frac{\dot{\chi}}{2}\widehat{g}_{,\varepsilon}^{\text{diag}} + \frac{\dot{\chi}^2}{2}\widehat{g}_{,\varepsilon\varepsilon}^{\text{diag}}. \tag{5.76}$$

Hereafter the terms proportional to $\ddot{\chi}$ are omitted owing to the assumed quasiclassical character of φ.

At the same time, one can make a gauge transformation of the function, defined by (5.71). Taking into account that the functions $\widehat{g}^R$ and $\widehat{g}^A$ transform in analogy to (5.74), one can demand the coincidence of the corresponding result with (5.76). This provides the transformation laws for functions f_1 and f_2 :

---

[6]At this stage it becomes clear, that the expression (5.74) is an equivalent form of the usual relation for the Green's function: $G \to G \exp[i\chi(t_1)/2 - i\chi(t_2)/2]$ at the gauge transformation. Expanding the exponent over the "fast" time $(t_1 - t_2)$ and finding the Fourier-transforms over the difference variable, one can obtain the expression (5.76) for the appropriate matrix component.

$$f_{1(2)} \rightarrow f_{1(2)} + \frac{(\dot{\chi}/2)\,(f_1 + f_2)_{,\varepsilon} + \left(\dot{\chi}^2/8\right)[(f_1 + f_2)_{,\varepsilon\varepsilon} + 2\,(N_{1,\varepsilon}/N_1)\,(f_1 + f_2)_{,\varepsilon}]}{2 + \dot{\chi} N_{1,\varepsilon}/N_1 + \dot{\chi}^2 N_{1,\varepsilon\varepsilon}/4N_1}$$

$$+(-)\frac{-(\dot{\chi}/2)\,(f_1 - f_2)_{,\varepsilon} + \left(\dot{\chi}^2/8\right)[(f_1 - f_2)_{,\varepsilon\varepsilon} + 2\,(\overline{N}_{1,\varepsilon}/\overline{N}_1)\,(f_1 - f_2)_{,\varepsilon}]}{2 - \dot{\chi}\overline{N}_{1,\varepsilon}/\overline{N}_1 + \dot{\chi}^2\overline{N}_{1,\varepsilon\varepsilon}/4\overline{N}_1}. \tag{5.77}$$

The functions N_1 and $\overline{N}_1$ in (5.77) are defined by the relations

$$N_1 = \frac{g^R - g^A}{2\pi i}, \quad \overline{N}_1 = \frac{\overline{g}^R - \overline{g}^A}{2\pi i}. \tag{5.78}$$

If $\varphi \rightarrow \varphi - \dot{\chi}/2$, we have in accordance with (5.71), (5.74) and (5.78)

$$N_1 \rightarrow N_1 + \frac{\dot{\chi}}{2}N_{1,\varepsilon} + \frac{\dot{\chi}^2}{8}N_{1,\varepsilon\varepsilon}, \tag{5.79}$$

$$\overline{N}_1 \rightarrow \overline{N}_1 - \frac{\dot{\chi}}{2}\overline{N}_{1,\varepsilon} + \frac{\dot{\chi}^2}{8}\overline{N}_{1,\varepsilon\varepsilon}. \tag{5.80}$$

Note that, owing to (5.77) and (5.79), the functions f_1 and f_2 (in analogy with N_1 and N_2) are functions of a general type. They have definite parity only in the absence of external fields.

5.3.5   Electron and Hole Distribution Functions

In the absence of external fields, as may be seen from the definition of these functions and (5.63), $f_1(\varepsilon)$ and $N_2(\varepsilon)$ are odd functions of ε, while $f_2(\varepsilon)$ and $N_1(\varepsilon)$ are even functions of ε. Introducing an arbitrary function n_ε (here $-\infty < \varepsilon < \infty$), we can write (making $\varphi = 0$)

$$f(\varepsilon_1) = a_1(n_\varepsilon + n_{-\varepsilon} - 1), \quad f_2(\varepsilon) = a_2(n_\varepsilon - n_{-\varepsilon}) \tag{5.81}$$

Because the function n_ε should be determined further from the kinetic equations, there is still an arbitrariness in the choice of coefficients a_1 and a_2. It is convenient to choose them in the form

$$a_1 = \text{sign}\varepsilon, \; a_2 = -u_\varepsilon^{-1}\text{sign}\varepsilon \tag{5.82}$$

where

$$u_\varepsilon = \frac{|\varepsilon|\theta(\varepsilon_\mathbf{p}^2 - |\Delta|^2)}{\sqrt{\varepsilon_\mathbf{p}^2 - |\Delta|^2}} \tag{5.83}$$

Then the expressions for $\widehat{g}^{R(A)}$ take the form

$$\widehat{g}_\varepsilon = -2\pi i \begin{pmatrix} u_\varepsilon \beta_\varepsilon + \alpha_\varepsilon & v_\varepsilon \beta_\varepsilon \\ -v_\varepsilon \beta_\varepsilon & -u_\varepsilon \beta_\varepsilon + \alpha_\varepsilon \end{pmatrix}, \tag{5.84}$$

$$(\widehat{g}^R - \widehat{g}^A)_\varepsilon = 2\pi i \begin{pmatrix} u_\varepsilon & v_\varepsilon \\ -v_\varepsilon & -u_\varepsilon \end{pmatrix}, \tag{5.85}$$

where

$$v_\varepsilon = \frac{|\Delta|\theta(\varepsilon_\mathbf{p}^2 - |\Delta|^2)}{\sqrt{\varepsilon_\mathbf{p}^2 - |\Delta|^2}} \text{sign}\varepsilon, \tag{5.86}$$

$$\beta_\varepsilon = (n_\varepsilon + n_{-\varepsilon} - 1)\theta(\varepsilon_\mathbf{p}^2 - |\Delta|^2)\text{sign}\varepsilon, \tag{5.87}$$

$$\alpha_\varepsilon = (n_\varepsilon - n_{-\varepsilon})\theta(\varepsilon_\mathbf{p}^2 - |\Delta|^2)\text{sign}\varepsilon . \tag{5.88}$$

The constant in the expression (5.68) may be chosen to be $-\pi^2$. Without a loss of generality, this ensures a limiting transition to expressions (5.66), (5.67) in the absence of imbalance, and simultaneously assigns to the function n_ε a transparent meaning of the energy distribution function in "pure" superconductors (this will be shown below). At the same time, it might be noted that the description of superconductor in terms of $\widehat{g}$-functions integrated over ξ-variable is also valid in cases when the concept of the energy spectrum turns deficient and ξ becomes a bad quantum number (e.g., in the case of superconductors containing very many impurities).

5.3.6 Kinetic Equations: Keldysh Option

As shown by Keldysh [10], in a nonequilibrium system, the Green's function technique allows us to formulate kinetic equations without integrating over energies. In such cases, the energy distribution function of excitations is connected to Green's functions, integrated over the frequency variable. If the energy spectrum is well-defined, these two methods are usually adequate.[7]

For the causal Green's function G_{11}^{--}, using the definition (5.53) and solving at $\varepsilon \ll \omega_D$ the Dyson equations by analogy to the normal metal case (cf. [12]), we have:

$$G_{11}^{--}(\varepsilon, \mathbf{p}) = \mathcal{U}_\mathbf{p}^2 \left(\frac{n_\mathbf{p}}{\varepsilon - \varepsilon_\mathbf{p} - i\delta} + \frac{1 - n_\mathbf{p}}{\varepsilon - \varepsilon_\mathbf{p} + i\delta} \right) + \mathcal{V}_\mathbf{p}^2 \left(\frac{n_{-\mathbf{p}}}{\varepsilon + \varepsilon_\mathbf{p} + i\delta} + \frac{1 - n_{-\mathbf{p}}}{\varepsilon + \varepsilon_\mathbf{p} - i\delta} \right), \tag{5.89}$$

---

[7]We have already noticed the advantage of energy integrated functions for "dirty" superconductors. At the same time, this technique fails when considering processes far from the Fermi surface, e.g., at the description of high-energy particle cascading in superconductors. In such situations usual (Keldysh') formulation of the kinetic scheme is preferable.

where the factors $\mathcal{U}_{\mathbf{p}}^2$ and $\mathcal{V}_{\mathbf{p}}^2$ are defined by the relation (1.23), $\varepsilon_{\mathbf{p}}-$ by (1.132), and $n_{\mathbf{p}}$ corresponds to the distribution function of electron-like ($\xi_{\mathbf{p}} > 0$) and hole-like ($\xi_{\mathbf{p}} < 0$) excitations.

Following Aronov and Gurevich [15], one can introduce a spectral representation for the causal function

$$G_{11}^{--}(\varepsilon, \mathbf{p}) = \int_{-\infty}^{\infty} \frac{d\varepsilon'}{2\pi i} \left[\frac{G_{11}^{-+}(\varepsilon', \mathbf{p})}{\varepsilon - \varepsilon' - i\delta} - \frac{G_{11}^{+-}(\varepsilon', \mathbf{p})}{\varepsilon - \varepsilon' + i\delta} \right]. \tag{5.90}$$

Comparing (5.89) and (5.90), one can find

$$G_{11}^{-+}(\varepsilon, \mathbf{p}) = 2\pi i \left[\mathcal{U}_{\mathbf{p}}^2 n_{\mathbf{p}} \delta(\varepsilon - \varepsilon_{\mathbf{p}}) + \mathcal{V}_{\mathbf{p}}^2 (1 - n_{-\mathbf{p}}) \delta(\varepsilon + \varepsilon_{\mathbf{p}}) \right], \tag{5.91}$$

$$G_{11}^{+-}(\varepsilon, \mathbf{p}) = -2\pi i \left[\mathcal{U}_{\mathbf{p}}^2 (1 - n_{\mathbf{p}}) \delta(\varepsilon - \varepsilon_{\mathbf{p}}) + \mathcal{V}_{\mathbf{p}}^2 n_{-\mathbf{p}} \delta(\varepsilon + \varepsilon_{\mathbf{p}}) \right]. \tag{5.92}$$

In the same manner

$$G_{12}^{--}(\varepsilon, \mathbf{p}) = \mathcal{U}_{\mathbf{p}} \mathcal{V}_{\mathbf{p}} \left(\frac{n_{\mathbf{p}}}{\varepsilon - \varepsilon_{\mathbf{p}} - i\delta} + \frac{1 - n_{\mathbf{p}}}{\varepsilon - \varepsilon_{\mathbf{p}} + i\delta} - \frac{n_{-\mathbf{p}}}{\varepsilon + \varepsilon_{\mathbf{p}} + i\delta} - \frac{1 - n_{-\mathbf{p}}}{\varepsilon + \varepsilon_{\mathbf{p}} - i\delta} \right), \tag{5.93}$$

$$G_{12}^{+-}(\varepsilon, \mathbf{p}) = -2\pi i \mathcal{U}_{\mathbf{p}} \mathcal{V}_{\mathbf{p}} \left[(1 - n_{\mathbf{p}}) \delta(\varepsilon - \varepsilon_{\mathbf{p}}) - n_{-\mathbf{p}} \delta(\varepsilon + \varepsilon_{\mathbf{p}}) \right], \tag{5.94}$$

$$G_{21}^{+-}(\varepsilon, \mathbf{p}) = G_{21}^{-+}(-\varepsilon, -\mathbf{p}). \tag{5.95}$$

Using these relations one can obtain the canonical forms of the collision integrals in superconductors [15] in a manner completely analogous to the case of a normal metal [12]. We will obtain the same kind of collision integrals (in $\varepsilon-$, rather than in $\xi-$ representation) in the next Chapter by employing the propagators integrated over energies. When both representations are applicable, these collision integrals are adequate to each other.

5.3.7 Expressions for Charge and Current

In general, the technique of the energy-integrated Green's functions is more convenient for those problems, where the kinetic processes occur in the vicinity of the Fermi-surface. In these cases it provides a powerful tool for the study of both pure and dirty superconductors. On the contrary, if the main processes occur in the regions remote from the Fermi surface, a straightforward application of this technique may lead to erroneous results. This difficulty may be overcome by properly accounting for the contribution that results from the equilibrium Green's function technique. Consider, for example, an expression for the electron charge in a superconductor. In the representation of the discrete imaginary frequencies, one can write (4.86) for the number of electrons in superconductors as

$$N_\omega(\mathbf{k}) = -Tr \left\{ T \sum_\varepsilon \int \frac{d^3\mathbf{p}}{(2\pi)^3} \mathfrak{G}_{\varepsilon\varepsilon-\omega}(\mathbf{p}, \mathbf{p} - \mathbf{k}) \right\}. \tag{5.96}$$

We will separate in this expression the contribution supplied by zero order Green's function $\mathfrak{G}^0_\varepsilon(\mathbf{p}) = (\xi - \varepsilon)^{-1}$, where $\xi = p^2/(2m) - \epsilon_F - e\varphi$ includes the quasiclassical scalar potential in the system. The deviation in the number of particles induced by the potential φ (in the first order in φ) has the form

$$\delta N(\varphi) = 2T \sum_\varepsilon \int \frac{d^3\mathbf{p}}{(2\pi)^3} \frac{1}{(\xi - \varepsilon)^2} e\varphi = -\frac{\partial N^{(0)}}{\partial \varepsilon_F} e\varphi = \frac{mp_F}{\pi^2} e\varphi, \tag{5.97}$$

where $N^{(0)}$ is the equilibrium electron density. Thus, the electron density in nonequilibrium superconductors is

$$N_\omega(\mathbf{k}) = N^{(0)}(2\pi)^4\delta(\omega)\delta(\mathbf{k}) - \frac{mp_F}{\pi^2}[e\varphi + \int_{-\infty}^\infty \frac{d\varepsilon}{4\pi i} \int \frac{d\Omega_\mathbf{p}}{4\pi} g'_{\varepsilon\varepsilon-\omega}] \tag{5.98}$$

$g'_{\varepsilon\varepsilon-\omega} = g_{\varepsilon\varepsilon-\omega} - 2\pi i \tanh(\varepsilon/2T)$ [the prime in (5.98) may be omitted, if the integration over ε is assumed in symmetrical limits]. From (5.98) the relation follows for a charge in nonequilibrium conditions:

$$\rho_\omega(\mathbf{k}) = e[N_\omega(\mathbf{k}) - N^{(0)}(2\pi)^4\delta(\omega)\delta(\mathbf{k})]. \tag{5.99}$$

The situation with the expression (4.85) for the electric current is analogous. The correct accounting of the contribution, supplied by the regions remote from the Fermi surface, results in the disappearance of the contribution from the second term in (4.85), which must be absent, if the $\hat{g}$-function technique [9] is used.

References

1. A.A. Abrikosov, L.P. Gor'kov, I.E. Dzyaloshinskiy, *Quantum Field Theoretical Methods in Statistical Physics*, 2nd edn. (Pergamon Press, Oxford, 1965), pp. 75–76
2. A.B. Migdal, Interaction between electrons and lattice vibrations in a normal metal. Sov. Phys. JETP **7**(6), 996–1001 (1958). [Zh. Eksp. i Teor. Phys. **34**(6), 1438–1446 (1958)]
3. A.S. Alexandrov, J. Ranninger, Bipolaronic superconductivity. Phys. Rev. B **24**(3), 1164–1169 (1981)
4. A.S. Alexandrov, N.F. Mott, *Polarons and Bipolarons* (World Scientific, Singapore, 1995), pp. 179–187
5. E.G. Brovman, Y.M. Kagan, Phonons in nontransition metals. Sov. Phys. Uspekhi **17** (2), 125–152 (1974). [Usp. Fiz. Nauk **112**(3), 369–426 (1974)]
6. G.M. Eliashberg, Temperature Green's function for electrons in a superconductor. Sov. Phys. JETP **12**(5), 1000–1002 (1960). [Zh. Eksp. i Teor. Phys. **39** [5(11)], 1437–1441 (1960)]
7. V.L. Ginzburg, D.A. Kirzhnits, *High-Temperature Superconductivity* (Consultants Bureau, New York, 1982), pp. 1–364

8. L.D. Landau, E.M. Lifshitz, *Fluid Mechanics* (Pergamon Press, Oxford, 1982), pp. 245–310
 9. G.M. Eliashberg, Inelastic electron collisions and nonequilibrium stationary states in superconductors. Sov. Phys. JETP **34**(3), 668–676 (1972). [Zh. Eksp. i Teor. Fiz. **61** [3(9)], 1254–1272 (1971)]
10. L.V. Keldysh, Diagram technique for nonequilibrium processes. Sov. Phys. JETP **20**(4), 1018–1026 (1965). [Zh. Eksp. i Teor. Phys. **47** [4(10)], 1515–1527 (1964)]
11. A.F. Volkov, S.M. Kogan, Collisionless relaxation of the energy gap in superconductors. Sov. Phys. JETP **38**(5), 1018–1021 (1974). [Zh. Eksp. i Teor. Phys. **65** [5(11)], 2038–2046 (1973)]
12. E.M. Lifshitz, L.P. Pitaevskii, *Physical Kinetics* (Pergamon Press, Oxford, 1981), pp. 391–412
13. G. Eilenberger, Transformation of Gor'kov equations for type II superconductors into transport-like equations. Z. Phys. **214**(2), 195–213 (1968)
14. A.I. Larkin, Y.N. Ovchinnikov, Nonlinear effects during the motion of vortices in superconductors. Sov. Phys. JETP **46**(1), 155–162 (1977). [Zh. Eksp. i Teor. Fiz. **73**[1(7)], 299–312 (1977)]
15. A.A. Aronov, V.L. Gurevich, Stability of nonequilibrium Fermi distributions with respect to Cooper pairing. Sov. Phys. JETP **38** (3), 550–556 (1974). [Zh. Eksp. i Teor. Phys. **65** 3(9), 1111–1124 (1973)]

Chapter 6
Electron and Phonon Collision Integrals

This Chapter is mainly devoted to the study of inelastic collision integrals of electrons with phonons, with each other, and with photons. Canonical forms of collision integrals will be derived. In the case of photon fields, for a large number of monochromatic photons, the electron-photon collision integral corresponds to the classical field terms derived by Eliashberg. Out of research curiosity, we derive the generalized collision integrals by the Keldysh's method as well. The results coincide with the results of the previous Chapter. An important feature of collision integrals is the nondiagonal channel of inelastic scattering. This yields the so-called branch imbalance in nonequilibrium superconductors and is specific for the superconducting state: no branch imbalance occurs in normal metals. In parallel with the kinetic equations for electrons, we introduce the kinetic equation for phonons. A coupled system of these equations, together with the self-consistency equation for the superconducting Cooper-pair condensate, serve as the basis for the next, final stage of the TDGL equations derivation.

6.1 Collision Integral Derivation

6.1.1 Spatially Homogeneous States

The generalized kinetic equations for integrated Green's functions $\widehat{g}_{\varepsilon\varepsilon-\omega}$ provide initial relations for constructing the canonical forms of collision integrals. As was shown in the preceding chapter, the matrix function $\widehat{g}_{\varepsilon\varepsilon-\omega}$ obeys (5.63), which in the spatially homogeneous case can be written as

$$\begin{pmatrix} \omega g & (2\varepsilon - \omega)f \\ (2\varepsilon - \omega)f^+ & -\omega\overline{g} \end{pmatrix} = \widehat{H}_1\widehat{g} - \widehat{g}\widehat{H}_1 + \widehat{I}, \tag{6.1}$$

© Springer Nature Switzerland AG 2020
A. Gulian, *Shortcut to Superconductivity*,
https://doi.org/10.1007/978-3-030-23486-7_6

where

$$\widehat{I} = \widehat{g}\widehat{\Sigma}^A - \widehat{\Sigma}^R\widehat{g} + \widehat{g}^R\widehat{\Sigma} - \widehat{\Sigma}\widehat{g}^A, \tag{6.2}$$

$$\widehat{g}^{R(A)} = \begin{pmatrix} g & f \\ -f^+ & g \end{pmatrix}^{R(A)}, \qquad \widehat{\Sigma}^{R(A)} = \begin{pmatrix} \Sigma_1 & \Sigma_2 \\ -\Sigma_2^+ & \Sigma_1 \end{pmatrix}^{R)A)}, \tag{6.3}$$

$$\widehat{H} = \begin{pmatrix} H_1 & 0 \\ 0 & H_1 \end{pmatrix}, \quad H_1 = -\frac{e}{c}\mathbf{v}\cdot\mathbf{A} + e\varphi, \quad \overline{H}_1 = \frac{e}{c}\mathbf{v}\cdot\mathbf{A} + e\varphi, \quad \mathbf{v} = \mathbf{v}_F \tag{6.4}$$

The retarded (advanced) functions in (6.1)–(6.3) are determined from the diagram expansion in which all the propagators and self-energies are retarded (advanced) (Sect. 5.2). For these functions equations of type (6.1) follow, where $\widehat{I}^{R(A)} = \widehat{g}^R\widehat{\Sigma}^{R(A)} - \widehat{\Sigma}^{R(A)}\widehat{g}^R$. The self-energy matrices $\widehat{\Sigma}^{R(A)}$ in (6.2) are additive functions[1]:

$$\widehat{\Sigma} = \widehat{\Sigma}^{(\mathrm{imp})} + \widehat{\Sigma}^{(\mathrm{e-ph})} + \widehat{\Sigma}^{(\mathrm{e-e})} + \widehat{\Sigma}^{(\mathrm{T})}..., \tag{6.5}$$

they correspond to the interaction of electrons with impurities, phonons, each other, tunneling, etc. Some of the self-energy parts will be examined in detail in subsequent sections.

6.1.2 Separation of Real and Virtual Processes

Separating in (6.2) the terms corresponding to the electron-phonon interaction, we will detach the virtual processes. Omitting the renormalization terms $\left(\Sigma_1^R + \Sigma_1^A\right)^{(\mathrm{e-ph})}$ and introducing a superconducting order parameter [2]

$$\Delta = \frac{1}{2}\left(\Sigma_1^R + \Sigma_1^A\right)^{(\mathrm{e-ph})}, \tag{6.6}$$

one finds for the 11-component of (6.2) the following expression

$$I_{\varepsilon\varepsilon-\omega} = \{-f\Delta^* + \Delta f_{\varepsilon\varepsilon-\omega}^+\} + \{-i(g\gamma + \gamma g) + i(-f\delta^+ + \delta f^+) + g^R\Sigma_1^{(\mathrm{e-ph})}$$
$$- \Sigma_1^{(\mathrm{e-ph})} g^A - f^R\Sigma_2^{+(\mathrm{e-ph})} + \Sigma_2^{(\mathrm{e-ph})} f^{+A}\}_{\varepsilon\varepsilon-\omega} + I'_{\varepsilon\varepsilon-\omega}, \tag{6.7}$$

where the quantities

$$2i\gamma_{\varepsilon\varepsilon-\omega} = \left(\Sigma_1^R - \Sigma_1^A\right)^{(\mathrm{e-ph})}, \quad 2i\delta_{\varepsilon\varepsilon-\omega} = \left(\Sigma_2^R - \Sigma_2^A\right)^{(\mathrm{e-ph})}, \tag{6.8}$$

---

[1]In principle, the interference between different physical processes, described by (6.5), is possible. Such interference was considered, e.g., by Reizer and Sergeev [1].

as well as $\Sigma_{1,2}^{(e-ph)}$, represent the real interactions between electrons and phonons, which are essential for the kinetics, and $I'_{\varepsilon\varepsilon-\omega}$ no longer contains $\Sigma^{(e-ph)}$ explicitly.

6.1.3 Nondiagonal Channel

The dissipation function γ in (6.7) (as well as δ and $\Sigma_{1,2}^{(e-ph)}$) has a characteristic magnitude of the order of the energy damping of electron excitations. In normal metals $\gamma \sim T^3/\omega_D^2$; in a superconducting state γ is even smaller, since a significant part of the electron-phonon interaction (the virtual processes) was already taken into account as being responsible for the superconducting transition. The γ function is less by orders of magnitude than the modulus of the order parameter Δ almost at all temperatures. Hence, before moving to the kinetic equation in (6.7), we must account exactly for the first expression in braces, using equations for the nondiagonal components of $\widehat{g}$-functions following from (6.1) (the nondiagonal channel, cf. [3]). From these equations it follows

$$(2\varepsilon - \omega)(f - f^+)_{\varepsilon\varepsilon-\omega} = \{i(f\overline{\gamma} + \overline{\gamma}f^+) - i(\gamma f + f^+\gamma) + i(\delta^+ g - g\delta)$$
$$+ i(\overline{g}\delta^+ - \delta\overline{g}) + (g\Delta - \Delta^* g) + (\overline{g}\Delta^* - \Delta\overline{g}) + f^R\overline{\Sigma}_1 + f^{+R}\Sigma_1 - \Sigma_1 f^A$$
$$- \overline{\Sigma}_1 f^{+A} + g^R\Sigma_2 - \Sigma_2\overline{g}^A + \overline{g}^R\Sigma_2^+ - \Sigma_2^+ g^A\}_{\varepsilon\varepsilon-\omega} + I''_{\varepsilon\varepsilon-\omega}. \qquad (6.9)$$

As in the derivation of relation (6.7), we have separated in the braces in (6.9) the virtual processes, which explicitly represent the electron-phonon interaction, while $I''_{\varepsilon\varepsilon-\omega}$ contains [in analogy to $I'_{\varepsilon\varepsilon-\omega}$ in (6.7)] all other processes. For the last quantity we have from (6.1):

$$I'' = g\Sigma_2^A + f\overline{\Sigma}_1^A - \Sigma_1^R f - \Sigma_2^R\overline{g} + g^R\Sigma_2 + f^R\overline{\Sigma}_1 - \Sigma_1 f^A - \Sigma_2\overline{g}^A$$
$$+ f^+\Sigma_1^A + \overline{g}\Sigma_2^{+A} - \Sigma_2^{+R}g - \overline{\Sigma}_1^R f^+ + f^{+R}\Sigma_1 + \overline{g}^R\Sigma_2^+ - -\Sigma_2^+ g^A - \overline{\Sigma}_1 f^{+A}. \qquad (6.10)$$

(Here all external and internal arguments are omitted; in this notation the order of the co-factors is important).

6.1.4 Impurities

Here we consider thin-film superconductors with a thickness on the order of superconducting correlation length. Such specimens always contain a number of electron elastic scattering centers (such as nonmagnetic impurities and lattice defects). If the number of these centers is sufficiently large, the superconducting films would be "dirty" and the mean free path of electrons would be shorter than the other lengths, which characterize their motion in superconductors. This circumstance allows one to make significant simplifications. In particular, one may ignore anisotropy effects, the non-locality of electrodynamics, the reflection of electrons from the boundaries, etc.

In the presence of impurities the self-energy matrix in (6.5) has the form (see Sect. 4.1)

$$\Sigma_{\varepsilon\varepsilon-\omega}^{(imp)R(A)} = \frac{1}{2\pi\tau} \int \frac{d\Omega_\mathbf{p}}{4\pi} \widehat{g}_{\varepsilon\varepsilon-\omega}^{R(A)}, \quad \Sigma_{\varepsilon\varepsilon-\omega}^{(imp)} = \frac{1}{2\pi\tau} \int \frac{d\Omega_\mathbf{p}}{4\pi} \widehat{g}_{\varepsilon\varepsilon-\omega}. \tag{6.11}$$

If paramagnetic impurities are also present, then (as follows from the analysis in Sect. 4.1) different factors $1/\tau_1$ and $1/\tau_2$ correspond to the functions g, $\overline{g}$ and f, f^+ respectively.

Having in mind the case of nonmagnetic impurities, in (6.1) we perform averaging over the angular variable, taking into account (6.7), (6.9) and (6.10). The self-energy parts, which correspond to the interaction of electrons with impurities, are eliminated owing to the isotropy of this interaction.

6.1.5   Effective Collision Integral

Before the derivation of the collision integrals in terms of the distribution function of electron (n_ε) and hole ($n_{-\varepsilon}$) type excitations, we examine the relation between $n_{\pm\varepsilon}$ and $\widehat{g}_\varepsilon$, where

$$\widehat{g}_{\varepsilon\varepsilon-\omega} = 2\pi\delta(\omega)\widehat{g}_\varepsilon. \tag{6.12}$$

The required relation was established in Sect. 5.2. Using (5.64)–(5.70), one finds

$$u(\varepsilon)\dot{n}_\varepsilon = -\frac{1}{8\pi i} \{(\dot{g}_\varepsilon - \dot{g}_{-\varepsilon}) + u(\varepsilon)(\dot{g}_\varepsilon - \dot{g}_{-\varepsilon})\} \operatorname{sign}\varepsilon, \tag{6.13}$$

where the dot designates the time derivative. Thus, the right side of (6.2) is expressed in terms of the 11-component of (6.1). Taking into account the nondiagonal channel, the effective collision integral becomes

$$I_{\text{eff}}(\varepsilon) = I_{\text{eff}}^{(e-ph)}(\varepsilon) + I_{\text{eff}}^{(e-e)}(\varepsilon) + I_{\text{eff}}^{(T)}(\varepsilon), \tag{6.14}$$

where the last two terms have the structure

$$I_{\text{eff}}(\varepsilon) = I_{\text{eff}}'(\varepsilon) - \frac{|\Delta|}{2\varepsilon} I_{\text{eff}}''(\varepsilon) = g\Sigma_1^A - f\Sigma_2^{+A} - \Sigma_1^R g + \Sigma_2^R f^+$$

$$+ g^R\Sigma_1 - f^R\Sigma_2^+ - \Sigma_1 g^A + \Sigma_2 f^{+A} - \frac{|\Delta|}{2\varepsilon}\{g\Sigma_2^A + f\overline{\Sigma}_1^A - \Sigma_1^R f$$

$$- \Sigma_2^R\overline{g} + g^R\Sigma_2 + f^R\overline{\Sigma}_1 - \Sigma_1 f^A - \Sigma_2\overline{g}^A + f^+\Sigma_1^A + \overline{g}\Sigma_2^{+A}$$

$$- \Sigma_2^{+R}g - \overline{\Sigma}_1^R f^+ + f^{+R}\Sigma_1 + \overline{g}^R\Sigma_2^+ - \Sigma_2^+ g^A - \overline{\Sigma}_1 f^{+A}\}. \tag{6.15}$$

A similar expression follows for $I_{\text{eff}}^{(e-ph)}$, which contains the quantities γ, δ, etc.

6.2 Inelastic Electron-Electron Collisions

To find the collision integral in canonical form [3, 4] we will use the general relations (6.13)–(6.15). First, it is necessary to specify the self-energy parts in them.

6.2.1 Diagram Evaluation of Electron-Electron Self-Energy

The diagrams corresponding to inter-collisions of two electrons are depicted in Fig. 6.1. The presence of pair condensate in the system, as usual, is responsible for the matrix structure of $\widehat{\Sigma}$. The contribution to Σ_1 from the first graph in Fig. 6.1 is shown in Fig. 6.2. In the representation of discrete imaginary frequencies, the elements of matrix $\widehat{\Sigma}$ may be written (omitting for a moment unessential indices) in the form

$$\Sigma_1(P, P - K) = T^2 \sum_{\varepsilon_1 \varepsilon_2} \int \int \frac{d^3 \mathbf{p}_1 \, d^3 \mathbf{p}_2}{(2\pi)^6} \left\{ A \mathfrak{G}_1 \mathfrak{G}_2 \overline{\mathfrak{G}}_3 - B \mathfrak{F}_1 \mathfrak{F}_2^+ \mathfrak{G}_3 \right\}, \quad (6.16)$$

$$\Sigma_2(P, P - K) = T^2 \sum_{\varepsilon_1 \varepsilon_2} \int \int \frac{d^3 \mathbf{p}_1 \, d^3 \mathbf{p}_2}{(2\pi)^6} \left\{ B \mathfrak{G}_1 \overline{\mathfrak{G}}_2 \mathfrak{F}_3 - A \mathfrak{F}_1 \mathfrak{F}_2 \mathfrak{F}_3^+ \right\}. \quad (6.17)$$

Here the 4-momentum variables of propagators are defined by the "decay" conservation laws $P = P_1 + P_2 + P_3$ and $K = K_1 + K_2 + K_3$. The quadratic forms A and B are related to the scattering amplitudes of two normal excitations on the Fermi surface. Using Fig. 6.1 the following expressions may be derived in Born's approximation:

$$A = -2|V_{\mathbf{p}-\mathbf{p}_2}|^2 + V_{\mathbf{p}-\mathbf{p}_2} V_{\mathbf{p}-\mathbf{p}_1}, \quad (6.18)$$

$$B = -2|V_{\mathbf{p}_1+\mathbf{p}_2}|^2 + V_{\mathbf{p}-\mathbf{p}_1} V_{\mathbf{p}-\mathbf{p}_2} + V_{\mathbf{p}_1+\mathbf{p}_2} V_{\mathbf{p}-\mathbf{p}_2} + V_{\mathbf{p}-\mathbf{p}_1} V_{\mathbf{p}_1+\mathbf{p}_2}, \quad (6.19)$$

where $V_{\mathbf{q}}$ is the interaction potential.

Fig. 6.1 Diagrams determining the self-energy functions (6.16) and (6.17). Wavy lines correspond to the electron-electron interaction potential

Fig. 6.2 Diagrams for 11-component of self-energy matrix (6.16) which corresponds to the first skeleton diagram in Fig. 6.1

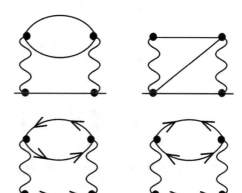

6.2.2 Analytical Continuation

From expressions (6.16) and (6.17), written in the discrete imaginary frequency representation, we move to the expressions on the real axis, using the Gor'kov-Eliashberg technique. To do this, consider the Nth order diagram of perturbation theory as the function of the complex variable ε at fixed imaginary frequencies of the field vertices. The analytical continuation over each of the frequencies must be made from the upper half-plane onto the real axis. The cuts lying between the lines $\mathrm{Im}\varepsilon = 0$ and $\mathrm{Im}(\varepsilon - \omega) = 0$ correspond to this diagram. Assuming that the cuts are

$$\mathrm{Im}(\varepsilon_1 - \omega_{1i}) = \mathrm{Im}(\varepsilon_2 - \omega_{2k}) = \mathrm{Im}(\varepsilon_3 - \omega_{3l}) = 0, \qquad (6.20)$$

we transform the sum over frequencies in (6.16) and (6.17) into a double integral. Because the directions of the arrows do not influence the procedure of analytical continuation, one can present (6.16) and (6.17) in the form (temporarily omitting unessential symbols)

$$\Sigma = T^2 \sum_{\varepsilon_1\varepsilon_2} \mathfrak{G}_{\varepsilon_1} \mathfrak{G}_{\varepsilon_2} \mathfrak{G}_{\varepsilon-\varepsilon_1-\varepsilon_2} = \oint\!\!\oint \frac{d\varepsilon_1 d\varepsilon_2}{(4\pi i)^2} \tanh\frac{\varepsilon_1}{2T} \tanh\frac{\varepsilon_2}{2T} \mathfrak{G}_{\varepsilon_1} \mathfrak{G}_{\varepsilon_2} \mathfrak{G}_{\varepsilon-\varepsilon_1-\varepsilon_2},$$
$$(6.21)$$

where the contours of integration enclose all the poles of hyperbolic tangents. Further step-by-step transformation of (6.21) to an integration over the real axis gives the result

$$\Sigma = \int\!\!\int \frac{dz_1 dz_2}{(4\pi i)^2} \left\{ \tanh\frac{z_1}{2T} \tanh\frac{z_2}{2T} \sum_{i,k} \delta_i\left(\mathfrak{G}_{z_1+\omega_{1i}}\right) \delta_k\left(\mathfrak{G}_{z_2+\omega_{2k}}\right) \mathfrak{G}_{\varepsilon-z_2-z_1-\omega_{1i}-\omega_{2k}} \right.$$

$$- \sum_{k,l} \mathfrak{G}_{\varepsilon+z_1-\omega_{2k}-\omega_{3l}} \delta_k\left(\mathfrak{G}_{z_2+\omega_{2k}}\right) \delta_l\left(\mathfrak{G}_{-z_2-z_1+\omega_{3l}}\right) \coth\frac{z_1}{2T} \tanh\frac{z_2}{2T}$$

$$+ \sum_{k,l} \mathfrak{G}_{\varepsilon+z_1-\omega_{2k}-\omega_{3l}} \delta_k\left(\mathfrak{G}_{z_2-z_1+\omega_{2k}}\right) \delta_l\left(\mathfrak{G}_{-z_2+\omega_{3l}}\right) \coth\frac{z_1}{2T} \tanh\frac{z_2}{2T}$$

$$\left. - \sum_{i,l} \delta_i\left(\mathfrak{G}_{z_1+\omega_{1i}}\right) \mathfrak{G}_{\varepsilon+z_2-z_1-\omega_{1i}-\omega_{3l}} \delta_l\left(\mathfrak{G}_{-z_2+\omega_{3l}}\right) \tanh\frac{z_1}{2T} \tanh\frac{z_2}{2T} \right\}. \qquad (6.22)$$

Here the external variable ε and the field frequencies remain imaginary. Continuing (6.22) over ε from the region $\mathrm{Im}(\varepsilon - \omega) > 0$ ($\mathrm{Im}\varepsilon < 0$) and next over all the frequencies onto the upper half-plane, one finds for $\Sigma^{R(A)}$ the expression

$$\Sigma^{R(A)} = \int_{-\infty}^{\infty} \frac{d\varepsilon_1 d\varepsilon_2}{(4\pi i)^2} \left\{ G_1 G_2 G_3^{R(A)} + G_1 G_2^{R(A)} G_3 + G_1^{R(A)} G_2 G_3 \right.$$
$$+ G_1^{R(A)} G_3^{R(A)} G_2^{R(A)} - G_1^{R(A)} G_2^{R(A)} G_3^{A(R)}$$
$$\left. - G_1^{R(A)} G_2^{A(R)} G_3^{R(A)} - G_1^{A(R)} G_2^{R(A)} G_3^{R(A)} \right\}. \qquad (6.23)$$

Using the definition of Σ in a form, analogous to (5.46)

$$\Sigma_{\varepsilon\varepsilon-\omega} = \sum_{N=0}^{\infty} \sum_{k=1}^{N+1} \delta_k \left(\Sigma_{\varepsilon\varepsilon-\omega}^{(N)}\right) \tanh \frac{\varepsilon - \Omega_k}{2T}, \tag{6.24}$$

where Ω_k is some combination of field vertex frequencies, one finds from (4.22):

$$\Sigma = \int_{-\infty}^{\infty} \frac{d\varepsilon_1 d\varepsilon_2}{(4\pi i)^2} \{G_1 G_2 G_3 + G_1(G^R - G^A)_2(G^R - G^A)_3 \\ + (G^R - G^A)_1 G_2(G^R - G^A)_3 + (G^R - G^A)_1(G^R - G^A)_2 G_3\}. \tag{6.25}$$

The expressions for the quantities $\Sigma^{R(A)}$ and Σ, which define the collision integral, follow from (6.16), (6.17) and (6.23), (6.25).

6.2.3 Transition to Energy-Integrated Propagators

Before writing down the corresponding results, we will integrate (6.16) and (6.17) over the variables $\xi = v_F(p - p_F)$; this is possible because the effective interaction is short range, so the amplitudes A and B depend on angle variables only. Hence one can write

$$\int\int \frac{d^3\mathbf{p}_1 d^3\mathbf{p}_2}{(2\pi)^6} AG_1 G_2 G_3 \\ = \left(\frac{mp_F}{2\pi^2}\right)^2 \int\int \frac{d\Omega_{\mathbf{p}_1} d\Omega_{\mathbf{p}_2}}{(4\pi)^2} A \int\int\int d\xi_1 d\xi_2 d\xi_3\, \delta(\xi_3 - v_F(p_3 - p_F))G_1 G_2 G_3. \tag{6.26}$$

The δ-function here restricts mainly the angle integration, requiring $p_3 = |\mathbf{p} - \mathbf{p}_1 - \mathbf{p}_2| \approx p_F$ and thus

$$\int\int \frac{d^3\mathbf{p}_1 d^3\mathbf{p}_2}{(2\pi)^6} AG_1 G_2 G_3 = \left(\frac{mp_F}{2\pi^2}\right)^2 \frac{1}{2\epsilon_F} \int\int \frac{d\Omega_{\mathbf{p}_1} d\Omega_{\mathbf{p}_2}}{(4\pi)^2} \delta\left(\frac{p_3}{p_F} - 1\right) Ag_1 g_2 g_3. \tag{6.27}$$

Gathering the results, we have

$$\Sigma_1^R - \Sigma_1^A = \widehat{L}\left[A\left\{g_1 g_2 \bar{g}_3\right\}^{(R-A)} - B\left\{f_1 f_2^+ g_3\right\}^{(R-A)}\right], \tag{6.28}$$

$$\Sigma_1 = \widehat{L}\left[A\left\{g_1 g_2 \bar{g}_3\right\} - B\left\{f_1 f_2^+ g_3\right\}\right], \tag{6.29}$$

$$\Sigma_2^R - \Sigma_2^A = \widehat{L}\left[B\left\{g_1 \bar{g}_2 f_3\right\}^{(R-A)} - A\left\{f_1 f_2 f_3^+\right\}^{(R-A)}\right], \tag{6.30}$$

$$\Sigma_2 = \widehat{L}\left[B\left\{g_1 \bar{g}_2 f_3\right\} - A\left\{f_1 f_2 f_3^+\right\}\right]. \tag{6.31}$$

The notations $\{\dots\}$ and $\{\dots\}^{(R-A)}$ are used:

$$\{g_1 g_2 g_3\}^{(R-A)} = g_1 g_2 \left(g_3^R - g_3^A\right) + g_1 \left(g_2^R - g_2^A\right) g_3 + \left(g_1^R - g_1^A\right) g_2 g_3$$
$$+ \left(g_1^R - g_1^A\right)\left(g_2^R - g_2^A\right)\left(g_3^R - g_3^A\right), \tag{6.32}$$

$$\{g_1 g_2 g_3\} = g_1 g_2 g_3 + g_1 \left(g_2^R - g_2^A\right)\left(g_3^R - g_3^A\right) + \left(g_1^R - g_1^A\right) g_2 \left(g_3^R - g_3^A\right)$$
$$+ \left(g_1^R - g_1^A\right)\left(g_2^R - g_2^A\right) g_3, \tag{6.33}$$

and the operator $\widehat{L}$ is defined as

$$\widehat{L} = \left(\frac{mp_F}{2\pi^2}\right)^2 \frac{1}{2\epsilon_F} \int_{-\infty}^{\infty}\int_{-\infty}^{\infty} \frac{d\varepsilon_1 d\varepsilon_2}{(4\pi i)^2} \iint \frac{d\Omega_{\mathbf{p}_1} d\Omega_{\mathbf{p}_2}}{(4\pi)^2} \delta\left(\frac{p_3}{p_F} - 1\right). \tag{6.34}$$

6.2.4 Derivation of the Canonical Form

Substituting expressions (6.16) and (6.17), subject to (6.32) and (6.33) and (5.81) to (5.85), into formula (6.15), we find

$$
\begin{aligned}
I_{\text{eff}}(\varepsilon) = 2\pi^4 \widehat{L}\{ &-A(-uu_1 u_2 u_3 \beta\beta_1\beta_2 - uu_2 u_3 \beta\beta_2\alpha_1 - uu_1 u_3 \beta\alpha_2\beta_1 \\
&- uu_3\beta\alpha_1\alpha_2 - uu_1 u_2 u_3 \beta\beta_1\beta_3 - uu_2 u_3\beta\beta_3\alpha_1 + uu_1 u_2\beta\beta_1\alpha_3 \\
&+ uu_2\beta\alpha_1\alpha_3 - uu_1 u_2 u_3\beta\beta_2\beta_3 - uu_1 u_3\beta\beta_3\alpha_2 + uu_1 u_2\beta\beta_2\alpha_3 \\
&+ uu_1\beta\alpha_2\alpha_3 - uu_1 u_2 u_3\beta - u_1 u_2 u_3\beta_1\beta_2\alpha - u_2 u_3\beta_2\alpha\alpha_1 \\
&- u_1 u_3\beta_1\alpha\alpha_2 - u_3\alpha\alpha_1\alpha_2 - u_1 u_2 u_3\beta_1\beta_3\alpha - u_2 u_3\beta_3\alpha\alpha_1 \\
&+ u_1 u_2\beta_1\alpha\alpha_3 + u_2\alpha\alpha_1\alpha_3 - u_1 u_2 u_3\beta_2\beta_3\alpha - u_1 u_3\beta_3\alpha\alpha_2 \\
&+ u_1 u_2\beta_2\alpha\alpha_3 + u_1\alpha\alpha_2\alpha_3 - u_1 u_2 u_3\alpha) + B(uu_3 v_1 v_2\beta\beta_1\beta_2 \\
&+ uu_3 v_1 v_2\beta\beta_1\beta_3 + uv_1 v_2\beta\beta_1\alpha_3 + uv_1 v_2 u_3\beta\beta_2\beta_3 + uv_1 v_2\beta\beta_2\alpha_3 \\
&+ uu_3 v_1 v_2\beta + u_3 v_1 v_2\beta_1\beta_2\alpha + u_3 v_1 v_2\beta_1\beta_3\alpha + v_1 v_2\beta_1\alpha\alpha_3 \\
&+ u_3 v_1 v_2\beta_2\beta_3\alpha + v_1 v_2\beta_2\alpha\alpha_3 + u_3 v_1 v_2\alpha) - B(u_1 u_2 v v_3\beta\beta_1\beta_2 \\
&+ u_2 v v_3\beta\beta_2\alpha_1 - u_1 v v_3\beta\beta_1\alpha_2 - v v_3\beta\alpha_1\alpha_2 + u_1 u_2 v v_3\beta\beta_1\beta_3 \\
&+ u_2 v v_3\beta\beta_3\alpha_1 + u_1 u_2 v v_3\beta\beta_2\beta_3 - u_1 v v_3\beta\beta_3\alpha_2 + u_1 u_2 v v_3\beta) \\
&- A(v v_1 v_2 v_3\beta\beta_1\beta_2 + v v_1 v_2 v_3\beta\beta_1\beta_3 + v v_1 v_2 v_3\beta\beta_2\beta_3 + v v_1 v_2 v_3\beta) \\
&- A(uu_1 u_2 u_3\beta_1\beta_2\beta_3 - uu_1 u_2\beta_1\beta_2\alpha_3 + uu_2 u_3\beta_2\beta_3\alpha_1 - uu_2\beta_2\alpha_1\alpha_3 \\
&+ uu_1 u_3\beta_1\beta_3\alpha_2 - uu_1\beta_1\alpha_2\alpha_3 + uu_3\beta_3\alpha_1\alpha_2 - u\alpha_1\alpha_2\alpha_3 \\
&+ uu_1 u_2 u_3\beta_1 + uu_2 u_3\alpha_1 + uu_1 u_2 u_3\beta_2 + uu_1 u_3\alpha_2 + uu_1 u_2 u_3\beta_3 \\
&- uu_1 u_2\alpha_3) - B(uu_3 v_1 v_2\beta_1\beta_2\beta_3 + uv_1 v_2\beta_1\beta_2\alpha_3 + uu_3 v_1 v_2\beta_1 \\
&+ uu_3 v_1 v_2\beta_2 + uu_3 v_1 v_2\beta_3 + uv_1 v_2\alpha_3) + B(u_1 u_2 v v_3\beta_1\beta_2\beta_3 \\
&+ u_2 v v_3\beta_2\beta_3\alpha_1 - u_1 v_3 v\beta_1\beta_3\alpha_2 - v v_3\beta_3\alpha_1\alpha_2 + u_1 u_2 v v_3\beta_1 \\
&+ u_2 v v_3\alpha_1 + u_1 u_2 v v_3\beta_2 - u_1 v v_3\alpha_2 + u_1 u_2 v v_3\beta_3) \\
&+ A(v v_1 v_2 v_3\beta_1\beta_2\beta_3 + v v_1 v_2 v_3\beta_1 + v v_1 v_2 v_3\beta_2 + v v_1 v_2 v_3\beta_3)
\end{aligned}
$$

$$- \frac{|\Delta|}{\varepsilon}[-A(-u_2 u_3 \beta \beta_2 \alpha_1 - u_1 u_3 v \beta \beta_1 \alpha_2 - u_2 u_3 v \beta \beta_3 \alpha_1 + u_1 u_2 v \beta \beta_1 \alpha_3$$
$$- u_1 u_3 v \beta \beta_3 \alpha_2) + B(v v_1 v_2 \beta \beta_1 \alpha_3 + v v_1 v_2 \beta \beta_2 \alpha_3) - B(v_3 \alpha \alpha_1 \alpha_2$$
$$- u_1 u_2 v_3 \beta_3 \alpha - u_2 v_3 \beta_3 \alpha \alpha_1 - u_1 u_2 v_3 \beta_2 \beta_3 \alpha - u_1 u_2 u_3 \beta_1 \beta_2 \alpha$$
$$- u_2 v_3 \beta_2 \alpha \alpha_1 + u_1 v_3 \beta_1 \alpha \alpha_2 + u_1 v_3 \beta_3 \alpha \alpha_2 - u_1 u_2 v_3 \alpha)$$
$$+ A(v_1 v_2 v_3 \beta_1 \beta_2 \alpha + v_1 v_2 v_3 \beta_1 \beta_3 \alpha + v_1 v_2 v_3 \beta_2 \beta_3 \alpha + v_1 v_2 v_3 \alpha)$$
$$- A(u_2 u_3 v \beta_2 \beta_3 \alpha_1 + u_1 u_3 v \beta_1 \beta_3 \alpha_2 - u_1 u_2 v \beta_1 \beta_2 \alpha_3 - v \alpha_1 \alpha_2 \alpha_3$$
$$+ v u_2 u_3 \alpha_1 + u_1 u_3 v \alpha_2 - u_1 u_2 v \alpha_3) - B(v v_1 v_2 \beta_1 \beta_2 \alpha_3 + v v_1 v_2 \alpha_3)]\}. \quad (6.35)$$

[Here $u = u(\varepsilon)$, $u_1 = u(\varepsilon_1)$, etc.] Reversing the sign of ε in this expression and substituting the values of $I(\varepsilon)$ and $I(-\varepsilon)$ into (6.13), we obtain, subject to relations (5.86) to (5.88), the following canonical form for the inelastic electron-electron collision integral:

$$J^{(e-e)}(n_{\pm \varepsilon}) = \frac{1}{16 \epsilon_F \sqrt{\varepsilon^2 - |\Delta|^2}} \int_{|\Delta|}^{\infty} \int_{|\Delta|}^{\infty} \int_{|\Delta|}^{\infty} \frac{d\varepsilon_1 d\varepsilon_2 d\varepsilon_3}{\sqrt{\varepsilon_1^2 - |\Delta|^2}\sqrt{\varepsilon_2^2 - |\Delta|^2}\sqrt{\varepsilon_3^2 - |\Delta|^2}}$$
$$\times \{E_1 \delta(\varepsilon - \varepsilon_1 - \varepsilon_2 - \varepsilon_3) + E_2 \delta(\varepsilon + \varepsilon_1 - \varepsilon_2 - \varepsilon_3) + E_3 \delta(\varepsilon + \varepsilon_2 + \varepsilon_3 - \varepsilon_1)\},$$
$$(6.36)$$

in which the factors E_i have a form

$$E_1 = M_1^1 \{[(1 - n_{\pm \varepsilon}) n_{\varepsilon_1} n_{\varepsilon_2} n_{\varepsilon_3} - n_{\pm \varepsilon}(1 - n_{\varepsilon_1})(1 - n_{\varepsilon_2})(1 - n_{\varepsilon_3})]$$
$$+ [(1 - n_{\pm \varepsilon}) n_{\varepsilon_1} n_{-\varepsilon_2} n_{-\varepsilon_3} - n_{\pm \varepsilon}(1 - n_{\varepsilon_1})(1 - n_{-\varepsilon_2})(1 - n_{-\varepsilon_3})]\}$$
$$+ M_1^2 \{[(1 - n_{\pm \varepsilon}) n_{-\varepsilon_1} n_{\varepsilon_2} n_{\varepsilon_3} - n_{\pm \varepsilon}(1 - n_{-\varepsilon_1})(1 - n_{\varepsilon_2})(1 - n_{\varepsilon_3})]$$
$$+ [(1 - n_{\pm \varepsilon}) n_{-\varepsilon_1} n_{-\varepsilon_2} n_{-\varepsilon_3} - n_{\pm \varepsilon}(1 - n_{-\varepsilon_1})(1 - n_{-\varepsilon_2})(1 - n_{-\varepsilon_3})]\}$$
$$+ 2M_1^3 [(1 - n_{\pm \varepsilon}) n_{\varepsilon_1} n_{-\varepsilon_2} n_{\varepsilon_3} - n_{\pm \varepsilon}(1 - n_{\varepsilon_1})(1 - n_{-\varepsilon_2})(1 - n_{\varepsilon_3})]$$
$$+ 2M_1^4 [(1 - n_{\pm \varepsilon}) n_{-\varepsilon_1} n_{-\varepsilon_2} n_{\varepsilon_3} - n_{\pm \varepsilon}(1 - n_{-\varepsilon_1})(1 - n_{-\varepsilon_2})(1 - n_{\varepsilon_3})], \quad (6.37)$$
$$E_2 = M_2^1 \{[n_{\varepsilon_2} n_{\varepsilon_3}(1 - n_{\pm \varepsilon})(1 - n_{\varepsilon_1}) - (1 - n_{\varepsilon_2})(1 - n_{\varepsilon_3}) n_{\pm \varepsilon} n_{\varepsilon_1}]$$
$$+ [n_{-\varepsilon_2} n_{-\varepsilon_3}(1 - n_{\pm \varepsilon})(1 - n_{\varepsilon_1}) - (1 - n_{-\varepsilon_2})(1 - n_{-\varepsilon_3}) n_{\pm \varepsilon} n_{\varepsilon_1}]\}$$
$$+ M_2^2 \{[n_{\varepsilon_2} n_{\varepsilon_3}(1 - n_{\pm \varepsilon})(1 - n_{-\varepsilon_1}) - (1 - n_{\varepsilon_2})(1 - n_{\varepsilon_3}) n_{\pm \varepsilon} n_{-\varepsilon_1}]$$
$$+ [n_{-\varepsilon_2} n_{-\varepsilon_3}(1 - n_{\pm \varepsilon})(1 - n_{-\varepsilon_1}) - (1 - n_{-\varepsilon_2})(1 - n_{-\varepsilon_3}) n_{\pm \varepsilon} n_{-\varepsilon_1}]\}$$
$$+ 2M_2^3 [n_{-\varepsilon_2} n_{\varepsilon_3}(1 - n_{\pm \varepsilon})(1 - n_{\varepsilon_1}) - (1 - n_{-\varepsilon_2})(1 - n_{\varepsilon_3}) n_{\pm \varepsilon} n_{\varepsilon_1}]$$
$$+ 2M_2^4 [n_{-\varepsilon_2} n_{\varepsilon_3}(1 - n_{\pm \varepsilon})(1 - n_{-\varepsilon_1}) - (1 - n_{-\varepsilon_2})(1 - n_{\varepsilon_3}) n_{\pm \varepsilon} n_{-\varepsilon_1}], \quad (6.38)$$
$$E_3 = M_3^1 \{[n_{\varepsilon_1}(1 - n_{\pm \varepsilon})(1 - n_{\varepsilon_2})(1 - n_{\varepsilon_3}) - (1 - n_{\varepsilon_1}) n_{\pm \varepsilon} n_{\varepsilon_2} n_{\varepsilon_3}]$$
$$+ [n_{\varepsilon_1}(1 - n_{\pm \varepsilon})(1 - n_{-\varepsilon_2})(1 - n_{-\varepsilon_3}) - (1 - n_{\varepsilon_1}) n_{\pm \varepsilon} n_{-\varepsilon_2} n_{-\varepsilon_3}]\}$$
$$+ M_3^2 \{[n_{-\varepsilon_1}(1 - n_{\pm \varepsilon})(1 - n_{\varepsilon_2})(1 - n_{\varepsilon_3}) - (1 - n_{-\varepsilon_1}) n_{\pm \varepsilon} n_{\varepsilon_2} n_{\varepsilon_3}]$$
$$+ [n_{-\varepsilon_1}(1 - n_{\pm \varepsilon})(1 - n_{-\varepsilon_2})(1 - n_{-\varepsilon_3}) - (1 - n_{-\varepsilon_1}) n_{\pm \varepsilon} n_{-\varepsilon_2} n_{-\varepsilon_3}]\}$$
$$+ 2M_3^3 [n_{\varepsilon_1}(1 - n_{\pm \varepsilon})(1 - n_{-\varepsilon_2})(1 - n_{\varepsilon_3}) - (1 - n_{\varepsilon_1}) n_{\pm \varepsilon} n_{-\varepsilon_2} n_{\varepsilon_3}]$$
$$+ 2M_3^4 [n_{-\varepsilon_1}(1 - n_{\pm \varepsilon})(1 - n_{-\varepsilon_2})(1 - n_{\varepsilon_3}) - (1 - n_{-\varepsilon_1}) n_{\pm \varepsilon} n_{-\varepsilon_2} n_{\varepsilon_3}]. \quad (6.39)$$

Coefficients M_i^j, entering (6.37) to (6.39), are given by the following relations:

$$M_1^1 = a(\varepsilon\varepsilon_1\varepsilon_2\varepsilon_3 - |\Delta|^4 - \varepsilon\varepsilon_1\sqrt{\varepsilon_2^2 - |\Delta|^2}\sqrt{\varepsilon_3^2 - |\Delta|^2}$$
$$\pm \sqrt{\varepsilon^2 - |\Delta|^2}\sqrt{\varepsilon_1^2 - |\Delta|^2}\varepsilon_2\varepsilon_3 \mp \sqrt{\varepsilon^2 - |\Delta|^2}\sqrt{\varepsilon_1^2 - |\Delta|^2}\sqrt{\varepsilon_2^2 - |\Delta|^2}$$
$$\times \sqrt{\varepsilon_3^2 - |\Delta|^2}) + b|\Delta|^2(\varepsilon\varepsilon_3 - \varepsilon_1\varepsilon_2 + \sqrt{\varepsilon_2^2 - |\Delta|^2}\sqrt{\varepsilon_3^2 - |\Delta|^2}$$
$$\pm \sqrt{\varepsilon^2 - |\Delta|^2}\sqrt{\varepsilon_1^2 - |\Delta|^2}), \tag{6.40}$$

$$M_1^2 = a(\varepsilon\varepsilon_1\varepsilon_2\varepsilon_3 - |\Delta|^4 - \varepsilon\varepsilon_1\sqrt{\varepsilon_2^2 - |\Delta|^2}\sqrt{\varepsilon_3^2 - |\Delta|^2}$$
$$\mp \sqrt{\varepsilon^2 - |\Delta|^2}\sqrt{\varepsilon_1^2 - |\Delta|^2}\varepsilon_2\varepsilon_3 \pm \sqrt{\varepsilon^2 - |\Delta|^2}\sqrt{\varepsilon_1^2 - |\Delta|^2}\sqrt{\varepsilon_2^2 - |\Delta|^2}$$
$$\times \sqrt{\varepsilon_3^2 - |\Delta|^2}) + b|\Delta|^2(\varepsilon\varepsilon_3 - \varepsilon_1\varepsilon_2 + \sqrt{\varepsilon_2^2 - |\Delta|^2}\sqrt{\varepsilon_3^2 - |\Delta|^2}$$
$$\mp \sqrt{\varepsilon^2 - |\Delta|^2}\sqrt{\varepsilon_1^2 - |\Delta|^2}), \tag{6.41}$$

$$M_1^3 = a(\varepsilon\varepsilon_1\varepsilon_2\varepsilon_3 - |\Delta|^4 + \varepsilon\varepsilon_1\sqrt{\varepsilon_1^2 - |\Delta|^2}\sqrt{\varepsilon_3^2 - |\Delta|^2}$$
$$\pm \sqrt{\varepsilon^2 - |\Delta|^2}\sqrt{\varepsilon_1^2 - |\Delta|^2}\varepsilon_2\varepsilon_3 \pm \sqrt{\varepsilon^2 - |\Delta|^2}\sqrt{\varepsilon_1^2 - |\Delta|^2}\sqrt{\varepsilon_2^2 - |\Delta|^2}$$
$$\times \sqrt{\varepsilon_3^2 - |\Delta|^2}) + b|\Delta|^2(\varepsilon\varepsilon_3 - \varepsilon_1\varepsilon_2 + \sqrt{\varepsilon_2^2 - |\Delta|^2}\sqrt{\varepsilon_3^2 - |\Delta|^2}$$
$$\pm \sqrt{\varepsilon^2 - |\Delta|^2}\sqrt{\varepsilon_1^2 - |\Delta|^2}), \tag{6.42}$$

$$M_1^4 = a(\varepsilon\varepsilon_1\varepsilon_2\varepsilon_3 - |\Delta|^4 + \varepsilon\varepsilon_1\sqrt{\varepsilon_2^2 - |\Delta|^2}\sqrt{\varepsilon_3^2 - |\Delta|^2}$$
$$\mp \sqrt{\varepsilon^2 - |\Delta|^2}\sqrt{\varepsilon_1^2 - |\Delta|^2}\varepsilon_2\varepsilon_3 \mp \sqrt{\varepsilon^2 - |\Delta|^2}\sqrt{\varepsilon_1^2 - |\Delta|^2}\sqrt{\varepsilon_2^2 - |\Delta|^2}$$
$$\times \sqrt{\varepsilon_3^2 - |\Delta|^2}) + b|\Delta|^2(\varepsilon\varepsilon_3 - \varepsilon_1\varepsilon_2 + \sqrt{\varepsilon_2^2 - |\Delta|^2}\sqrt{\varepsilon_3^2 - |\Delta|^2}$$
$$\mp \sqrt{\varepsilon^2 - |\Delta|^2}\sqrt{\varepsilon_1^2 - |\Delta|^2}). \tag{6.43}$$

The quantities M_2^j and M_3^j are defined by the expressions

$$M_2^j = -M_1^j(\varepsilon, -\varepsilon_1, \varepsilon_2, \varepsilon_3) - M_1^j(\varepsilon, \varepsilon_2, -\varepsilon_1, \varepsilon_3) - M_1^j(\varepsilon, \varepsilon_3, \varepsilon_2, -\varepsilon_1), \tag{6.44}$$
$$M_3^j = M_1^j(\varepsilon, \varepsilon_1, -\varepsilon_2, -\varepsilon_3) + M_1^j(\varepsilon, -\varepsilon_2, \varepsilon_1, -\varepsilon_3) + M_1^j(\varepsilon, -\varepsilon_3, -\varepsilon_2, \varepsilon_1). \tag{6.45}$$

Factors a and b, entering (6.40) to (6.43), are numbers (of an order of unity) and are connected with A and B by relations of the type

$$a = -2\pi \left(\frac{mp_F}{2\pi^2}\right)^2 \int\int\int \frac{d\Omega_p d\Omega_{p_1} d\Omega_{p_2}}{(4\pi)^2} \delta\left(\frac{|\mathbf{p} - \mathbf{p}_1 - \mathbf{p}_2|}{p_F} - 1\right) A. \tag{6.46}$$

6.2.5 Essence of Elementary Acts

The meaning of the elementary acts, described by the collision operator (6.36), is quite transparent. Consider, for example, the term in E_1 that is proportional to M_1^1. With a positive sign of ε, the first component in this term describes the merger of three electron excitations into a single electron-type excitation. With a negative sign of ε, three merging electron excitations create an excitation on the hole branch. Thus in the first case the difference in the number of electrons and holes changes by 2, while in the second case it changes by 4. Analogous processes are described by other items in this term. It vanishes in the case of a normal metal ($M_1^1 = 0$ when $|\Delta| = 0$); hence the channel of homogeneous relaxation of electron-hole imbalance is closed in a normal metal [5].

Note that the collision integral (6.36) has obtained such a transparent meaning, owing to the specific selection of the form of the functions $n_{\pm\varepsilon}$ in the expressions for $\widehat{g}^{(R,A)}$ (5.81) to (5.88).

6.3 Kinetic Equation for Phonons

6.3.1 Application of Keldysh Technique

The phonon Green-Keldysh function [6] is introduced in the usual manner[2]:

$$\mathcal{D}_\nu^{ik}(1, 2) = -i\langle T\hat{\varphi}_\nu(1i)\hat{\varphi}_\nu(2k)\rangle, \qquad 1 \equiv X \equiv (r, t) \tag{6.47}$$

The Keldysh indices i, k are the signs minus or plus, according to the location of the time coordinate on each of the two time axes ($\infty, +\infty$ and $+\infty, -\infty$). Recall, that the time on the second axis (with the sign plus) is greater than any time on the first axis (with the sign minus), and the T-ordering on the second axis proceeds in the reversed order. The free phonon field operators are real ($\hat{\varphi}^\dagger = \hat{\varphi}$, see Sect. 5.1):

$$\hat{\varphi}(x) = \frac{1}{\sqrt{V_0}} \sum_k \sqrt{\frac{\omega_0(\mathbf{k})}{2}} \left\{ b_\mathbf{k} e^{i(\mathbf{k}\cdot\mathbf{r}-\omega_0(\mathbf{k})t)} + b_\mathbf{k}^\dagger e^{-i(\mathbf{k}\cdot\mathbf{r}-\omega_0(\mathbf{k})t)} \right\}. \tag{6.48}$$

Here V_0 is the volume of the system; $\omega_0(\mathbf{k})$ is the dispersion of phonons in normal metal; $b_\mathbf{k}^\dagger$ and $b_\mathbf{k}$ are the phonon creation and annihilation operators.

The "bare" Green-Keldysh functions, defined by (6.47) and (6.48), may be easily found in the homogeneous and stationary cases. For instance, the expression for $\mathcal{D}_0^{-+}$ is:

---

[2]We omit below the index ν of phonon polarization. It may be restored in the final expressions. As was mentioned in Sect. 4.3, in the isotropic model of metals the electrons interact only with longitudinal acoustic phonons. Such interaction is implied in this Chapter.

$$\mathcal{D}_0^{-+}(\mathbf{r}, t) = -\frac{i}{2} \int \frac{d^3 k}{(2\pi)^3} \omega_0(\mathbf{k}) e^{i\mathbf{k}\cdot\mathbf{r}} \left[N_{\mathbf{k}} e^{-i\omega_0(\mathbf{k})t} + (1 + N_{-\mathbf{k}}) e^{i\omega_0(\mathbf{k})t} \right], \quad (6.49)$$

where $N_{\mathbf{k}} = \langle b_{\mathbf{k}}^{\dagger} b_{\mathbf{k}} \rangle$ is the nonequilibrium phonon distribution function [6].

In addition, we introduce an operator $\mathcal{D}_{01(2)}^{-1}$, which acts on the first (second) argument of the phonon propagator (u is the phonon's velocity):

$$\mathcal{D}_{01(2)}^{-1} = \frac{\partial^2}{\partial t_{1(2)}^2} - u^2 \nabla_{1(2)}^2, \quad (6.50)$$

where

$$\mathcal{D}_{01(2)}^{-1} \mathcal{D}_0^{ik}(1, 2) = u^2 \hat{\sigma}_z^{ik} \delta(t_1 - t_2) \nabla_{1(2)} \Delta(\mathbf{r}_1 - \mathbf{r}_2), \quad (6.51)$$

and $\hat{\sigma}_z$ is the third of the Pauli-matrices $\hat{\sigma}_x, \hat{\sigma}_y, \hat{\sigma}_z$.

In the general case the phonon function obeys the Dyson equation

$$\widehat{\mathcal{D}}(1, 2) = \widehat{\mathcal{D}}_0(1, 2) + \int \widehat{\mathcal{D}}_0(1, 4) \widehat{\Pi}(4, 3) \widehat{\mathcal{D}}(3, 2) d^4 x_3 d^4 x_4, \quad (6.52)$$

or

$$\widehat{\mathcal{D}}(1, 2) = \widehat{\mathcal{D}}_0(1, 2) + \int \widehat{\mathcal{D}}(1, 3) \widehat{\Pi}(3, 4) \widehat{\mathcal{D}}_0(4, 2) d^4 x_3 d^4 x_4, \quad (6.53)$$

where all the functions are matrices in Keldysh indices. Note that owing to their definition (6.47), the Green-Keldysh functions are linear dependent ($\mathcal{D}^{--} + \mathcal{D}^{++} - \mathcal{D}^{-+} - \mathcal{D}^{+-} = 0$) and, consequently, the polarization operators are also linear dependent: ($\Pi^{--} + \Pi^{++} + \Pi^{-+} + \Pi^{+-} = 0$).

The electron Green-Keldysh-Nambu function is defined analogously:

$$G_{\mu\nu}^{ik}(1, 2) = -i \langle T \psi_\mu(1i) \psi_\nu^{\dagger}(2k) \rangle \quad (6.54)$$

as mentioned in Sect. 5.2. Here $\mu = 1, 2$ and $\nu = 1, 2$ are the Nambu indices of the field operators

$$\psi_1(1i) \equiv \psi_\uparrow(1i), \qquad \psi_2(1i) \equiv \psi_\downarrow^{\dagger}(1i). \quad (6.55)$$

The Green's function thus introduced (in absence of interactions, which depend explicitly on spin variables) has the symmetry property

$$\left(G_{\mu\nu}^{ik} \right)^* = (-1)^{\mu+\nu} G_{\overline{\mu}\overline{\nu}}^{\overline{ik}}, \quad (6.56)$$

where the bar above the index means its reversion [i.e., $\overline{1} = 2$; $\overline{(-)} = (+)$]. From (6.54) and (6.56) it follows that:

$$G_{\mu\nu}^{ik}(1, 2) = -\left(G_{\nu\mu}^{ik}(2, 1)\right)^* = (-1)^{\mu+\nu+1}G_{\overline{\nu\mu}}^{\overline{ik}} \quad (\text{for } i \neq k), \quad (6.57)$$

$$G_{\mu\nu}^{ii}(1, 2) = -\left(G_{\nu\mu}^{ii}(2, 1)\right)^* = (-1)^{\mu+\nu+1}G_{\overline{\nu\mu}}^{ii}. \quad (6.58)$$

The functions G, G^R, and G^A are defined according to the relations (5.55):

$$G + G^R + G^A = 2G^{--}, \quad (6.59)$$

$$G + G^R - G^A = 2G^{+-}, \quad (6.60)$$

$$G - G^R + G^A = 2G^{+-}, \quad (6.61)$$

$$G - G^R - G^A = 2G^{++}, \quad (6.62)$$

from which [taking into account (6.57) and (6.58)] equalities follow:

$$G_{\mu\nu}(1, 2) = -G_{\nu\mu}^*(2, 1) = (-1)^{\mu+\nu+1}G_{\overline{\nu\mu}}(2, 1), \quad (6.63)$$

$$G_{\mu\nu}^R(1, 2) = G_{\nu\mu}^{A*}(2, 1) = (-1)^{\mu+\nu+1}G_{\overline{\nu\mu}}^A(2, 1), \quad (6.64)$$

$$G_{\mu\nu}^A(1, 2) = G_{\nu\mu}^R(2, 1) = (-1)^{\mu+\nu+1}G_{\overline{\nu\mu}}^R(2, 1). \quad (6.65)$$

6.3.2 Quasiclassical Approximation

In homogeneous and stationary cases, the Green-Keldysh functions depend on the difference of space-time coordinate. If the evolution of the phonon system is taking place sufficiently slow, one can assume that all the quantities depend only weakly on the summary variable $(1 + 2)$ and are the functions mainly of the difference variable $(1 - 2)$. Separating these variables [we use the notations of the type $1 \equiv x_1 \equiv (\mathbf{r}_1, t_1)$, etc.]

$$\widehat{\mathcal{D}}(x_1, x_2) = \widehat{\mathcal{D}}\left(\frac{x_1 + x_2 + (x_1 - x_2)}{2}; \frac{x_1 + x_2 - (x_1 - x_2)}{2}\right), \quad (6.66)$$

one can perform the Fourier-transformation over the difference variables $\mathbf{R} = \mathbf{r}_1 - \mathbf{r}_2$ and $\Theta = t_1 - t_2$:

$$\mathcal{D}^{ik}(\mathbf{q}, \omega; \mathbf{r}, t) = \int \mathcal{D}^{ik}(\mathbf{r}, t; R, \Theta)e^{i\mathbf{q}\cdot\mathbf{R}+i\omega\Theta}d^3\mathbf{R}d\Theta, \quad (6.67)$$

where, obviously, $\mathbf{r} = (\mathbf{r}_1 + \mathbf{r}_2)/2$, $t = (t_1 + t_2)/2$. Acting by the operator $\mathcal{D}_{01}^{-1}$ on (6.52) and by $\mathcal{D}_{02}^{-1}$ on (6.53), and subtracting, we obtain the result for the $(-+)$-component:

$$\left(\mathcal{D}_{02}^{-1} - \mathcal{D}_{01}^{-1}\right)\mathcal{D}^{-+}(x_1, x_2) = -\int d^4x_3 d^4x_4 \{[\mathcal{D}^{--}(1, 3)\Pi^{-+}(3, 4)$$

$$+ \mathcal{D}^{-+}(1, 3)\Pi^{++}(3, 4)]\delta(t_2 - t_4)\nabla_2^2\delta(\mathbf{r}_4 - \mathbf{r}_2) + [\Pi^{--}(4, 3)\mathcal{D}^{-+}(3, 2)$$

$$+ \Pi^{-+}(4, 3)\mathcal{D}^{++}(3, 2)]\delta(t_1 - t_4)\nabla_1^2\delta(\mathbf{r}_1 - \mathbf{r}_4)\}. \qquad (6.68)$$

Consider first the right side of (6.68) and transform it with the help of quasiclassical conditions. For the phonon system, these conditions mean that the quantities characterizing its evolution in time (Δt) and space ($\Delta \mathbf{r}$), must be large in comparison with the characteristic phonon reciprocal frequencies $\omega(\mathbf{q})^{-1}$ and wave numbers q^{-1} ($\hbar = 1$), i.e.:

$$\omega(q)\Delta t \gg 1, \qquad q\Delta r \gg 1. \qquad (6.69)$$

This is a good approximation when the perturbation of the phonon system is caused by the superconducting electron system. The condition (6.69) permits us to simplify in the usual manner (cf., e.g., [6]) the left side of (6.68). Taking into account that the operator $\left(\mathcal{D}_{02}^{-1} - \mathcal{D}_{01}^{-1}\right)$ in the left side of (6.68) may be presented in the form

$$\mathcal{D}_{02}^{-1} - \mathcal{D}_{01}^{-1} = -\frac{\partial^2}{\partial t \partial \Theta} + u^2 \frac{\partial^2}{\partial \mathbf{r} \partial \mathbf{R}}, \qquad (6.70)$$

and carrying out the Fourier-transformation of (6.68), we obtain the expression

$$2\left(i\omega \frac{\partial \mathcal{D}^{-+}}{\partial t} + iu^2 \mathbf{q} \cdot \frac{\partial}{\partial \mathbf{r}}\mathcal{D}^{-+}\right) = -(\mathbf{u} \cdot \mathbf{q})^2 \left(\Pi^{-+}\mathcal{D}^{+-} - \mathcal{D}^{-+}\Pi^{+-}\right). \qquad (6.71)$$

[the arguments of all the functions in parentheses are ($\mathbf{q}, \omega; \mathbf{r}, t$); we have used here the linear interdependence of $\mathcal{D}^{ik}$ and also of Π^{ik}, mentioned earlier].

6.3.3 Phonon Distribution Function

To obtain the kinetic equation in terms of the distribution function $N(\mathbf{q}, \mathbf{r}, t)$, it is necessary to find a relation between functions N and $\mathcal{D}$. In the quasiclassical case, such a relation may be found rather simply. As noted in Sect. 5.1, the superconducting transition negligibly influences [because $\Delta\omega_0(\mathbf{q})/\omega_0(\mathbf{q}) \sim 10^{-4}$] the bare phonon spectrum: $\omega_0(\mathbf{q}) \approx \omega(\mathbf{q})$. Implying the quasiclassical condition for the phonons, we assume that $N(\mathbf{q}, \mathbf{r}, t)$ and $\widehat{\mathcal{D}}(\omega, \mathbf{q}, \mathbf{r}, t)$ obey the relation

$$\mathcal{D}^{-+}(\omega, \mathbf{q}, \mathbf{r}, t) = -\frac{i}{2}\omega(\mathbf{q})\{[1 + N(-\mathbf{q}; \mathbf{r}, t)]\delta[\omega + \omega(\mathbf{q})]$$

$$+ N(\mathbf{q}, \mathbf{r}, t)\delta[\omega - \omega(\mathbf{q})]\}. \qquad (6.72)$$

For acoustic phonons (only these are important in the isotropic case) and for momenta that are small compared with the extreme value in the crystal, the following relation is valid:

$$\mathbf{q} = \frac{\partial \omega(\mathbf{q})}{\partial \mathbf{u}} \approx \frac{\omega(q)}{u^2}\mathbf{u}. \tag{6.73}$$

Using also the property

$$\mathcal{D}^{+-}(\omega, \mathbf{q}; \mathbf{r}, t) = \mathcal{D}^{-+}(-\omega, -\mathbf{q}; \mathbf{r}, t) \tag{6.74}$$

[which follows from (6.47)], substituting (6.72) to (6.74) into (6.71) and integrating over ω within the limits $(0, \infty)$, one obtains the following kinetic equation:

$$\frac{\partial N(\mathbf{q}, \mathbf{r}, t)}{\partial t} + \mathbf{u} \cdot \frac{\partial N(\mathbf{q}, \mathbf{r}, t)}{\partial \mathbf{r}} = I(N), \tag{6.75}$$

where the quantity

$$I(N) = i\frac{\omega(\mathbf{q})}{2}\left\{\Pi^{-+}(\mathbf{q}, \omega(\mathbf{q}); \mathbf{r}, t)\left[1 + N(\mathbf{q}, \mathbf{r}, t)\right] - \Pi^{+-}(\mathbf{q}, \omega(\mathbf{q}); \mathbf{r}, t) N(\mathbf{q}, \mathbf{r}, t)\right\} \tag{6.76}$$

is the inequilibrium source.

Note that expression (6.76) may also be presented in a somewhat different form, if one makes the standard transformation (cf. [6]) from the Π^{ik} matrix [in analogy with (5.55)] to the linearly independent functions Π, Π^R, Π^A:

$$I(N) = \frac{i\omega(\mathbf{q})}{2}\left\{\left(\Pi^R - \Pi^A\right)N - \frac{1}{2}\left[\Pi - (\Pi^R - \Pi^A)\right]\right\}. \tag{6.77}$$

Now we specify the polarization operators in (6.77).

6.3.4 Polarization Operators in Keldysh's Technique

In the Keldysh-Nambu technique, the polarization operator is represented by a diagram expansion:

$$\tag{6.78}$$

Regular Feynman rules [7] are applied; the only difference is that all the quantities, including the vertices, are matrices in Keldysh-Nambu indices. Since the transition to the superconducting state (as well as the interaction with an external electromagnetic field) affects only a minor smeared region ($\sim T/\epsilon_F$, $|\Delta|/\epsilon_F$) at the Fermi surface, we can set (as in Sect. 5.1) the total vertex Γ in the polarization operator equal to $\Gamma_0 \sim g$ with adiabatic accuracy $\sim u/v_F$. The electron functions $G_{\mu\nu}^{ik}$ [the bold lines in the l.h.s. of (6.78)] are considered as exact functions [they contain electron interactions between themselves, with the external field, phonons, impurities, etc., symbolically depicted by the wavy lines in (6.78); wavy lines on the bottom parts in the r.h.s. are also assumed but not shown]. Because in this technique the vertices Γ have the matrix structure [8]:

$$\Gamma_{\alpha\beta}^{ij,k} \propto \widehat{\sigma}_z^{ij} \delta_{jk} \widehat{\sigma}_z^{\alpha\beta}, \tag{6.79}$$

we obtain (g is the electron-phonon interaction constant)

$$\Pi^{kk'} = -g^2(-1)^{k+k'} \left[G_{11}^{kk'} G_{11}^{k'k} - G_{12}^{kk'} G_{21}^{k'k} - G_{21}^{kk'} G_{12}^{k'k} + G_{22}^{kk'} G_{22}^{k'k} \right]. \tag{6.80}$$

However, it is more convenient to deal with linearly-independent quantities G, G^R and G^A (5.55). Moving simultaneously to Π, Π^R, and Π^A and omitting components like $G_{11}^A(1,2)G_{11}^A(2,1)$ (which are identically zero), we obtain

$$\Pi^{A(R)}(1,2) = -\frac{1}{2}g^2[G_{11}^{A(R)}(1,2)G_{11}(2,1) + G_{11}G_{11}^{R(A)} + G_{22}^{A(R)}G_{22}$$
$$+ G_{22}G_{22}^{R(A)} - G_{12}^{A(R)}G_{21} - G_{12}G_{21}^{R(A)} - G_{21}^{A(R)}G_{12} - G_{21}G_{12}^{R(A)}], \tag{6.81}$$

and correspondingly

$$\Pi(1,2) = -\frac{i}{2}g^2[G_{11}(1,2)G_{11}(2,1) + G_{22}G_{22} - G_{21}G_{12} - G_{12}G_{21} + G_{11}^A G_{11}^R$$
$$+ G_{11}^R G_{11}^A + G_{22}^A G_{22}^R + G_{22}^R G_{22}^A - G_{12}^A G_{21}^R - G_{12}^R G_{21}^A - G_{21}^A G_{12}^R - G_{21}^R G_{12}^A]. \tag{6.82}$$

Expressions (6.81) and (6.82) allow one to obtain the collision integral for the phonon kinetic equation in a superconducting system. All the influence of the electromagnetic field is contained in Green's functions for electrons, which are exact and also account for the impurities and other fields acting on the electron system.

As required by the kinetic equation, written in the form of (6.75), we should move to the (x, p)-representation. It is clear that the polarization operators can be expressed in terms of the energy-integrated Green's functions. Before making this transformation, we will derive the expressions for the polarization operators in (6.77), using the analytical continuation technique.

6.3.5 Polarization Operators: Analytical Continuation Technique

In a discrete imaginary frequency representation $[\varepsilon_n = (2n + 1)\pi T i, \omega_m = 2m\pi T i]$ we have the following expression for the polarization operator

$$\Pi_{\omega\omega-\omega'}(\mathbf{p}, \mathbf{p} - \mathbf{p}') = g^2 T \sum_{\varepsilon_1} \int \frac{d^3\mathbf{p}_1}{(2\pi)^3} \{ \mathfrak{G}_{\varepsilon_1} \overline{\mathfrak{G}}_{\omega-\varepsilon_1} + \overline{\mathfrak{G}}_{\varepsilon_1} \mathfrak{G}_{\omega-\varepsilon_1}$$

$$- \mathfrak{F}_{\varepsilon_1} \mathfrak{F}^+_{\omega-\varepsilon_1} - \mathfrak{F}^+_{\varepsilon_1} \mathfrak{F}_{\omega-\varepsilon_1} \}. \tag{6.83}$$

For brevity we omit the second arguments of Green's functions ($\mathfrak{G}_{\varepsilon_1\varepsilon_2}$, etc.), which may be reconstructed from the "decay" conservation law for internal variables:

$$\varepsilon_1 + \varepsilon_2 = \omega, \qquad \omega_1 + \omega_2 = \omega'. \tag{6.84}$$

Rule (6.84) is responsible for the appearance of $\overline{\mathfrak{G}}$-functions in (6.83), which differs from $\mathfrak{G}$ by the reversed directions of the arrows in the diagrams. In addition, the $\mathfrak{F}\mathfrak{F}^+$ pair in (6.83) is accompanied (as in Sect. 3.4.1) by a change in the diagram's sign. Starting the analytic continuation of the polarization operator, we consider each component in (6.83) as the infinite sum of the diagrams of various orders in the external field. The entire procedure is analogous to that used earlier in deriving the analytically continued self-energy parts of electron-electron collisions. The only difference is that the external frequencies here are Bose frequencies (and, naturally, there are two electron lines). Since the directions of arrows in the diagrams do not influence the analytic continuation process, we will consider only the expression

$$\Pi_{\omega\omega-\omega'} = T \sum_{\varepsilon_1} \mathfrak{G}_{\varepsilon_1\varepsilon_1-\omega} \mathfrak{G}_{\omega-\varepsilon_1\omega-\varepsilon_1+\omega_1-\omega'}. \tag{6.85}$$

For simplicity of notation, we omit the signs of internal frequency integration and summation at the vertices of interaction with the external field and phonons. Remember that expression (6.85) corresponds to the diagram series of perturbation theory and contains the sum of the various diagrams up to the infinite order; Green's functions for electrons in the polarization loops contain the phonon self-energy insertions that in turn contain field vertices of an arbitrary order. The diagram of a specific order in the external field (considered as a function of the complex variable ω) has a cut on the line $\text{Im}(\omega - \Omega_m) = 0$ for fixed imaginary frequencies of the field vertices, which goes between the uppermost ant lowermost banks:

$$0 \leq \text{Im}\omega \leq \text{Im}\omega'. \tag{6.86}$$

As in the case of the electron-electron self-energy parts, the quantities Ω_m represent certain combinations of the field vertex frequencies; here the set of these combinations and their total number depend on the distribution of vertices along the electron lines. Assuming that the cuts with $\mathrm{Im}(\varepsilon_1 - \omega_{1i}) = 0$, $\mathrm{Im}(\varepsilon_2 - \omega_{2k}) = 0$ correspond to these lines, we transform the frequency sum in (6.85) to the contour integral

$$\pi_{\omega\omega-\omega'} = \oint_C \frac{\mathrm{d}z}{4\pi i}\, \tanh \frac{z}{2T} \mathfrak{G}_z \mathfrak{G}_{\omega-z'}, \tag{6.87}$$

where C is the contour shown in Fig. 4.1. Deforming the contour C to C', which goes along the banks of the cut, and noting that for the diagrams of any order the integrals along the big arcs vanish, when the corresponding radii tend to infinity, after some straightforward calculations we obtain:

$$\pi_{\omega\omega-\omega'} = \sum_{i,k} \int \frac{\mathrm{d}z}{4\pi i}\{\delta_i\left(\mathfrak{G}_{z+\omega_{1i}}\right) \mathfrak{G}_{\omega-z-\omega_{1i}} \tanh \frac{z}{2T}$$
$$- \mathfrak{G}_{z+\omega-\omega_{2k}}\delta_k\left(\mathfrak{G}_{-z+\omega_{2k}}\right) \tanh \frac{z}{2T}\}, \tag{6.88}$$

where $\delta_{i,k}(\mathfrak{G})$ is the jump in the Green's function at the corresponding cut line. The external variable ω and the field frequencies remain imaginary. Their combination determines the set of cuts for the given diagram. Assuming in all diagrams $\omega > \omega'$ (the upper bank of the uppermost cut line), shifting the integration variable in all diagram expressions (as was done in Sects. 4.3 and 6.2) and summing over all orders of the perturbation series, we obtain

$$\pi^R_{\omega\omega-\omega'} = \int_{-\infty}^{\infty} \frac{\mathrm{d}z}{4\pi i}\left(G_z G^R_{\omega-z} + G^R_z G_{\omega-z}\right), \tag{6.89}$$

where the G-function is determined as

$$G_{\varepsilon\varepsilon-\omega} = \sum_{N=0}^{\infty} \sum_{i=1}^{N+1} \delta_i\left(\mathfrak{G}^{(N)}\right) \tanh \frac{\varepsilon - \omega_{1i}}{2T}, \tag{6.90}$$

while the functions $G^{R(A)}$ are determined directly from the diagram expansion (or the Dyson equation), where all the Green's functions for electrons are retarded (advanced) and their entire set $\{G, G^R, G^A\}$ coincides with the functions figuring in Sect. 4.3. The expression for $\omega < 0$ also follows from (6.88) for $\omega \leq 0$ (the lower bank of the lowermost cut line):

$$\pi^A_{\omega\omega-\omega'} = \int_{-\infty}^{\infty} \frac{\mathrm{d}z}{4\pi i}\left(G_z G^A_{\omega-z} + G^A_z G_{\omega-z}\right). \tag{6.91}$$

Using for $\Pi_{\omega\omega-\omega'}$ an expression analogous to (6.90), but with the external frequencies representing now the Bose field, i.e.:

$$\pi_{\omega\omega-\omega'} = \sum_{N=0}^{\infty} \sum_{k=1}^{N+1} \delta_k \left(\Pi^{(N)}\right) \coth \frac{\omega - \Omega_k}{2T}, \tag{6.92}$$

we obtain after the substitution of (6.88) into (6.92):

$$\pi_{\omega\omega-\omega'} = \sum_{i,k} \left\{ \int_{-\infty}^{\infty} \left[\delta_i \left(\mathfrak{G}_{z+\omega_{1i}}\right) \left(\mathfrak{G}_{\omega-z-\omega_{1i}}\right) \tanh \frac{z}{2T} \coth \frac{\omega - \omega_{ik} - \omega_{2k}}{2T} \right. \right.$$
$$\left. \left. - \delta_i \left(\mathfrak{G}_{z+\omega-\omega_{2k}}\right) \delta_k \left(\mathfrak{G}_{z+\omega_{2k}}\right) \tanh \frac{z}{2T} \coth \frac{\omega - \omega_{2k} - \omega_{1i}}{2T} \right] \right\}. \tag{6.93}$$

Shifting the integration variable in each of the terms in the first integral $z + \omega_{1i} \to z$ and $z + \omega - \omega_{2k} \to z$ in the second integral, and using the identity (5.98), one gets

$$\pi_{\omega\omega-\omega'} = \sum_{i,k} \int \frac{dz}{4\pi i} \delta_i \left(\mathfrak{G}_z\right) \delta_k \left(\mathfrak{G}_{\omega-z}\right) \left[1 + \tanh \frac{z - \omega_{1i}}{2T} \tanh \frac{\omega - z - \omega_{2k}}{2T} \right]. \tag{6.94}$$

Summing in (6.94) all orders of perturbation theory and accounting for the definition of (6.89), we finally obtain

$$\pi_{\omega\omega-\omega'} = \int \frac{dz}{4\pi i} \left[G_z G_{\omega-z} + \left(G_z^R - G_z^A\right) \left(G_{\omega-z}^R - G_{\omega-z}^A\right) \right]. \tag{6.95}$$

Thus, for polarization operator (6.83), the complete set of functions Π, Π^R, Π^A is found, which describes the nonequilibrium case. One can show that they are identical to those obtained earlier on the basis of the Keldysh technique.

6.3.6 Equivalence of Keldysh and Eliashberg Approaches

Consider for this purpose anyone of the components in $\Pi_{\omega\omega-\omega'}$, which follows from (6.83), for example,

$$\Pi_{\omega\omega-\omega'}^1 (\mathbf{p}, \mathbf{p} - \mathbf{p}') = \frac{g^2}{2i} \left(\int \frac{d\varepsilon_1}{2\pi} G_{\varepsilon_1} G_{\omega-\varepsilon_1} \right) \equiv \frac{g^2}{2i} \int \frac{d\varepsilon_1 d\omega_1 d^3 p_1 d^3 k}{(2\pi)^8}$$
$$\times G_{\varepsilon_1 \varepsilon_1 - \omega} (\mathbf{p}_1, \mathbf{p}_1 - \mathbf{k}) G_{\omega-\varepsilon_1, \omega-\varepsilon_1+\omega_1-\omega'} (\mathbf{p} - \mathbf{p}_1 \mathbf{p} - \mathbf{p}_1 + \mathbf{k} - \mathbf{p}'), \tag{6.96}$$

and make a Fourier transformation to the spatial and temporal variables. As a result, one obtains

$$\Pi^1 (x_1, x_2) = -\frac{ig^2}{2} G(x_1 x_2) G(x_1 x_2). \tag{6.97}$$

Now we write down all the terms $\Pi(x_1, x_2)$ obtained from the analytical continuation. Taking into account (6.83) and (6.97) one finds

$$\Pi(1, 2) = -\frac{ig^2}{2}\{G(1, 2)\overline{G}(1, 2) + \overline{G}G - FF^+ - F^+F$$
$$+ (\overline{G}^R - \overline{G}^A)(G^R - G^A) + (G^R - G^A)(\overline{G}^R - \overline{G}^A)$$
$$- (F^R - F^A)(F^{+R} - F^{+A}) - (F^{+R} - F^{+A})(F^R - F^A)\}. \quad (6.98)$$

To compare this with result (6.82) obtained by the Keldysh technique, we must make the substitution $(1, 2) \rightarrow (2, 1)$ in the second multiplier of each of the components either in braces in (6.98) or in brackets in (6.82). In the latter case, this should be done using the relations (6.63) to (6.65). We use the first possibility, noting that the Eliashberg functions have the properties

$$G^R(1, 2) = \overline{G}^A(2, 1), \ G(1, 2) = \overline{G}(21), \ F^R(1, 2) = F^A(2, 1), \ F(1, 2) = F(2, 1).$$
$$(6.99)$$

[in the absence of the spin-dependent interactions $G_{\alpha\beta} = \delta_{\alpha\beta}G$, $F_{\alpha\beta} = i(\widehat{\sigma}_y)_{\alpha\beta}F$]. After removing the parentheses certain components vanish [for example, $\overline{G}^A(1, 2)\overline{G}^A(2, 1) \equiv 0$], so (6.98) can be reduced to the form

$$\Pi = \frac{ig^2}{2}\{G(1, 2)(2, 1) + \overline{G}G - FF^+ - F^+F + \overline{G}^R\overline{G}^A + \overline{G}^A\overline{G}^R + G^RG^A$$
$$+ G^AG^R - F^RF^{+A} - F^AF^{+R} - F^{+R}F^A - F^{+A}F^R\}. \quad (6.100)$$

Comparing this expression with (6.82), we see that they coincide, if the functions $G^{R(A)}, \overline{G}^{R(A)}, F^{R(A)}$, and $F^{+R(A)}$ are replaced by $G_{11}^{R(A)}, G_{22}^{R(A)}, G_{12}^{R(A)}$, and $G_{21}^{R(A)}$. Because these functions coincide up to the sign,[3] the polarization operators (which are quadratic in Green's functions) coincide identically.

6.3.7   Transition to Energy-Integrated Propagators

Consider now an arbitrary component [e.g., the first one in the expression for $\Pi_{\omega\omega-\omega'}(\mathbf{p}, \mathbf{p} - \mathbf{p}')$], which follows from (6.83), taking into account (6.94). In this expression we can move from the integration over $d^3\mathbf{p}_1$ to angle and energy integrations, based on the relation

$$\frac{d^3\mathbf{p}_1}{(2\pi)^3} \approx \frac{mp_F}{2\pi^2}\frac{d\Omega_{\mathbf{p}_1}}{4\pi}d\xi_1. \quad (6.101)$$

---

[3] These functions coincide for nondiagonal components of $\widehat{G}$-matrix and differ in sign for diagonal ones. The reason is the sign difference between "bare" propagators of the normal state, mentioned in Sect. 4.3.2.

Using the auxiliary δ-function: $\delta(\xi_2 - \xi_1 - \mathbf{q} \cdot \mathbf{p}_1/m)$, we may integrate with respect to the variable ξ_2. This makes it possible to express the quantity

$$\frac{mp_F}{2\pi^2} \int \frac{d\Omega_{\mathbf{p}_1}}{4\pi} \int \int d\xi_1 d\xi_2 \, \delta(\xi_2 - \xi_1 + \mathbf{q} \cdot \mathbf{p}_1/m) G_1 \overline{G}_2 \tag{6.102}$$

in terms of energy-integrated functions, determined by a relation of the type

$$g_{\varepsilon\varepsilon-\omega}(\mathbf{p}, \mathbf{k}) = \int_{-\infty}^{\infty} d\xi \, G_{\varepsilon\varepsilon-\omega}(\mathbf{p}, \mathbf{p} - \mathbf{k}), \tag{6.103}$$

since the δ-function in (6.102) restricts mainly the angular integration and hence it may be factored out of the ξ-integral. Thus we have

$$\int \frac{d^3\mathbf{p}_1}{(2\pi)^3} G_1 \overline{G}_2 = \frac{mp_F}{2\pi^2} \frac{1}{2qv_F} \int \frac{d\Omega_{\mathbf{p}_1}}{4\pi} \delta\left(\frac{\mathbf{q} \cdot \mathbf{p}_1}{qp_1}\right) g_1 \overline{g}_2. \tag{6.104}$$

To shorten the notations we introduce the operator

$$\hat{M} = \frac{g^2}{2i} \frac{mp_F}{2\pi^2} \frac{1}{2qv_F} \int \frac{d\varepsilon_1 d\omega_1 d^3\mathbf{k} \, d\Omega_{\mathbf{p}_1}}{(2\pi)^5} \delta\left(\frac{\mathbf{q} \cdot \mathbf{p}_1}{qp_1}\right). \tag{6.105}$$

and the convention

$$[A, B]_+ = A_{\varepsilon_1\varepsilon_1-\omega_1}(\mathbf{p}_1, \mathbf{k}) B_{\omega-\varepsilon_1\omega-\varepsilon_1+\omega_1-\omega'}(\mathbf{p} - \mathbf{p}_1, \mathbf{p}' - \mathbf{k}) + BA, \tag{6.106}$$

obtaining thus the final expressions for the Fourier components of the quantities, which define the inequilibrium source in the phonon kinetic equation:

$$\Pi_{\omega\omega-\omega'} = \hat{M}\{[g^A\overline{g}]_+ - [f, f^+]_+ + [g^R - g^A, \overline{g}^R - \overline{g}^A]_+$$
$$- [f^R - f^A, f^{+R} - f^{+A}]_+\}, \tag{6.107}$$
$$(\Pi^R - \Pi^A)_{\omega\omega-\omega'} = M\{[g, \overline{g}^R - \overline{g}^A]_+ + [\overline{g}^A, g^R - g^A]_+$$
$$- [f, f^{+R}]_+ - [f^+, f^R - f^A]_+\}. \tag{6.108}$$

Before bringing the equation for phonons to the canonical form, we will obtain the expression for the collision integral of electrons with phonons.

6.4 Inelastic Electron-Phonon Collisions

The self-energy functions for an electron-phonon interaction were derived in Sect. 5.2 assuming an equilibrium phonon distribution. We will consider now the general case when the phonon system is not in equilibrium.

Fig. 6.3 Self-energy
function of the
electron-phonon interaction

6.4.1   Electron-Phonon Self-Energy Parts

In the representation of discrete imaginary frequencies $[P = \{\varepsilon, \mathbf{p}\}, \; K = \{\omega, \mathbf{k}\},$ $\omega = 2m\pi T i, \; \varepsilon = (2n + 1)\pi T i, \; m$ and n are integers], we have:

$$\widehat{\Sigma}(P, P - K) = T \sum_{\varepsilon'} \frac{d^3 \mathbf{p}'}{(2\pi)^3} \mathfrak{D}(P' - P, P' - P - K' + K) \widehat{\mathfrak{G}}(P', P' - K'),$$

(6.109)

which corresponds to the diagram of Fig. 6.3. The functions $\widehat{\mathfrak{G}}$ (as well as $\mathfrak{D}$) in (6.109) are assumed to be complete, including both external field and the self-energy parts and polarization operators. Assuming that initially in the absence of an external field the system is in equilibrium, we expand $\widehat{\mathfrak{G}}$ and $\mathfrak{D}$ in a power series over the field and consider the analytical structure of the Nth order diagram as the function of the complex variable ε at fixed imaginary frequencies ω_l ($\omega_l = 2m_l \pi T i,$ $\sum_l \omega_l = \omega$). The manifold of cuts in the object under consideration consists of the cuts of the internal $\mathfrak{G}$-function and the $\mathfrak{D}$-function. We denote this manifold by Ω_n. These cuts may be considered as situated on the lines $\mathrm{Im}(\varepsilon - \Omega_n) = 0$ in the complex plane ε between the uppermost line $\mathrm{Im}(\varepsilon - \omega) = 0$ and the abscissa (as was assumed earlier in accordance with the causality principle). The combinations of ω_l, which constitute Ω_n, are defined by the distribution of the field vertices over the internal lines of the diagram $\widehat{\Sigma}^{(N)}$. Let us assume that manifolds of cuts $\mathrm{Im}(\varepsilon' - \omega_{1i}) = 0$ and $\mathrm{Im}(\varepsilon' - \varepsilon - \omega_{2k}) = 0$ correspond to $\widehat{\mathfrak{G}}$ and $\mathfrak{D}$-functions. Replacing in (6.109) the summation over ε' by contour integration and shifting as usual the integration contour to the banks of the cuts, we find the resulting expression

$$\widehat{\Sigma}^{(N)} = \sum_{i,k} \int \frac{dz}{4\pi i} \left\{ \tanh \frac{z}{2T} \delta_i \left(\widehat{\mathfrak{G}}_{z+\omega_{1i}} \right) \mathfrak{D}_{z-\varepsilon+\omega_{1i}} \right.$$
$$\left. + \widehat{\mathfrak{G}}_{z+\omega_{2k}+\varepsilon} \coth \frac{z}{2T} \delta_k \left(\mathfrak{D}_{z+\omega_{2k}} \right) \right\},$$

(6.110)

where $\delta_i \left(\mathfrak{G}_{z+\omega_{1i}} \right)$ and $\delta_k \left(\mathfrak{D}_{z+\omega_{2k}} \right)$ are the jumps in Green's functions on the corresponding cuts (hereafter for brevity we omit second indices of Green's functions). Continuing now analytically in (6.110) over ε from the upper bank of the uppermost cut (the lower bank of the lowermost cut), we obtain (after returning to real $\omega_{i,k}$, shifting the integration variable, summing over all the orders of perturbation theory, and integrating over the energy ξ) the expression

$$\widehat{\Sigma}^{R(A)} = \int_{-\infty}^{\infty} \frac{d\varepsilon'}{4\pi i} \left\{ \widehat{g}_{\varepsilon'} D_{\varepsilon'-\varepsilon}^{A(R)} + D_{\varepsilon'-\varepsilon} \widehat{g}_{\varepsilon'}^{R(A)} \right\},$$

(6.111)

in which the $\mathcal{D}$-function is defined as

$$\mathcal{D}_\omega = \sum_{N=0}^{\infty} \sum_{k=1}^{N+1} \coth \frac{\omega - \omega_k}{2T} \delta_k \left(\mathfrak{D}^{(N)}\right). \tag{6.112}$$

Introducing $\widehat{\Sigma}_{\varepsilon\varepsilon-\omega}$ as

$$\widehat{\Sigma}_{\varepsilon\varepsilon-\omega} = \sum_{N=0}^{\infty} \sum_{k=1}^{N+1} \delta_k \left(\widehat{\Sigma}_{\varepsilon\varepsilon-\omega}^N\right) \tanh \frac{\varepsilon - \Omega_k}{2T}, \tag{6.113}$$

we obtain, starting from (6.110), the expressions for the matrix elements of $\widehat{\Sigma}$, which may be presented in the form (omitting for simplicity the second arguments)

$$\Sigma_1 = \int_{-\infty}^{\infty} \frac{d\varepsilon'}{4\pi i} \int \frac{d\Omega_{\mathbf{p}'}}{4\pi} \left[\mathcal{D}_{\varepsilon'-\varepsilon} g_{\varepsilon'} - (g^R - g^A)_{\varepsilon'} (D^R - D^A)_{\varepsilon'-\varepsilon}\right], \tag{6.114}$$

$$\Sigma_1^R - \Sigma_1^A = 2i\gamma = \int_{-\infty}^{\infty} \frac{d\varepsilon'}{4\pi i} \int \frac{d\Omega_{\mathbf{p}'}}{4\pi} \left[\mathcal{D}_{\varepsilon'-\varepsilon} (g^R - g^A)_{\varepsilon'} - g_{\varepsilon'} (^R - D^A)_{\varepsilon'-\varepsilon}\right],$$
$$\tag{6.115}$$

$$\Sigma_2 = \Sigma_1 \left(g^{R(A)} \to f^{R(A)}\right), \quad \delta = \gamma \left(g^{R(A)} \to f^{R(A)}\right). \tag{6.116}$$

In the diagonal over frequencies (quasiclassical) approximation, the phonon propagators may be expressed through the function $N_{\omega_{\mathbf{q}}}$:

$$\mathcal{D}_{\varepsilon'-\varepsilon} = \left(1 + 2N_{\omega_{\mathbf{p}'-\mathbf{p}}}\right) \operatorname{sign}(\varepsilon' - \varepsilon)(D^R - D^A)_{\varepsilon'-\varepsilon}, \tag{6.117}$$

$$\mathcal{D}_{\varepsilon'-\varepsilon}^{R(A)} = \lambda \frac{2\omega_{\mathbf{p}'-\mathbf{p}}^2}{\omega_{\mathbf{p}'-\mathbf{p}}^2 - \left(\varepsilon' - \varepsilon \, {}^+_{(-)} \, i\delta\right)^2}. \tag{6.118}$$

6.4.2 Canonical Form for Electron-Phonon Collisions

Using the relations (6.114) to (6.118), (5.83) to (5.88), (6.7), (6.9), and (6.13), one arrives at the following form of the electron-phonon collision integral:

$$J^{(\text{e-ph})}(n_{\pm\varepsilon}) = \frac{\pi\lambda}{4(up_F)^2} \int_0^{\infty} \omega_q^2 d\omega_q \int_{|\Delta|}^{\infty} d\varepsilon' [p_1 \delta(\varepsilon' - \varepsilon - \omega_q)$$
$$+ p_2 \delta(\varepsilon - \varepsilon' - \omega_q) + p_3 \delta(\varepsilon + \varepsilon' - \omega_q)], \tag{6.119}$$

where the factors p_{1-3} are

$$p_1 = (u_\varepsilon u_{\varepsilon'} - v_\varepsilon v_{\varepsilon'} \pm 1) \left[n_{\varepsilon'}(1 - n_{\pm\varepsilon})(1 + N_{\omega_q}) - n_{\pm\varepsilon}(1 - n_{\varepsilon'})N_{\omega_q} \right]$$
$$+ (u_\varepsilon u_{\varepsilon'} - v_\varepsilon v_{\varepsilon'} \mp 1) \left[n_{-\varepsilon'}(1 - n_{\pm\varepsilon})(1 + N_{\omega_q}) - n_{\pm\varepsilon}(1 - n_{-\varepsilon'})N_{\omega_q} \right], \quad (6.120)$$

$$p_2 = (u_\varepsilon u_{\varepsilon'} - v_\varepsilon v_{\varepsilon'} \pm 1) \left[n_{\varepsilon'}(1 - n_{\pm\varepsilon})N_{\omega_q} - n_{\pm\varepsilon}(1 - n_{\varepsilon'})(1 + N_{\omega_q}) \right]$$
$$+ (u_\varepsilon u_{\varepsilon'} - v_\varepsilon v_{\varepsilon'} \mp 1) \left[n_{-\varepsilon'}(1 - n_{\pm\varepsilon})N_{\omega_q} - n_{\pm\varepsilon}(1 - n_{-\varepsilon'})(1 + N_{\omega_q}) \right], \quad (6.121)$$

$$p_3 = (u_\varepsilon u_{\varepsilon'} + v_\varepsilon v_{\varepsilon'} \mp 1) \left[(1 - n_{\pm\varepsilon})(1 - n_{\varepsilon'})N_{\omega_q} - n_{\pm\varepsilon}n_{\varepsilon'}(1 + N_{\omega_q}) \right]$$
$$+ (u_\varepsilon u_{\varepsilon'} + v_\varepsilon v_{\varepsilon'} \pm 1) \left[(1 - n_{\pm\varepsilon})(1 - n'_{-\varepsilon})N_{\omega_q} - n_{\pm\varepsilon}n_{-\varepsilon'}(1 + N_{\omega_q}) \right]. \quad (6.122)$$

Expression (6.119) describes, besides the energy relaxation of electrons, inelastic collision processes that produce the relaxation of electron-hole population imbalance in superconductors. The situation here is fully analogous to that discussed in Sect. 6.2 and requires no further comments.

Having ascertained that the function $N_{\omega q}$, introduced by (6.117), plays the role of a phonon nonequilibrium distribution function, we will now obtain the kinetic equation for this quantity. We start from expression (6.112) for $\mathcal{D}_{\omega\omega-\omega'}$ and (6.92) for a polarization operator $\Pi_{\omega\omega-\omega'}$. Separating the anomalous parts $\mathcal{D}^{(a)}$ and $\Pi^{(a)}$, one obtains

$$\mathcal{D}_{\omega\omega-\omega'} = \mathcal{D}^R_{\omega\omega-\omega'} \coth \frac{\omega - \omega'}{2T} - \mathcal{D}^A_{\omega\omega-\omega'} \coth \frac{\omega}{2T} + \mathcal{D}^{(a)}_{\omega\omega-\omega'}, \quad (6.123)$$

$$\Pi_{\omega\omega-\omega'} = \Pi^R_{\omega\omega-\omega'} \coth \frac{\omega - \omega'}{2T} - \Pi^A_{\omega\omega-\omega'} \coth \frac{\omega}{2T} + \Pi^{(a)}_{\omega\omega-\omega'}. \quad (6.124)$$

Regular functions in (6.123) and (6.124) (for example, the advanced function $\mathcal{D}^A$) can be determined by the diagram expansion, in which all the functions (propagators and polarization operators) are advanced ones. Separating in the diagram expansion for $\mathcal{D}^{(a)}_{\omega\omega-\omega'}$ the left free line $\mathcal{D}^R_\omega$, we have the equation

$$\left(\mathcal{D}^0_\omega \right)^{-1} \mathcal{D}^{(a)}_{\omega\omega-\omega'} = \left\{ \Pi^R \mathcal{D}^{(a)} + \Pi^{(a)} \mathcal{D}^A \right\}_{\omega\omega-\omega'}, \quad (6.125)$$

where the following notation is used

$$\{AB\}_{\omega\omega-\omega'} = \int \frac{d\omega_1}{2\pi} A_{\omega\omega-\omega_1} B_{\omega-\omega_1\omega-\omega'} \quad (6.126)$$

(an integration over internal momentum or a coordinate variable is also assumed). Separating the right free line $\mathcal{D}^A_{\omega\omega-\omega'}$ in the same as above manner, subtracting the result from (6.125), and using formulae (6.123) and (6.124) together with the expression for regular $\mathcal{D}^{R,A}$ functions, one obtains the relation

$$\left[\left(\mathcal{D}^0_\omega \right)^{-1} - \left(\mathcal{D}^0_{\omega-\omega'} \right)^{-1} \right] \mathcal{D}_{\omega\omega-\omega'} = \left\{ \Pi^R \mathcal{D} - \Pi \mathcal{D}^A - \mathcal{D}\Pi^A - \mathcal{D}^R \Pi \right\}_{\omega\omega-\omega'}, \quad (6.127)$$

which is the desired general form of the kinetic equation for phonons. Expressions (6.75) and (6.77) follow from (6.127) after integration over the positive half-axis ω in quasiclassical limit.

Note that at this stage the "bath" temperature T, which enters into imaginary frequency variables of initial equations, is eliminated both from (6.127) and from (6.114) to (6.116). The situation here is fully equivalent to that obtained by the Keldysh technique. In the technique of analytical continuation, the bath temperature plays a role of equilibrium density matrix in Keldysh's method—this matrix is also eliminated from the final expressions.

6.4.3 Canonical Form for Phonon-Electron Collisions

The canonical form of the phonon-electron collision integral follows from (6.75), (6.77), (6.107), (6.108), and (5.83) to (5.88):

$$I\left(N_{\omega_q}\right) = \frac{\pi\lambda\,\omega_D}{8\,\epsilon_F} \int_\Delta^\infty \int_\Delta^\infty d\varepsilon d\varepsilon' \left\{\delta(\varepsilon + \varepsilon' - \omega_q)s_1 + 2\delta(\varepsilon - \varepsilon' - \omega_q)s_2\right\},$$

(6.128)

$$
\begin{aligned}
s_1 = &(u_\varepsilon u_{\varepsilon'} + v_\varepsilon v_{\varepsilon'} + 1)\left\{\left[(N_{\omega_q} + 1)n_\varepsilon n_{-\varepsilon'} - N_{\omega_q}(1 - n_\varepsilon)(1 - n_{-\varepsilon'})\right]\right. \\
&+ \left.\left[(N_{\omega_q} + 1)n_{-\varepsilon}n_{\varepsilon'} - N_{\omega_q}(1 - n_{-\varepsilon})(1 - n_{\varepsilon'})\right]\right\} \\
&+ (u_\varepsilon u_{\varepsilon'} + v_\varepsilon v_{\varepsilon'} - 1)\left\{\left[(N_{\omega_q} + 1)n_\varepsilon n_{\varepsilon'} - N_{\omega_q}(1 - n_\varepsilon)(1 - n_{\varepsilon'})\right]\right. \\
&+ \left.\left[(N_{\omega_q} + 1)n_{-\varepsilon}n_{\varepsilon'} - N_{\omega_q}(1 - n_{-\varepsilon})(1 - n_{-\varepsilon'})\right]\right\}, \qquad (6.129)
\end{aligned}
$$

$$
\begin{aligned}
s_2 = &(u_\varepsilon u_{\varepsilon'} - v_\varepsilon v_{\varepsilon'} - 1)\left\{\left[(N_{\omega_q} + 1)n_\varepsilon(1 - n_{-\varepsilon'}) - N_{\omega_q}(1 - n_\varepsilon)n_{-\varepsilon'}\right]\right. \\
&+ \left.\left[(N_{\omega_q} + 1)n_{-\varepsilon}(1 - n_{\varepsilon'}) - N_{\omega_q}(1 - n_{-\varepsilon})n_{\varepsilon'}\right]\right\} \\
&+ (u_\varepsilon u_{\varepsilon'} - v_\varepsilon v_{\varepsilon'} + 1)\left\{\left[(N_{\omega_q} + 1)n_\varepsilon(1 - n_{\varepsilon'}) - N_{\omega_q}(1 - n_\varepsilon)n_{\varepsilon'}\right]\right. \\
&+ \left.\left[(N_{\omega_q} + 1)n_{-\varepsilon}(1 - n_{\varepsilon'}) - N_{\omega_q}(1 - n_{-\varepsilon})n_{-\varepsilon'}\right]\right\}. \qquad (6.130)
\end{aligned}
$$

6.5 Seminar 3. Cooling by Heating

The effects which we will consider now have no direct relationship to TDGL derivation. However, they are typical for nonequilibrium superconductivity. Learning about them will facilitate better understanding of the apparatus in use for our general task.

6.5.1 Gap Enhancement

Let us consider a stationary state caused by the action of a high-frequency electromagnetic field on a thin superconducting film. Our direct interest is in determining the distribution function of electrons, which we will assume to be symmetric: $n_\varepsilon = n_{-\varepsilon}$.

This assumption is justified if the frequency of the field is small: $\omega \lesssim 2\Delta$ (we will assume Δ real). The stationary solution for n_ε can then be found from the kinetic equation

$$0 = J^{\text{(e-phonon)}}(n_\varepsilon) + J^{\text{(e-photon)}}(n_\varepsilon). \tag{6.131}$$

Here the first collision integral, of electrons with phonons, is given by (6.119)–(6.122), while $J^{\text{(e-photon)}}(n_\varepsilon)$ must be derived. For that, let us first simplify $J^{\text{(e-phonon)}}(n_\varepsilon)$ using the condition $n_\varepsilon = n_{-\varepsilon}$:

$$J^{\text{(e-phonon)}}(n_\varepsilon) = \frac{\pi\lambda}{2(up_F)^2} \int_0^\infty \omega_q^2 d\omega_q \int_{|\Delta|}^\infty d\varepsilon' \frac{\varepsilon\varepsilon'}{\sqrt{\varepsilon^2 - \Delta^2}\sqrt{\varepsilon'^2 - \Delta^2}}$$

$$\left\{ \left(1 - \frac{\Delta^2}{\varepsilon\varepsilon'}\right) \left[n_{\varepsilon'}(1 - n_\varepsilon)(1 + N_{\omega_q}) - n_\varepsilon(1 - n_{\varepsilon'})N_{\omega_q}\right] \delta(\varepsilon' - \varepsilon - \omega_q) \right.$$

$$+ \left(1 - \frac{\Delta^2}{\varepsilon\varepsilon'}\right) \left[n_{\varepsilon'}(1 - n_\varepsilon)N_{\omega_q} - n_\varepsilon(1 - n_{\varepsilon'})(1 + N_{\omega_q})\right] \delta(\varepsilon - \varepsilon' - \omega_q)$$

$$\left. + \left(1 + \frac{\Delta^2}{\varepsilon\varepsilon'}\right) \left[(1 - n_\varepsilon)(1 - n_{\varepsilon'})N_{\omega_q} - n_\varepsilon n_{\varepsilon'}(1 + N_{\omega_q})\right] \delta(\varepsilon + \varepsilon' - \omega_q) \right\} \tag{6.132}$$

In this equation, N_{ω_q} is the phonon distribution function, which we will assume to be an equilibrium one, at the phonon heatbath temperature T:

$$N_{\omega_q} \approx N_{\omega_q}^{(0)} = \frac{1}{e^{\frac{\omega_q}{T}} - 1} \tag{6.133}$$

This assumption means that the phonon heatbath model is valid. Practically, a phonon heatbath can be reached in thin enough films. The collision integral of electrons with photons will have the same structure as (6.132) if we describe photons in Fock's representation where photons are represented by their occupation numbers N_ω^p. The difference will be in coherence factors. The coherence factors $\left(1 \pm \Delta^2/\varepsilon\varepsilon'\right)$ in (6.132) correspond to a longitudinal acoustic phonon field. The photon field is transverse, so the coherence factors will have the opposite internal sign: $\left(1 \mp \Delta^2/\varepsilon\varepsilon'\right)$, see [9] for details. Also the interaction constant λ should be replaced by the fine structure constant e^2/hc. The overall coefficient will be different from that of (6.132). We will call it Q_0. Then

$$J^{\text{(e-photon)}}(n_\varepsilon) = Q_0 \int_0^\infty \omega^2 d\omega \int_{|\Delta|}^\infty d\varepsilon' \frac{\varepsilon\varepsilon'}{\sqrt{\varepsilon^2 - \Delta^2}\sqrt{\varepsilon'^2 - \Delta^2}}$$

$$\left\{ \left(1 + \frac{\Delta^2}{\varepsilon\varepsilon'}\right) \left[n_{\varepsilon'}(1 - n_\varepsilon)(1 + N_\omega^p) - n_\varepsilon(1 - n_{\varepsilon'})N_\omega^p\right] \delta(\varepsilon' - \varepsilon - \omega) \right.$$

$$+ \left(1 + \frac{\Delta^2}{\varepsilon\varepsilon'}\right) \left[n_{\varepsilon'}(1 - n_\varepsilon)N_\omega^p - n_\varepsilon(1 - n_{\varepsilon'})(1 + N_\omega^p)\right] \delta(\varepsilon - \varepsilon' - \omega)$$

$$+ \left(1 - \frac{\Delta^2}{\varepsilon\varepsilon'}\right)\left[(1 - n_\varepsilon)(1 - n_{\varepsilon'})N_\omega^P - n_\varepsilon n_{\varepsilon'}(1 + N_\omega^P)\right]\delta(\varepsilon + \varepsilon' - \omega)\Big\}. \quad (6.134)$$

We are interested in a classical limit of this quantum expression, which is the case when the occupation numbers are large: $N_\omega^P \gg 1$. Then (6.134) can be transformed into

$$J^{(e-\text{photon})}(n_\varepsilon) = Q_0 \int_0^\infty N_\omega^P \omega^2 d\omega \int_{|\Delta|}^\infty d\varepsilon' \frac{\varepsilon\varepsilon'}{\sqrt{\varepsilon^2 - \Delta^2}\sqrt{\varepsilon'^2 - \Delta^2}}$$

$$\Big\{\left(1 + \frac{\Delta^2}{\varepsilon\varepsilon'}\right)(n_{\varepsilon'} - n_\varepsilon)\,\delta(\varepsilon' - \varepsilon - \omega)$$

$$+ \left(1 + \frac{\Delta^2}{\varepsilon\varepsilon'}\right)(n_{\varepsilon'} - n_\varepsilon)\,\delta(\varepsilon - \varepsilon' - \omega)$$

$$+ \left(1 - \frac{\Delta^2}{\varepsilon\varepsilon'}\right)(1 - n_\varepsilon - n_{\varepsilon'})\delta(\varepsilon + \varepsilon' - \omega)\Big\}. \quad (6.135)$$

We will next assume that the electromagnetic field is monochromatic: $N_\omega^P = N^0 \times \delta(\omega - \omega_0)$. That will allow immediate integration in (6.135), so that it will acquire the form

$$J^{(e-\text{photon})}(n_\varepsilon) = Q_0 \int_{|\Delta|}^\infty d\varepsilon' \frac{\varepsilon\varepsilon'}{\sqrt{\varepsilon^2 - \Delta^2}\sqrt{\varepsilon'^2 - \Delta^2}}$$

$$\Big\{\left(1 + \frac{\Delta^2}{\varepsilon\varepsilon'}\right)(n_{\varepsilon'} - n_\varepsilon)\,\delta(\varepsilon' - \varepsilon - \omega_0)$$

$$+ \left(1 + \frac{\Delta^2}{\varepsilon\varepsilon'}\right)(n_{\varepsilon'} - n_\varepsilon)\,\delta(\varepsilon - \varepsilon' - \omega_0)$$

$$+ \left(1 - \frac{\Delta^2}{\varepsilon\varepsilon'}\right)(1 - n_\varepsilon - n_{\varepsilon'})\delta(\varepsilon + \varepsilon' - \omega_0)\Big\}. \quad (6.136)$$

Here, the factor Q_0 contains both the occupation number of photons, and their density of states $\propto \omega_0^2$; it is proportional to the intensity of the classical electromagnetic field $|E_{\omega_0}|^2$. Using the remnant δ-functions, we can now perform the integration in (6.136), and obtain:

$$J^{(e-\text{photon})}(n_\varepsilon) = Q_0 \frac{\varepsilon}{\sqrt{\varepsilon^2 - \Delta^2}}\theta(\varepsilon - \Delta)$$

$$\Big\{\frac{\varepsilon + \omega_0}{\sqrt{(\varepsilon + \omega_0)^2 - \Delta^2}}\left(1 + \frac{\Delta^2}{\varepsilon(\varepsilon + \omega_0)}\right)(n_{\varepsilon + \omega_0} - n_\varepsilon)$$

$$+ \frac{\varepsilon - \omega_0}{\sqrt{(\varepsilon - \omega_0)^2 - \Delta^2}}\left(1 + \frac{\Delta^2}{\varepsilon(\varepsilon - \omega_0)}\right)(n_{\varepsilon - \omega_0} - n_\varepsilon)\,\theta(\varepsilon - \omega_0 - \Delta)$$

$$+ \frac{\omega_0 - \varepsilon}{\sqrt{(\omega_0 - \varepsilon)^2 - \Delta^2}} \left(1 - \frac{\Delta^2}{\varepsilon(\omega_0 - \varepsilon)} \right) (1 - n_\varepsilon - n_{\omega_0 - \varepsilon}) \theta(\omega_0 - \varepsilon - \Delta) \Bigg\},$$

$$(6.137)$$

or, after simple transformations,

$$J^{(e-photon)}(n_\varepsilon) = Q_0 \Bigg\{ -\frac{\left[\varepsilon(\varepsilon + \omega_0) + \Delta^2 \right] \theta(\varepsilon - \Delta)}{\sqrt{\left[(\varepsilon + \omega_0)^2 - \Delta^2 \right] (\varepsilon^2 - \Delta^2)}} \left(n_\varepsilon - n_{\varepsilon + \omega_0} \right)$$

$$+ \frac{\left[(\varepsilon - \omega_0)\varepsilon + \Delta^2 \right] \theta(\varepsilon - \omega_0 - \Delta)}{\sqrt{\left[(\varepsilon - \omega_0)^2 - \Delta^2 \right] (\varepsilon^2 - \Delta^2)}} \left(n_{\varepsilon - \omega_0} - n_\varepsilon \right)$$

$$+ \frac{\left[\varepsilon(\omega_0 - \varepsilon) - \Delta^2 \right] \theta(\omega_0 - \varepsilon - \Delta)\theta(\varepsilon - \Delta)}{\sqrt{(\omega_0 - \varepsilon)^2 - \Delta^2}} (1 - n_\varepsilon - n_{\omega_0 - \varepsilon}) \Bigg\}. \quad (6.138)$$

This so-called "Eliashberg field term" with $Q_0 = 2D(e/c)^2 A_{\omega_0} A_{-\omega_0}$ [2] is the driving term in (6.131). Here D is the diffusion coefficient $D = v_F^2 \tau_{imp}/3$, where τ_{imp} is the scattering time of electrons on impurities.[4] It is responsible for breaking the Cooper pairs and generating single-electron quasiparticles from the condensate (in case of $\omega_0 > 2\Delta$) as well as the nonequilibrium redistribution of existing quasiparticles. In contrast, the electron-phonon term in (6.131) is responsible for the relaxation of the excited quasiparticles towards the thermal equilibrium described by the function $n_\varepsilon^0 = 1/\left[\exp(\varepsilon/T) + 1\right]$. In the relaxation-time approximation this term can be written as

$$J^{(e-phonon)}(n_\varepsilon) \approx -2\gamma \frac{\varepsilon}{\sqrt{\varepsilon^2 - \Delta^2}} \delta n_\varepsilon \quad (6.139)$$

where γ is the damping coefficient proportional to the average number of phonons and electron-phonon coupling: it has the same value as in the normal-metal state: $\sim \max\left(T^3/\omega_D^2, \; T^2/\varepsilon_F \right)$; and $\delta n_\varepsilon = n_\varepsilon - n_\varepsilon^0$. We will analyze the action of (6.138) for $\omega_0 \ll 2\Delta$, so that only the first two terms in braces of (6.138) are non-zero:

$$J^{(e-photon)}(n_\varepsilon) = Q_0 \left[U_{\varepsilon\varepsilon - \omega_0}(n_{\varepsilon - \omega_0} - n_\varepsilon) - U_{\varepsilon + \omega_0 \varepsilon}(n_\varepsilon - n_{\varepsilon + \omega_0}) \right] \quad (6.140)$$

where we denoted

$$U_{\varepsilon\varepsilon - \omega_0} = \frac{\left[(\varepsilon - \omega_0)\varepsilon + \Delta^2 \right] \theta(\varepsilon - \omega_0 - \Delta)}{\sqrt{\left[(\varepsilon - \omega_0)^2 - \Delta^2 \right] (\varepsilon^2 - \Delta^2)}}. \quad (6.141)$$

---

[4]We should mention that in absence of impurities, or, in a more general sense, of scattering centers, free electrons cannot absorb single photons.

Fig. 6.4 Nonequilibrium deviation of electron quasiparticles δn_ε. Parameters are the same as in Fig. 6.5 below

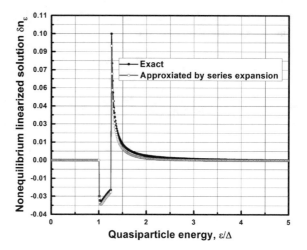

The dimensionless ratio Q_0/γ which characterizes the level on external electromagnetic pumping, may be both smaller and larger than unity. In case $Q_0/\gamma \ll 1$ we can consider the linearized solution of (6.131):

$$\delta n_\varepsilon = \frac{Q_0}{2\gamma} \left[U_{\varepsilon\varepsilon-\omega_0} (n^0_{\varepsilon-\omega_0} - n^0_\varepsilon) - U_{\varepsilon+\omega_0\varepsilon} (n^0_\varepsilon - n^0_{\varepsilon+\omega_0}) \right] \frac{\sqrt{\varepsilon - \Delta}}{\varepsilon} \qquad (6.142)$$

This solution is plotted in Fig. 6.4.

To analyze it analytically it is convenient to perform a series expansion in (6.142) and represent it as

$$\delta n_\varepsilon = \frac{Q_0}{2\gamma} \left\{ -\omega_0 \frac{\partial n^0_\varepsilon}{\partial \varepsilon} \left(U_{\varepsilon\varepsilon-\omega_0} - U_{\varepsilon+\omega_0\varepsilon} \right) \right.$$
$$\left. + \frac{\omega_0^2}{2} \frac{\partial^2 n^0_\varepsilon}{\partial \varepsilon^2} \left(U_{\varepsilon\varepsilon-\omega_0} + U_{\varepsilon+\omega_0\varepsilon} \right) \right\}, \qquad (6.143)$$

Both the first and second derivatives of the electron distribution function are changing in the range of ε of the order of T. Meanwhile, the first term in the braces of (6.143) is nonzero in the range $\varepsilon \sim \Delta$. If $T \lesssim T_c$, then $\Delta \ll T$, and in the vicinity of the gap edge, the solution (6.143) can be approximated by

$$\delta n_\varepsilon = \frac{Q_0}{2\gamma} \left[\frac{\omega_0}{4T} \cosh^{-2} \left(\frac{\varepsilon}{2T} \right) \left(U_{\varepsilon\varepsilon-\omega_0} - U_{\varepsilon+\omega_0\varepsilon} \right) \right] \frac{\sqrt{\varepsilon^2 - \Delta^2}}{\varepsilon}. \qquad (6.144)$$

This solution (also plotted in Fig. 6.4) should be substituted into the self-consistency equation

$$1 = \lambda \int_\Delta^{\omega_D} \frac{d\varepsilon}{\sqrt{\varepsilon^2 - \Delta^2}} (1 - 2n_\varepsilon). \qquad (6.145)$$

Its equilibrium solution at $n_\varepsilon = n_\varepsilon^0$ will be denoted as Δ_0. At $\delta n_\varepsilon = n_\varepsilon - n_\varepsilon^0 \neq 0$, one can transform (6.145) into

$$\delta\Delta = -\frac{T^2}{\Delta}\left(\frac{8\pi^2}{7\zeta(3)}\right)\int_\Delta^{\omega_D}\frac{d\varepsilon}{\sqrt{\varepsilon^2 - \Delta^2}}\delta n_\varepsilon. \tag{6.146}$$

Substituting (6.144) into (6.146), we find

$$\delta\Delta = a_0 T\left(\frac{Q_0\omega_0^2}{2\gamma\Delta^2}\right), \tag{6.147}$$

where

$$a_0 = \frac{8\pi^2}{7\zeta(3)}\ln\frac{8\Delta}{\omega_0} \sim 1. \tag{6.148}$$

Since $Q_0 \leq \gamma$ for the application of our approach, and also $\omega_0^2 \ll \Delta^2$, we see that $\delta\Delta \ll T$, as one should expect. However, the most important feature of solution (6.147) is its positive sign: the energy gap, and superconductivity itself are enhanced by the action of the (weak) electromagnetic field.

Indeed, the enhanced values of the critical currents were revealed by initial experimental measurements [10, 11] performed on microbridges and attempts were made to explain them by involving spatial inhomogeneities. It was not until 1970 that Eliashberg [12] recognized that this enhancement is caused by the effective cooling of electrons by high frequency electromagnetic fields. This specific mechanism which we described was further elaborated for cases when the "heating" energy was supplied by electromagnetic [13–18] and acoustic [19] fields, as well as the tunneling process [20–23]. Experiments [19, 24–30] confirmed these predictions. Interestingly, the "gap enhancement" effect should be accompanied by a "phonon deficit effect" [31], which will constitute the second half of our Seminar.

6.5.2 Negative Phonon Fluxes

Let us now calculate the phonon fluxes, which correspond to the same physical "phonon heat-bath" model in which the electronic distribution has been derived above. As we already mentioned, in the nonequilibrium "dressed" phonon Green's function, the polarization operators will differ from the "bare" equilibrium values, and contain information on the phonon fluxes, which leave the superconductor. After transferring to representation, that information is encoded now in the collision integrals (6.128)–(6.130). Let us rewrite these equations for the symmetric case: $n_\varepsilon = n_{-\varepsilon}$

$$I\left(N_{\omega_q}\right) = \frac{\pi\lambda}{2}\frac{\omega_D}{\epsilon_F}\int_{\Delta}^{\infty}\int_{\Delta}^{\infty}d\varepsilon d\varepsilon'\left\{\delta(\varepsilon+\varepsilon'-\omega_q)s_1 + 2\delta(\varepsilon-\varepsilon'-\omega_q)s_2\right\},$$

$$(6.149)$$

$$s_1 = (u_\varepsilon u_{\varepsilon'} + v_\varepsilon v_{\varepsilon'})\left[(N_{\omega_q}+1)n_\varepsilon n_{\varepsilon'} - N_{\omega_q}(1-n_\varepsilon)(1-n_{\varepsilon'})\right]$$

$$s_2 = (u_\varepsilon u_{\varepsilon'} - v_\varepsilon v_{\varepsilon'})\left[(N_{\omega_q}+1)n_\varepsilon(1-n_{\varepsilon'}) - N_{\omega_q}(1-n_\varepsilon)n_{-\varepsilon'}\right]$$

To find the phonon fluxes leaving the superconductor in conditions of free phonon exchange with the external world (thermostat) we substitute into (6.149) the *equilibrium* phonon distribution function (6.133), and the function (6.142) for n_ε. As we will see, the collision integral then is not zero: $I\left(N_{\omega_q}^{(0)}\right) \neq 0$. Since the canonical collision integrals define influx and outflux of the particles per unit time,

$$\frac{dN_{\omega_q}}{dt} = I(N_{\omega_q}^{(0)}) = \frac{\partial N_{\omega_q}}{\partial t} + \frac{\partial \mathbf{r}}{\partial t}\frac{\partial N_{\omega_q}}{\partial \mathbf{r}} \neq 0. \qquad (6.150)$$

In a stationary regime $\partial N_{\omega_q}/\partial t = 0$, and $\mathbf{v}\cdot\nabla N_{\omega_q}$ corresponds to the phonon flux from the superconducting volume. We are interested in the collisional integral of these phonons, $I(N_{\omega_q}^{(0)})$, which describes the outcome of phonons leaving the superconductor at a constant rate. As soon as it is obtained, the number of phonons leaving the superconductor per unit time can be computed as the integral of the distribution function times the density of states over their frequencies:

$$N = \int_0^{\infty} d\omega_q\left[\omega_q^2 I(N_{\omega_q}^{(0)})\right]. \qquad (6.151)$$

Alternatively, one can find the energy outflux:

$$E = \int_0^{\infty} d\omega_q\left[\omega_q^3 I(N_{\omega_q}^{(0)})\right]. \qquad (6.152)$$

Thus, the quantity of primary interest is $I(N_{\omega_q}^{(0)})$. To compute this function, it is convenient to integrate over one of the variables using $\delta-$functions. We will then have:

$$I\left(N_{\omega_q}^{(0)}\right) = \frac{\pi\lambda}{2}\frac{\omega_D}{\epsilon_F}\{\theta(\omega_q - 2\Delta)\int_{\Delta}^{\omega_q - \Delta}d\varepsilon\,\frac{\varepsilon(\omega_q - \varepsilon)}{\sqrt{\varepsilon^2 - \Delta^2}\sqrt{(\omega_q - \varepsilon)^2 - \Delta^2}}$$

$$\times\left[1 + \frac{\Delta^2}{\varepsilon(\omega_q - \varepsilon)}\right][(N_{\omega_q}^{(0)}+1)n_\varepsilon n_{\omega_q - \varepsilon} - N_{\omega_q}^{(0)}(1-n_\varepsilon)(1-n_{\omega_q-\varepsilon})]$$

$$+2\int_{\Delta}^{\infty}d\varepsilon'\,\frac{(\varepsilon'+\omega_q)\varepsilon'}{\sqrt{(\varepsilon'+\omega_q)^2 - \Delta^2}\sqrt{\varepsilon'^2 - \Delta^2}}\left[1 - \frac{\Delta^2}{(\varepsilon'+\omega_q)\varepsilon'}\right]$$

$$\times[(N_{\omega_q}^{(0)}+1)n_{(\varepsilon'+\omega_q)}(1-n_{\varepsilon'}) - N_{\omega_q}^{(0)}(1-n_{(\varepsilon'+\omega_q)})n_{\varepsilon'}]\} \qquad (6.153)$$

$$U1(x) := \begin{vmatrix} \dfrac{\left[x \cdot (x - \omega 0) + \Delta^2\right]}{\sqrt{\left[(x - \omega 0)^2 - \Delta^2 + \gamma \cdot \Delta\right]\left(x^2 - \Delta^2\right)}} & \text{if } x > \omega 0 + \Delta \\ 0 & \text{otherwise} \end{vmatrix}$$

$$E(x) := \begin{vmatrix} \dfrac{\sqrt{x^2 - \Delta^2}}{x} & \text{if } x > \Delta \\ 0 & \text{otherwise} \end{vmatrix}$$

$$U2(x) := \begin{vmatrix} \dfrac{\left[x \cdot (x + \omega 0) + \Delta^2\right]}{\sqrt{\left[(x + \omega 0)^2 - \Delta^2\right]\left(x^2 - \Delta^2\right)}} & \text{if } x > \Delta \\ 0 & \text{otherwise} \end{vmatrix}$$

$$N0(x) := \dfrac{1}{\exp\left(\dfrac{x}{T}\right) - 1}$$

$$\delta n(x) := \dfrac{Q0}{2\gamma} \cdot \left[U1(x)(n0(x - \omega 0) - n0(x)) - U2(x) \cdot (n0(x) - n0(x + \omega 0))\right] \cdot E(x)$$

$$n0(x) := \dfrac{1}{\exp\left(\dfrac{|x|}{T}\right) + 1}$$

$$L1(x,q) := \dfrac{x \cdot (q - x) + \Delta^2}{\sqrt{x^2 - \Delta^2} \cdot \sqrt{(q - x)^2 - \Delta^2}}$$

$$L2(x,q) := \dfrac{x \cdot (q + x) - \Delta^2}{\sqrt{x^2 - \Delta^2} \cdot \sqrt{(q + x)^2 - \Delta^2}}$$

$$n(x) := n0(x) + \delta n(x)$$

$$Irec(q) := \begin{vmatrix} \displaystyle\int_{\Delta}^{q-\Delta} L1(x,q) \cdot \left[(N0(q) + 1) \cdot n(q - x) \cdot n(x) - N0(q) \cdot (1 - n(q - x)) \cdot (1 - n(x))\right] dx & \text{if } q > 2\Delta \\ 0 & \text{otherwise} \end{vmatrix}$$

$$Irel(q) := 2 \cdot \int_{\Delta}^{Inf} L2(x,q) \cdot \left[(N0(q) + 1) \cdot n(q + x) \cdot (1 - n(x)) - N0(q) \cdot (1 - n(q + x)) \cdot n(x)\right] dx$$

$$\omega q := q \qquad\qquad \Delta \equiv 1 \qquad \omega 0 \equiv 0.5 \qquad Inf \equiv 20 \qquad T \equiv 1$$

$$I(\omega q) := Irec(\omega q) + Irel(\omega q) \qquad Q0 \equiv 0.001 \qquad \gamma \equiv 0.005 \qquad q \equiv 0, 0.01 .. Inf$$

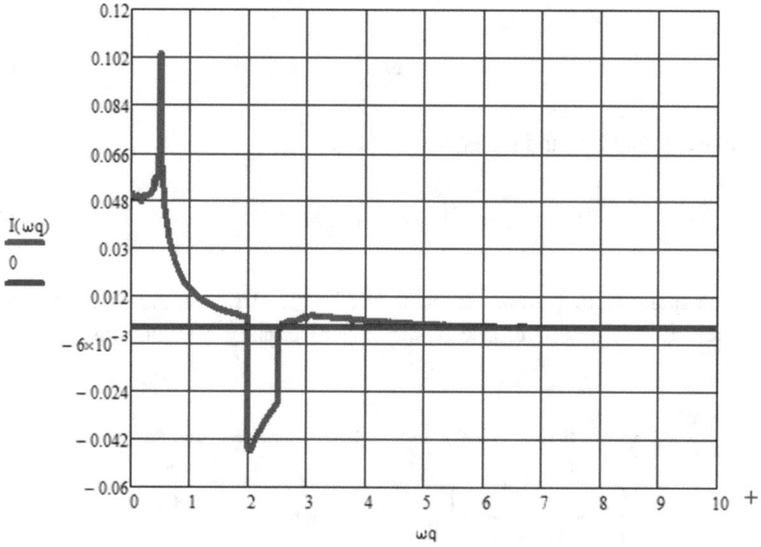

Fig. 6.5 MathCAD code with a plot of the result of numeric computation of (6.153)

This expression where the function n_ε is given by (6.142) can be easily evaluated numerically. It also can be evaluated analytically using the linearized approximation (6.144) for n_ε. We will describe the results of the numeric approach. The readers interested in the analytical approach are referred to [32]. The numerical computation of (6.153) is easy to perform by any numerical solver. We used MathCAD, and the code and results are shown in Fig. 6.5. As can be noticed, the phonon source is becoming negative in a narrow range above the gap-edge 2Δ. Negative phonon source means negative phonon fluxes in the outlined range of phonon frequencies: $2\Delta \leq \omega_q \leq 2\Delta + \omega_0$. This becomes possible because of the violation of detailed equilibrium by the external field action: absorbing photons are enforcing the drift of existing thermal excitations away from the Fermi surface/gap edge. As a result, phonons with energy slightly above the threshold value will break the Cooper pairs more effectively than the reciprocal process of recuperation of broken Cooper pairs takes place with the emission of phonons at the same frequency. Thus, the number of phonons at the frequency range $2\Delta \leq \omega_q \leq 2\Delta + \omega_0$ becomes less than in equilibrium, rising the influx of external phonons influx to eliminate the phonon deficit. This "phonon deficit effect" was proposed for cryogenic cooling. If one will compute the total energy balance of the phonon exchange with the thermostat, it is positive, as should be expected from the energy conservation law. However, by phonon filtering one can prohibit propagation of higher-energy phonons, thus making the net cooling effect via this mechanism potentially achievable in more complex systems. Details on the suggested design of the "phonon deficit"-based coolers could be found in [33].

References

1. M.Yu. Reizer, A.V. Sergeev, Electron-phonon interaction in impurity-containing metals and superconductors. Sov. Phys. JETP **63**(3), 616–624 (1986) [Zh. Eksp. i Teor. Fiz. **90**(3), 1056–1070 (1986)]
2. G.M. Eliashberg, Inelastic electron collisions and nonequilibrium stationary states in superconductors. Sov. Phys. JETP **34**(3), 668–676 (1972) [Zh. Eksp. i Teor. Fiz. **61** [3(9)], 1254–1272 (1971)]
3. I.E. Bulyzhenkov, B.I. Ivlev, Nonequilibrium phenomena in junctions of superconductors. Sov. Phys. JETP **47**(1), 115–120 (1978) [Zh. Eksp. i Teor. Fiz. **74**(1), 224–235 (1978)]
4. A.M. Gulian, G.F. Zharkov, Electron and phonon kinetics in a nonequilibrium Josephson junction. Sov. Phys. JETP **62**(1), 89–97 (1985) [Zh. Eksp. i Teor. Fiz. **89**(3), 1056–1070 (1985)]
5. M. Tinkham, Tunneling generation, relaxation and tunneling detection of hole-electron imbalance in superconductors. Phys. Rev. B **6**(5), 1747–1756 (1972)
6. E.M. Lifshitz, L.P. Pitaevskii, *Physical Kinetics* (Pergamon Press, Oxford, 1981), pp. 391–412
7. A.A. Abrikosov, L.P. Gor'kov, I.E. Dzyaloshinskii, *Quantum Field Theoretical Methods in Statistical Physics*, 2nd edn. (Pergamon Press, Oxford, 1965), pp. 63–70
8. L.V. Keldysh, Diagram technique for nonequilibrium processes. Sov. Phys. JETP **20**(4), 1018–1026 (1965) [Zh. Eksp. i Teor. Phys. **47** [4(10)], 1515–1527 (1964)]
9. J.R. Schrieffer, *Theory of Superconductivity*, 2nd edn. (CRC Press, Boca Raton, FL, 2018), pp. 74–77
10. A.F.G. Wyatt, V.M. Dmitriev, W.S. Moore, F.W. Sheard, Microwave-enhanced critical supercurrents in constricted tin films. Phys. Rev. Lett. **16**(25), 1166–1169 (1966)

11. A.H. Dayem, J.J. Wiegand, Behaviour of thin-film superconducting bridges in a microwave field. Phys. Rev. **155**(2), 419–428 (1967)
12. G.M. Eliashberg, Film superconductivity stimulated by a high-frequency field. JETP Lett. **11**(3), 114–117 (1970) [Pis'ma v Zh. Eksp. i Teor. Fiz. **11**(3), 186–188 (1970)]
13. B.I. Ivlev, G.M. Eliashberg, Influence of nonequilibrium excitations on the properties of superconducting films in a high-frequency field. JETP Lett. **13**(8), 333–336 (1971) [Pis 'ma v Zh. Eksp. i Teor. Fiz. **13**(8), 464–468 (1971)]
14. B.I. Ivlev, S.G. Lisitsyn, G.M. Eliashberg, Nonequilibrium excitations in superconductors in high-frequency fields. J. Low Temp. Phys. **10**(3/4), 449–468 (1973)
15. J.J. Chang, D.J. Scalapino, Gap enhancement in superconducting thin films due to microwave irradiation. J. Low Temp. Phys. **29**(5/6), 447–485 (1977)
16. L.G. Aslamazov, V.I. Gavrilov, On the micropower-induced enhancement of the critical temperature of a superconductor. Sov. J. Low Temp. Phys. **6**(7), 877–881 (1980) [Fiz. Nizk. Temp. **6**(7), 877–881 (1980)]
17. A.M. Gulian, G.F. Zharkov, Dependence of a superconducting gap on temperature in an UHF field. JETP Lett. **33**(9), 454–458 (1981) [Pis 'ma Zh. Eksp. Teor. Fiz. **33**(9), 471–474 (1981)]
18. A.M. Gulyan, V.E. Mkrtchyan, Stimulation of superconductivity by electromagnetic radiation. Sov. Phys. Lebedev Inst. Reports **2**, 40–44 (1989) [Kratk. Soobsch. po Fiz. FIAN **2**, 31–34 (1989)]
19. T.J. Tredwell, E.H. Jacobsen, Phonon-induced enhancement of the superconducting energy gap. Phys. Rev. Lett. **35**(4), 244–247 (1975)
20. A.G. Aronov, V.L. Gurevich, The tunneling of excitations from a superconductor and the increase of T_c. Sov. Phys. JETP **36**(5), 957–963 (1972) [Zh. Eksp. i Teor. Fiz. **63**(5), 1809–1821 (1972)]
21. S.A. Peskovatskii, V.P. Seminozhenko, Stimulation of superconductivity by constant tunnel currents. Sov. J. Low Temp. Phys. **2**(7), 464–465 (1976) [Fiz. Niz. Temp. **2**(7), 943–945 (1976)]
22. J.J. Chang, Gap enhancement in superconducting thin films due to quasiparticle tunnel injection. Phys. Rev. B **17**(5), 2137–2140 (1978)
23. C.C. Chi, J. Clarke, Enhancement of the energy gap in superconducting aluminum by tunneling extraction of quasiparticles. Phys. Rev. B **20**(11), 4465–4473 (1979)
24. Yu.I. Latyshev, F.Ya. Nad', Mechanism of superconductivity stimulated by microwave radiation. Sov. Phys. JETP **44**(6), 1136–1141 (1976) [Zh. Eksp. i Teor. Fiz. **71**(6) 2158–2167 (1976)]
25. T.M. Klapwijk, J.E. Mooij, Microwave-enhanced superconductivity in aluminum films. Physica B + C **81**, 132–136 (1976)
26. K.E. Gray, Enhancement of superconductivity by quasiparticle tunneling. Solid State Comm. **26**, 633–635 (1978)
27. J.A. Pals, Microwave-enhanced critical currents in superconducting Al strips with local injection of electrons. Phys. Lett. A **61**(4), 275–277 (1977)
28. J.A. Pals, J. Dobben, Observation of order-parameter enhancement by microwave irradiation in a superconducting aluminum cylinder. Phys. Rev. Lett. **44**(7), 1143–1146 (1980)
29. Yu.I. Latyshev, F.Ya. Nad', Frequency dependence of superconductivity stimulated by a high-frequency field. JETP Lett. **19**(12), 380–382 (1974) [Pis 'ma v Zh. Eksp. i Teor. Fiz. **19**(12), 737–741 (1974)]
30. T.M. Kommers, J. Clarke, Measurement of microwave-enhanced energy gap in superconducting aluminum by tunneling. Phys. Rev. Lett. **38**(19), 1091–1094 (1977)
31. A.M. Gulian, G.F. Zharkov, The "phonon deficit" effect in superconductors induced by UHF radiation. Phys. Lett. A **80**(1), 79–80 (1980)
32. A.M. Gulian, G.F. Zharkov, *Nonequilibrium Electrons and Phonons in Superconductors* (Kluwer Academic/Plenum Publishers, New York, 1999), pp. 137–150
33. A.M. Gulian, G.G. Melkonyan, Cooling by heating: deep cryogenic refrigeration by photons based on the phonon deficit effect in superconductors, in *Recent Advancement in Superconductivity Research*, ed. by C.B. Taylor (Nova Science Publishers Inc, New York, 2013), pp. 29–54

Chapter 7
Time-Dependent Ginzburg–Landau (TDGL) Equations

The Initial microscopic derivation of TDGL equations was successfully performed by Eliashberg and Gor'kov. However, their theory, elucidated in Chap. 4 corresponds to the so-called gapless regime, where depairing factors (like magnetic impurities) squeeze the gap to zero, while leaving the Cooper condensate alive. Its quantum behavior is governed by this set of "gapless" TDGL equations. However, later developments demonstrated that as soon as the electron-phonon interaction smears, to a certain extent, the BCS-peculiarity in the electronic density of states (which occurs at a local equilibrium between Cooper pairs, electrons and phonons), the kinetic equations and the self-consistency equation converge into a closed system of TDGL equations, closely resembling the gapless TDGL equations. Using the tools developed in previous chapters, we demonstrate in this Chapter how the set of TDGL equations in the "local-equilibrium approximation" comes out. An interesting difference between gapless and local-equilibrium approximations is the presence of interference current in the latter, more general, case. Another important feature of finite-gap TDGL equations is related to the fact that the electron-phonon system is no longer decoupled as it was in the gapless case. This makes the class of physical systems which can be described by the resultant TDGL equations much more broad.

7.1 Order Parameter, Electron Excitations, and Phonons

The external fields acting on a superconductor may lead to nonstationary phenomena that have to be described by dynamic equations. However, as was shown in the previous chapters, the set of nonstationary equations in the general case is very complicated and in addition to the equations for the main parameters, characterizing superconductivity (such as $|\Delta|$, μ, $\mathbf{Q}$), it includes generalized kinetic equations for distribution functions (see Sect. 5.3). In the vicinity of the critical temperature (in analogy with the stationary case, Sect. 3.3), one can simplify the general time-dependent equations by

© Springer Nature Switzerland AG 2020
A. Gulian, *Shortcut to Superconductivity*,
https://doi.org/10.1007/978-3-030-23486-7_7

considering the gapless case (Sect. 4.2). For finite gap superconductors, the attempt
to simplify the general scheme encounters serious difficulties connected with the
non-local kernels of the integral equations, governing the order parameter. To derive
the equations for such superconductors, one needs to account simultaneously for the
condensate, the excitations, and the interaction between them. The success achieved
in this direction [1–6] is due to progress in the kinetic description of single-particle
excitations in nonequilibrium superconductors (see the review articles [7–10]). The
dynamic equations for the order parameter were obtained in their most complete form
by Watts-Tobin et al. [6]. But in some respects the theory still had some deficiencies,
which we have tried to fix up [11].

In many situations, the possible deviation of the phonon system from equilibrium
should be taken into account. The role of phonons in the problem considered is
twofold. First, the nonequilibrium in the phonon system may be essential for the
dynamics of the order parameter. Second, the time variations of the order parameter
modulus might lead to excess phonon generation and to phonon exchange between
a superconductor and its environment.

7.1.1 Basic Kinetic Equations

We will use here the generalized kinetic equations [12, 13] for energy-integrated
Green-Gor'kov functions. As was shown in Chap. 5, these equations are still valid
also in the case where the phonon system is not at equilibrium. In a real-time approximation,[1] the equations may be written in a very compact form:

$$
i\mathbf{v}\cdot\frac{\partial \breve{g}}{\partial \mathbf{r}} + i\breve{\sigma}_z\frac{\partial \breve{g}}{\partial t_1} + i\frac{\partial \breve{g}}{\partial t_2}\breve{\sigma}_z = \breve{H}(t_1)\breve{g} - \breve{g}\breve{H}(t_2)
$$

$$
+ \int_{-\infty}^{\infty} dt_3 \left\{ \breve{\Sigma}(t_1 t_3)\breve{g}(t_3 t_2) - \breve{g}(t_1 t_3)\breve{\Sigma}(t_3 t_2) \right\}. \tag{7.1}
$$

Here

$$
\breve{g} = \breve{g}(t_1, t_2, \mathbf{r}, \mathbf{p}_F), \quad \breve{g} = \begin{pmatrix} \widehat{g}^R & \widehat{g} \\ \widehat{0} & \widehat{g}^A \end{pmatrix}, \quad \widehat{g}^{(R,A)} = \begin{pmatrix} g & f \\ -f^+ & g \end{pmatrix}^{(R,A)}, \tag{7.2}
$$

$$
\breve{\Sigma} = \breve{\Sigma}(t_1, t_2, \mathbf{r}, \mathbf{p}_F), \quad \breve{\Sigma} = \begin{pmatrix} \widehat{\Sigma}^R & \widehat{\Sigma} \\ \widehat{0} & \widehat{\Sigma}^A \end{pmatrix}, \quad \widehat{\Sigma}^{(R,A)} = \begin{pmatrix} \Sigma_1 & \Sigma_2 \\ -\Sigma_2^+ & \Sigma_1 \end{pmatrix}^{(R,A)}, \tag{7.3}
$$

$$
\breve{H}_1(t) = \mathbf{v}\mathbf{A}(t)\breve{\sigma}_z - \breve{1}\cdot\varphi(t), \quad \breve{1} = \begin{pmatrix} \widehat{1} & \widehat{0} \\ \widehat{0} & \widehat{1} \end{pmatrix}, \quad \breve{\sigma}_z = \begin{pmatrix} \widehat{\sigma}_z & \widehat{0} \\ \widehat{0} & \widehat{\sigma}_z \end{pmatrix}, \tag{7.4}
$$

---

[1]In (7.1) the integration over the intrinsic coordinate (or, depending on representation, over the
momentum variable) is assumed.

where $\mathbf{p}_F = m\mathbf{v}_F = m\mathbf{v}$ is the Fermi momentum, and $\mathbf{r}$ is the quasiclassical coordinate, the Fourier-transform of which is denoted by $\mathbf{k}$. In (7.1) $(\mathbf{A}, \varphi)$ are the electromagnetic field potentials $(e = \hbar = c = 1)$.

7.1.2 Normalization Condition

Equation (7.1) must be supplemented by the normalization condition, which allows us to select the necessary solution of homogeneous (relative to the $\breve{g}$-functions) equations (7.1):

$$\int_{-\infty}^{\infty} dt_3\, \breve{g}(t_1, t_3, \mathbf{r}, \mathbf{p}_F)\breve{g}(t_3, t_2, \mathbf{r}, \mathbf{p}_F) = -\pi^2 \breve{1} \cdot \delta(t_1 - t_2). \tag{7.5}$$

We wrote this condition in Sect. 5.3 (see 5.68). It may be proven in the following way (see, e.g., [5]). Equation (7.1) may be presented in the form

$$\left[\breve{Z} * \breve{g} \right]_{-} = \left[\breve{Z} * \breve{g} - \breve{g} * \breve{Z} \right] = 0, \tag{7.6}$$

where the operator $\breve{Z}$, as follows from (7.6) and (7.1), is

$$\breve{Z} = i\breve{\sigma}_z \frac{\partial}{\partial t_1} \delta(t_1 - t_2) - \breve{H}(t_1) + \breve{1} i \mathbf{v} \cdot \frac{\partial}{\partial \mathbf{r}} - \breve{\Sigma}. \tag{7.7}$$

Since the convolution $*$ is commutative [which follows directly from its definition in (5.70)], it is easy to see that the condition

$$\breve{g} * \breve{g} = \text{const} \cdot \breve{1} \tag{7.8}$$

is compatible with the equation

$$\left[\breve{Z}*, \breve{g} * \breve{g} \right]_{-} = 0, \tag{7.9}$$

which follows from (7.6). The value of a constant in (7.8) can be obtained by considering (7.8) either in a superconducting region that is in an equilibrium state, or in a normal area, where $|\Delta| = 0$. The latter option is simpler, and it is possible to calculate the constant immediately. Larkin and Ovchinnikov [13] introduced the normalization in which const = 1. Because a particular value of this constant is of no importance, we will retain the normalization const $= -\pi^2$, used earlier.

7.1.3 Definition of Order Parameter

We will now use the results obtained in Chaps. 5 and 6. The self-energy function Σ is an additive quantity that contains certain terms corresponding to the interaction of electrons with impurities, with phonons, with each other, and so on. The nonequilibrium order parameter Δ in a weak-coupling limit $\lambda \ll 1$ (λ is the dimensionless electron-phonon coupling parameter) is defined by the formula

$$\Delta = \frac{1}{2} \left(\Sigma_2^R + \Sigma_2^A \right)^{(e-ph)}, \tag{7.10}$$

where the self-energy function representing the interaction of electrons with phonons is (see Sect. 6.4):

$$\left(\Sigma_2^{R(A)} \right)^{(e-ph)} = \int_{-\infty}^{\infty} \frac{d\varepsilon'}{4\pi i} \int \frac{d\Omega_{\mathbf{p}'}}{4\pi} \left\{ f_{\varepsilon'} D_{\varepsilon'-\varepsilon}^{A(R)} + D_{\varepsilon'-\varepsilon} f_{\varepsilon'}^{R(A)} \right\}. \tag{7.11}$$

The phonon propagator is expressed in terms of the nonequilibrium phonon distribution function $N_{\omega_{\mathbf{q}}}$ by the relations:

$$D_{\varepsilon'-\varepsilon} = \left(1 + 2N_{\omega_{\mathbf{p}-\mathbf{p}'}} \right) \operatorname{sign}(\varepsilon' - \varepsilon) \left(D^R - D^A \right)_{\varepsilon'-\varepsilon}, \tag{7.12}$$

$$D_{\varepsilon'-\varepsilon}^{R(A)} = \lambda \frac{2\omega_{\mathbf{p}'-\mathbf{p}}^2}{\omega_{\mathbf{p}'-\mathbf{p}}^2 - (\varepsilon' - \varepsilon \, {}^{+}_{(-)} i\delta)^2}. \tag{7.13}$$

When phonons are in equilibrium, the contribution of the second term in (7.11) is small by the parameter $(T/\omega_D)^2$, so to find Δ one can use the simplified equation that follows from (7.10) to (7.13):

$$\Delta_\omega(\mathbf{k}) = \lambda \int_{-\omega_D}^{\omega_D} \frac{d\varepsilon}{4\pi i} \int \frac{d\Omega_{\mathbf{p}}}{4\pi} f_{\varepsilon\varepsilon-\omega}(\mathbf{p}, \mathbf{k}). \tag{7.14}$$

If the phonon distribution function $N_{\omega_{\mathbf{q}}}$ is localized at energies $\omega_{\mathbf{q}} \ll \omega_D$ and has no singularities as a function of a real argument $\omega_{\mathbf{q}}$ (this will be assumed further), (7.14) may be applied to the situations with nonequilibrium phonons.

7.1.4 Nondiagonal Collision Channel

To obtain the propagator $f_{\varepsilon\varepsilon-\omega}(\mathbf{p}, \mathbf{k})$, one can use the equation that follows from (7.1) for the nondiagonal "Keldysh" component. Separating in (7.1) the virtual processes (see Sect. 6.1), which lead to (7.14), and ignoring the renormalization terms, one finds the expression for the $\widehat{I}$-matrix (6.2):

$$\widehat{I} = \begin{pmatrix} (-f\Delta^* + \Delta f^+) + K_{11} & (g\Delta - \Delta\overline{g}) + K_{12} \\ (-\overline{g}\Delta^* - \Delta^* g) + K_{21} & (-f^+\Delta - \Delta^* f) + K_{22} \end{pmatrix}, \tag{7.15}$$

where the coefficients K_{ij} are connected with the self-energy functions [the definition of these quantities follows from comparison of (7.15) with (6.2)]. Taking into account the nondiagonal channel in the kinetic equation for the electron-hole distribution function n_ε [14, 15], we get the canonical form of the collision integral. We recall here that the general expression for the $\widehat{g}$-function, which satisfies the normalization condition (7.8), was discussed in Sect. 5.3, where the functions f_1 and f_2 where introduced [see (5.72)]. These functions connect Green's functions with the distribution functions of electron-like (n_ε) and hole-like ($n_{-\varepsilon}$) excitations.

7.1.5 Spectral Functions R_1, R_2, N_1, and N_2

According to (5.77), (5.79), and (5.80), the functions f_1 and f_2 (as well as N_1, $\overline{N}_1$) are of general type, i.e., they have definite ε-parity only in absence of an external electromagnetic field. In the latter case they are equal to

$$N_1 = -\overline{N}_1 = \mathrm{Re}\left(\frac{\varepsilon + i\gamma}{\sqrt{(\varepsilon + i\gamma)^2 - |\Delta|^2}}\right), \tag{7.16}$$

$$f_1 = \mathrm{sign}\varepsilon\,(1 - n_\varepsilon - n_{-\varepsilon}), \tag{7.17}$$

$$f_2 = -\frac{\mathrm{sign}\varepsilon}{N_1}(n_\varepsilon - n_{-\varepsilon}), \tag{7.18}$$

where γ is the energy damping of electrons. Introducing also the functions

$$R_1 = \mathrm{Im}\left(\frac{\varepsilon + i\gamma}{\sqrt{(\varepsilon + i\gamma)^2 - |\Delta|^2}}\right), \tag{7.19}$$

$$R_2 = \mathrm{Re}\left(\frac{|\Delta|}{\sqrt{(\varepsilon + i\gamma)^2 - |\Delta|^2}}\right), \tag{7.20}$$

$$N_2 = -\mathrm{Im}\left(\frac{|\Delta|}{\sqrt{(\varepsilon + i\gamma)^2 - |\Delta|^2}}\right), \tag{7.21}$$

we can express the $\widehat{g}$-function as

$$\begin{pmatrix} g & f \\ -f^+ & g \end{pmatrix} = 2\pi i \left\{ \begin{pmatrix} N_1 & R_2 e^{i\theta} \\ -R_2 e^{-i\theta} & \overline{N}_1 \end{pmatrix} f_1 + \begin{pmatrix} N_1 & i N_2 e^{i\theta} \\ -i N_2 e^{-i\theta} & -\overline{N}_1 \end{pmatrix} f_2 \right.$$

$$\left. + \frac{1}{2} \left[\frac{\partial f_1}{\partial \varepsilon} \frac{\partial}{\partial t} \begin{pmatrix} R_1 & -N_2 e^{i\theta} \\ N_2 e^{-i\theta} & -R_1 \end{pmatrix} - \frac{\partial f_1}{\partial t} \frac{\partial}{\partial \varepsilon} \begin{pmatrix} R_1 & -N_2 e^{i\theta} \\ N_2 e^{-i\theta} & -R_1 \end{pmatrix} \right] \right\}. \quad (7.22)$$

In expression (7.22), only the lowest convolution corrections are kept (the contribution from the f_2-function is negligible).

7.1.6 Gap-Control Term

Separating in (7.14) the equilibrium part and making standard calculations (see Sect. 3.4) we get an equation for the order parameter near T_c:

$$-\frac{\pi}{8 T_c} \left[\frac{\partial}{\partial t} - D(\nabla - 2i\mathbf{A})^2 \right] \Delta + \left[\frac{T_c - T}{T_c} - \frac{7\zeta(3)}{8(\pi T_c)^2} |\Delta|^2 \right] \Delta + \varkappa(t) = 0,$$
$$(7.23)$$

where $D = l v_F / 3$ is the diffusion constant and $\varkappa(t)$ is the so-called "gap-control" term:

$$\varkappa(\mathbf{r}, t) = \int_{-\infty}^{\infty} \frac{d\varepsilon}{4\pi i} \left\{ [f_1(\varepsilon) - f_1^0(\varepsilon)](f^R - f^A)_\varepsilon - f_2(\varepsilon)(f^R + f^A)_\varepsilon \right\}, \quad (7.24)$$

where $f_1^0(\varepsilon) = \tanh(\varepsilon/2T)$. The nonequilibrium functions f_1 and f_2 should be found from the kinetic equation (7.1), where one can assume the phonon system to be initially in equilibrium. Note that the terms generated by $N_{1,\varepsilon}$ make insignificant contributions to (7.23) because the function $N_{1,\varepsilon}$ is non-zero at $\varepsilon \sim |\Delta|$. Only the values of $\varepsilon \sim T$ play a major role in the integrand of (7.24). For this reason, one can neglect the terms proportional to $N_{1,\varepsilon}/N_1$ in (5.77), which then take the form

$$f_1 \to f_1 + \frac{1}{2} \dot{\chi} f_{2,\varepsilon} + \frac{1}{8} \dot{\chi}^2 f_{1,\varepsilon\varepsilon}, \quad f_2 \to f_2 + \frac{1}{2} \dot{\chi} f_{1,\varepsilon} + \frac{1}{8} \dot{\chi}^2 f_{2,\varepsilon\varepsilon}. \quad (7.25)$$

From the kinetic equations for f_1 and f_2 in the absence of the potential φ in the local equilibrium approximation, it follows that

$$f_1 = f_1^0 - f_{1,\varepsilon}^0 \frac{R_2}{N_1} \tau_\varepsilon \frac{\partial |\Delta|}{\partial t}, \quad f_2 = \frac{N_2 \tau_\varepsilon |\Delta|}{N_1 + 2\tau_\varepsilon |\Delta| N_2} \dot{\theta} f_{1,\varepsilon}^0. \quad (7.26)$$

We will briefly follow the derivation procedure of these relations to clarify the essence of local equilibrium approximation.

7.1.7 Local-Equilibrium Approximation

If the characteristic frequencies and gradients of the electron system perturbation obey the relations[2]

$$(Dk^2, \omega) \ll \gamma, \tag{7.27}$$

where γ is the damping caused by inelastic processes in the electron system, then in the kinetic equations [(7.1) and (5.63)] that define the functions f_1 and f_2, one can neglect the left hand sides and the terms connected with the Hamiltonian $\widehat{H}_1$. This means that the functions f_1 and f_2 do not depend explicitly on the space coordinate $\mathbf{r}$ and time t. Only implicit dependence on $\mathbf{r}$ and t remains through the parameter $\Delta(\mathbf{r}, t)$, which enters into (7.1). This means that owing to effective inelastic collisions, the behavior of single-particle electron excitations in an external field is fully determined by the evolution of the order parameter that governs the formation of the distribution function n_ε (and does not depend, e.g., on the diffusion mechanism). In other words, local equilibrium between the system of single-particle excitations and the pair-condensate is taking place.

7.1.8 Determination of f_1-Function

In this approximation from the diagonal components of (5.63), the equation for the function f_1 follows:

$$0 = \left\{ -f\Delta^* + \Delta f^+ \right\}_{\varepsilon\varepsilon-\omega} + \left\{ -f^+\Delta + \Delta^* f \right\}_{\varepsilon\varepsilon-\omega} + K_{11} + K_{22}$$
$$\approx -\Delta\omega f_{,\varepsilon}^+ - \Delta^*\omega f_{,\varepsilon} + K_{11} + K_{22} \tag{7.28}$$

[the series expansion of functions $f_{\varepsilon-\omega}$ and $f_{\varepsilon-\omega}^+$ in (7.28) may be restricted to the first terms owing to the quasiclassical conditions]. Inserting (7.22) into (7.28) and omitting convolution corrections, one finds

$$0 = f_{1,\varepsilon}^0 R_2 \frac{\partial |\Delta|}{\partial t} + \frac{1}{4\pi}(K_{11} + K_{22}), \tag{7.29}$$

where the transformation rule ($i\omega \doteq \partial/\partial t$) is used and the inequalities

$$(f_1 - f_1^0)_{,\varepsilon} \ll f_{1,\varepsilon}^0, \qquad f_{2,\varepsilon} \ll f_{1,\varepsilon}^0. \tag{7.30}$$

have been taken into account. The functions $K_{11} + K_{22}$ are expressed through the collision operators $J(n_{\pm\varepsilon})$, obtained in Chap. 6:

---

[2]We assume also the quasiclassical character of external fields $\mathbf{A}(\mathbf{r}, t)$ and $\varphi(\mathbf{r}, t)$.

$$K_{11} + K_{22} = 4\pi \mathrm{sign}\varepsilon \left[J(n_\varepsilon) + J(n_{-\varepsilon}) \right]. \tag{7.31}$$

Furthermore, we will assume that the electron-phonon collisions provide the most effective channel of inelastic relaxation and represent $J(n_{\pm\varepsilon}) \approx J(n_{\pm\varepsilon})^{(\mathrm{e-ph})}$ in the form

$$J(n_{\pm\varepsilon})^{(\mathrm{e-ph})} = \frac{\pi\lambda}{4(up_F)^2} \int_0^\infty \omega_q^2 d\omega_q \int_{|\Delta|}^\infty d\varepsilon' \{ p_1(n_{\pm\varepsilon})\delta(\varepsilon' - \varepsilon - \omega)$$
$$+ p_2(n_{\pm\varepsilon})\delta(\varepsilon - \varepsilon' - \omega) + p_3(n_{\pm\varepsilon})\delta(\varepsilon + \varepsilon' - \omega) \}, \tag{7.32}$$

where

$$p_1(n_{\pm\varepsilon}) = (u_\varepsilon u_{\varepsilon'} - v_\varepsilon v_{\varepsilon'} \pm 1) \left[n_{\varepsilon'}(1 - n_{\pm\varepsilon})(1 + N_{\omega_q}) - n_{\pm\varepsilon}(1 - n_{\varepsilon'})N_{\omega_q} \right]$$
$$+ (u_\varepsilon u_{\varepsilon'} - v_\varepsilon v_{\varepsilon'} \mp 1) \left[n_{-\varepsilon'}(1 - n_{\pm\varepsilon})(1 + N_{\omega_q}) - n_{\pm\varepsilon}(1 - n_{-\varepsilon'})N_{\omega_q} \right], \tag{7.33}$$
$$p_2(n_{\pm\varepsilon}) = p_1(N_{\omega_q} \leftrightarrow N_{\omega_q} + 1), \tag{7.34}$$
$$p_3(n_{\pm\varepsilon}) = (u_\varepsilon u_{\varepsilon'} + v_\varepsilon v_{\varepsilon'} \mp 1) \left[(1 - n_{\pm\varepsilon})(1 - n_{\varepsilon'})N_{\omega_q} - n_{\pm\varepsilon}n_{\varepsilon'}(1 + N_{\omega_q}) \right]$$
$$+ (u_\varepsilon u_{\varepsilon'} + v_\varepsilon v_{\varepsilon'} \pm 1) \left[(1 - n_{\pm\varepsilon})(1 - n_{-\varepsilon'})N_{\omega_q} - n_{\pm\varepsilon}n_{-\varepsilon'}(1 + N_{\omega_q}) \right]. \tag{7.35}$$

In (7.33)–(7.35), the function N_{ω_q} is the distribution function of phonons, which as yet is assumed to be an equilibrium one:

$$N_{\omega_q} = N_{\omega_q}^0 = \exp \left[(\omega_q/T) - 1 \right]^{-1}. \tag{7.36}$$

In the vicinity of critical temperature, where $T \gg |\Delta|$, for the collision integral the relaxation-time approximation may be used[3]

$$J(n_{\pm\varepsilon})^{(\mathrm{e-ph})} \approx -2\gamma u_\varepsilon(n_{\pm\varepsilon} - n_\varepsilon^0), \tag{7.37}$$

where

$$n_\varepsilon^0 = \exp \left(\frac{|\varepsilon|}{T} + 1 \right)^{-1}, \qquad \gamma = (2\tau_\varepsilon^{-1}) \approx \frac{7\pi\lambda\zeta(3)T^3}{(up_F)^2}. \tag{7.38}$$

Using (7.31), (7.32), and (7.37), we find from (7.29) the first expression in (7.26).

---

[3]This opportunity emerges because the perturbation of the distribution function n_ε is localized in the energy range smaller than the temperature diffusion scale. Due to this, the term, containing nonequilibrium distribution function n_ε in the integral form, is smaller than the "free" term. We stress this circumstance, because it remains valid also in derivation of the function f_2(see below). However, sometimes the τ-approximation is criticized and, moreover, negated (see, e.g., [6]) as violating a condition, related to the particle number conservation. In our calculation scheme the missing term automatically appears from the gauge-transformation rules for the functions f_1 and f_2, which were established in Sect. 5.3.

7.1.9 Determination of f_2-Function

Let us now determine the function f_2. The nondiagonal elements of (5.63) and (7.15) are essential, because the first term (proportional to f_1) in Green's functions f and f^+ (7.22) does not contribute to the equation

$$0 = -2f_\varepsilon \Delta^* - 2\Delta f_\varepsilon^+ - \Delta\omega f_{,\varepsilon}^+ + \Delta^*\omega f_{,\varepsilon} + K_{11} - K_{22}. \tag{7.39}$$

Accounting for this, one finds

$$0 = 2|\Delta|N_2 \left(f_2 - \frac{1}{2}\dot\theta f_{1,\varepsilon}^0 \right) + \frac{1}{4\pi}\left(K_{11} - K_{22} + \left[-\frac{\Delta^*}{\varepsilon}K_{12} + \frac{\Delta}{\varepsilon}K_{21} \right] \right). \tag{7.40}$$

The same approximations are used here as in deriving (7.28). Using the relation

$$K_{11} - K_{22} - \left[\frac{\Delta^*}{\varepsilon}K_{12} + \frac{\Delta}{\varepsilon}K_{21} \right] = \frac{4\pi}{N_1}\text{sign}\varepsilon \{J(n_\varepsilon) - J(n_{-\varepsilon})\}, \tag{7.41}$$

one obtains from (7.37), (7.40), and (7.41) the second expression in (7.26).

The potential φ may be restored now in (7.26) with the help of (7.25), where one should make $\dot\chi/2 = -\varphi$. Omitting the term proportional to $\dot\theta\varphi$ [its contribution to (7.26) is small], one finds

$$f_1 = f_1^0 - f_{1,\varepsilon}^0 \frac{R_2}{N_1}\tau_\varepsilon \frac{\partial|\Delta|}{\partial t} + \frac{\varphi^2}{2}f_{1,\varepsilon\varepsilon}^0. \tag{7.42}$$

As for the function f_2, the term quadratic in φ may be omitted—it is proportional to $f_{1,\varepsilon\varepsilon\varepsilon}^0$. The linear term, which takes into account the transformation rule for θ, gives the equation

$$f_2 = -\frac{\varphi N_1 f_{1,\varepsilon}^0 - \tau_\varepsilon|\Delta|N_2\dot\theta f_{1,\varepsilon}^0}{N_1 + 2\tau_\varepsilon|\Delta|N_2}. \tag{7.43}$$

7.1.10 Order Parameter Equation

Utilizing (7.42) and (7.43), the equation for an order parameter [(7.23) and (7.24)] takes the final form

$$-\frac{\pi}{8T_c}\frac{1}{\sqrt{1+(2\tau_\varepsilon|\Delta|)^2}}\left[\frac{\partial}{\partial t} + 2i\varphi + 2\tau_\varepsilon^2\frac{\partial|\Delta|^2}{\partial t} \right]\Delta$$

$$+ \frac{\pi}{8T_c}\left[D(\nabla - 2i\mathbf{A})^2 \right]\Delta + \left[\frac{T_c - T}{T_c} - 7\zeta(3)\frac{(|\Delta|^2 + 2\mu^2)}{8(\pi T_c)^2} + P(|\Delta|) \right]\Delta = 0. \tag{7.44}$$

The function $P(|\Delta|)$ in (7.44) is due to the contribution of nonequilibrium phonon sub-system.

7.1.11 Contribution of Nonequilibrium Phonons

We will now trace the origin of $P(|\Delta|)$. If the phonons are shifted from equilibrium, the collision integral (7.37) acquires the contribution

$$\delta J(n_{\pm\varepsilon})^{(\text{e-ph})} = u_\varepsilon \Gamma(\varepsilon), \tag{7.45}$$

as follows from (7.32). The factor $\Gamma(\varepsilon)$ is the functional, and is linked with the deviation of the phonon distribution function from the equilibrium $\delta N_{\omega_q} = N_{\omega_q} - N^0_{\omega_q}$:

$$\Gamma(\varepsilon) = \frac{\pi\lambda}{2(up_F)^2} \int_0^\infty \omega_q^2 d\omega_q \int_{|\Delta|}^\infty d\varepsilon' \delta(\varepsilon' + \varepsilon - \omega_q)(u_\varepsilon u_{\varepsilon'} + v_\varepsilon v_{\varepsilon'})$$
$$\times (1 - n_\varepsilon - n_{\varepsilon'})\delta N_{\omega_q}. \tag{7.46}$$

This leads to the redefinition of the function f_1 (7.42), which now has the form

$$f_1 = f_1^0 - f_1^0 \frac{R_2}{N_1}\tau_\varepsilon \frac{\partial|\Delta|}{\partial t} + \frac{\varphi^2}{2}f_{1,\varepsilon\varepsilon}^0 - 2\text{sign}\varepsilon\, \tau_\varepsilon \Gamma(\varepsilon). \tag{7.47}$$

The function f_2 (7.43) remains unchanged. Substituting (7.47) into (7.24), one finds for $P(|\Delta|)$ a form

$$P(\Delta) = -2\tau_\varepsilon \,\text{Re} \int_0^\infty d\varepsilon \frac{\Gamma(\varepsilon)}{\sqrt{(\varepsilon + i\gamma)^2 - |\Delta|^2}}. \tag{7.48}$$

In section (7.1.13) we will discuss the contribution to (7.44), introduced by $P(|\Delta|)$.

7.1.12 Galayko's μ^2-Term

Another peculiarity of (7.44), compared with [2–6], is the additional term,[4] which is proportional to μ^2. Such a term was obtained by Galayko [16] in a static limit of the dynamic equations. The presence of the nonlinear μ term in (7.44) is principally

---

[4]This term is presented in a form, which guarantees the gauge-invariance of (7.44), i.e., we have replaced φ^2 by μ^2. We have resorted to this procedure, because at the derivation of (7.44) the higher time-derivatives were not kept. At more consecutive calculations the term φ^2 might be replaced, for instance, by the operator $[-1/4(\partial/\partial t + 2i\varphi)^2]$.

important. If $\mu = 0$, the relation for the gap follows from (7.44):

$$|\Delta| = \Delta_{\text{BCS}} = \Delta_0, \quad \Delta_0 = T_c \left[\frac{8\pi^2}{7\zeta(3)} \left(1 - \frac{T}{T_c} \right) \right]^{1/2}. \tag{7.49}$$

However, if $\mu \neq 0$, the expression for the gap $|\Delta|$ in a spatially homogeneous and stationary case is found from the equation

$$|\Delta| = \sqrt{\Delta_0^2 - 2\mu^2}. \tag{7.50}$$

Hence the initial static pattern cannot exist at $\mu \geq \Delta_0/\sqrt{2}$ (this was first pointed out by Galayko [17]).

Based on the assumption that the behavior of superconductors in a nonstationary steady-state has a close analogy with the usual thermodynamics (this principle was discussed in [18] for superconductors; see also [19] for more general cases), one can write the free energy functional of the Ginzburg–Landau type for the (7.44) discarding the first (dynamic) term. Considering μ as a parameter in this functional, it is easy to verify that the absolute minimum of the functional is obtained at $\mu = 0$. Thus in thermodynamic equilibrium, the value of μ vanishes.

7.1.13 Phonons and Order Parameter Dynamics

Now we return to the definition (7.44) of the function $P(|\Delta|)$, which contains the nonequilibrium addition $\delta N_{\omega_{\mathbf{q}}}$ to the phonon distribution function. Substituting the value $\delta N_{\omega_{\mathbf{q}}} \sim N_{\omega_{\mathbf{q}}}^0$ into (7.44), one would find the value of $P(|\Delta|)$ to be an order of unity that greatly exceeds all other terms in (7.44). In reality, however, the value of $\delta N_{\omega_{\mathbf{q}}}$ must be determined from the phonon kinetic equation, which has the form

$$\frac{\mathrm{d}}{\mathrm{d}t}(\delta N_{\omega_{\mathbf{q}}}) = I(N_{\omega_{\mathbf{q}}}) + L(N_{\omega_{\mathbf{q}}}), \tag{7.51}$$

where $I(N_{\omega_{\mathbf{q}}})$ is the phonon-electron collision integral, and $L(N_{\omega_{\mathbf{q}}})$ is the operator describing the phonon exchange of a superconductor with its environment (the heat-bath). In the simplest approximation [20, 21], the latter may be defined as

$$L(N_{\omega_q}) \approx -\frac{\delta N_{\omega_{\mathbf{q}}}}{\tau_{\text{es}}}, \tag{7.52}$$

where $\tau_{\text{es}} \sim d/u$ is the phonon escape time (into the heat-bath), and d is the characteristic dimension of the superconductor. The inelastic collision integral $I(N_{\omega_{\mathbf{q}}}) = I(N_{\omega_{\mathbf{q}}})^{(\text{ph}-\text{e})}$ was derived in Sect. 6.4 in the form

$$I(N_{\omega_{\mathbf{q}}})^{(\text{ph}-\text{e})} = \frac{\pi\lambda}{8}\frac{\omega_D}{\epsilon_F}\int_{|\Delta|}^{\infty}\int_{|\Delta|}^{\infty} d\varepsilon d\varepsilon'\left\{\delta(\varepsilon + \varepsilon' - \omega_{\mathbf{q}})s_1 + 2\delta(\varepsilon - \varepsilon' - \omega_{\mathbf{q}})s_2\right\}.$$

(7.53)

Moving in this expression to the functions f_1 and f_2 in the local equilibrium approximation (7.47) and (7.43) [omitting the term with φ^2 in (7.47)], and expressing u_ε and v_ε through N_1 (7.16) and R_2 (7.20), respectively, one arrives at

$$I(N_{\omega_{\mathbf{q}}})^{(\text{ph}-\text{e})} \approx \frac{\pi\lambda}{2}\frac{\omega_D}{\epsilon_F}\left\{2N_{\omega_{\mathbf{q}}}^0\left[\frac{\partial|\Delta|}{\partial t}\frac{\tau_\varepsilon}{T}\eta_1 + \eta_3\right] - \eta_2\delta N_{\omega_{\mathbf{q}}}\right\},$$

(7.54)

where the functions η_1, η_2, and η_3 are defined by relations

$$\eta_1 = \frac{1}{4}\int_0^{\infty} d\varepsilon \frac{P(\varepsilon)R_2(\varepsilon)}{N_1(\varepsilon)\cosh^2(\varepsilon/2T)}$$
$$- \frac{1}{4}\int_0^{\infty} d\varepsilon\, Q(\varepsilon)\left\{\frac{R_2(\varepsilon + \omega_q)}{N_1(\varepsilon + \omega_q)\cosh^2[(\varepsilon + \omega_q)/2T]} - \frac{R_2(\varepsilon)}{N_1(\varepsilon)\cosh^2(\varepsilon/2T)}\right\},\quad (7.55)$$

$$\eta_2 = \int_0^{\infty} d\varepsilon\, P(\varepsilon)\tanh\frac{\varepsilon}{2T} + \int_0^{\infty} d\varepsilon\, Q(\varepsilon)\left(\tanh\frac{\varepsilon + \omega_q}{2T} - \tanh\frac{\varepsilon}{2T}\right),\quad (7.56)$$

$$\eta_3 = \tau_\varepsilon\left\{\int_0^{\infty} d\varepsilon\, P(\varepsilon)\Gamma(\varepsilon) + \int_0^{\infty} d\varepsilon\, Q(\varepsilon)\left[\Gamma(\varepsilon + \omega_{\mathbf{q}}) - \Gamma(\varepsilon)\right]\right\}\quad (7.57)$$

with

$$P(\varepsilon) = N_1(\varepsilon)N_1(\omega_{\mathbf{q}} - \varepsilon) + R_2(\varepsilon)R_2(\omega_{\mathbf{q}} - \varepsilon),\quad (7.58)$$
$$Q(\varepsilon) = N_1(\varepsilon)N_1(\omega_{\mathbf{q}} + \varepsilon) - R_2(\varepsilon)R_2(\omega_{\mathbf{q}} + \varepsilon).\quad (7.59)$$

These relations subject to (7.46) allow one to find $\delta N_{\omega_{\mathbf{q}}}(t)$ and to study the interplay between the dynamics of the order parameter and nonequilibrium phonons. We will define a "generalized local equilibrium approximation" (between the pair condensate, electron excitations, and phonons) as the approximation in which (besides the fulfillment of the conditions of the local equilibrium approximation) the characteristic frequencies (and wave vectors) of variations of $N_{\omega_{\mathbf{q}}}$ are small compared with $\lambda\omega_D T/\epsilon_F$, so the left side of (7.51) may be neglected. In this case the function $\delta N_{\omega_{\mathbf{q}}}$ depends on $\mathbf{r}$ and t implicitly, through $\Delta(\mathbf{r}, t)$. From (7.51) to (7.59) it follows that

$$\delta N = \left\{\pi\lambda\frac{\omega_D}{\epsilon_F}N_{\omega_{\mathbf{q}}}^0\left[\frac{\partial|\Delta|}{\partial t}\frac{\tau_\varepsilon}{T}\eta_1 + \eta_3\right]\right\}\Big/\left(\eta_2\frac{\pi\lambda\omega_D}{2\epsilon_F} + \tau_{es}^{-1}\right).$$

(7.60)

The solution of the integral equation (7.60) may be sought in the form

$$\delta N_{\omega_{\mathbf{q}}} = K(\omega_q)\frac{\partial|\Delta|}{\partial t}.$$

(7.61)

In doing this, (7.60) transforms into the equation

$$K(\omega_{\mathbf{q}}) = \left\{ \pi\lambda \frac{\omega_D}{\epsilon_F} N^0_{\omega_{\mathbf{q}}} \left[\frac{\tau_\varepsilon}{T} \eta_1 + \eta_4 \right] \right\} / + \left(\eta_2 \frac{\pi\lambda\omega_D}{2\epsilon_F} + \tau^{-1}_{es} \right), \qquad (7.62)$$

where $\eta_4 = \eta_3[\delta N_{\omega_{\mathbf{q}}} \to K(\omega_{\mathbf{q}})]$. The function $K(\omega_{\mathbf{q}})$ depends on $|\Delta|$, γ, and τ_{es} parametrically and may be found from (7.62) by numerical methods. Rough estimates based on (7.62) show that

$$\delta N_{\omega_{\mathbf{q}}} \sim \frac{|\Delta|}{\gamma T} N^0_{\omega_{\mathbf{q}}}. \qquad (7.63)$$

Substituting (7.63) into (7.46) and (7.48), one can see that the quantity $P(|\Delta|)$ in (7.44) significantly renormalizes the term, which is proportional to $\partial|\Delta|/\partial t$ (this term changes by its order of magnitude at $\tau_{es} \to \infty$). The accurate evaluation of this term is outside the scope of this analysis. A detailed investigation is necessary for situations where the conditions of the generalized local equilibrium approximation are violated; in those cases, the value in (7.63) may turn out to be underestimated.

Consider now the limiting case $\tau_{es} \to 0$, when according to (7.51) and (7.52) $\delta N_{\omega_{\mathbf{q}}} \to 0$. This condition is fulfilled when $d < \xi_0$; for example, in the case of a superconducting film or filament (see the discussion in Sect. 5.2). The phonon radiation from the superconductor into the surrounding medium (the heat-bath) is then determined by (7.52).

According to the results of our Seminar 3, the intensity of the phonon flux emitted by the volume V in a spectral range $d\omega_{\mathbf{q}}$ is

$$dW_{\omega_{\mathbf{q}}} = I(N^0_{\omega_{\mathbf{q}}})^{(ph-e)} \rho(\omega_{\mathbf{q}}) d\omega_{\mathbf{q}}, \qquad (7.64)$$

where $\rho(\omega_{\mathbf{q}}) = V\omega^3_{\mathbf{q}}/(2\pi^2 u^3)$. Using expression (7.52) for $I(N_{\omega_{\mathbf{q}}})^{(ph-e)}$, one obtains

$$dW_{\omega_{\mathbf{q}}} = \frac{\pi\lambda}{2} \frac{\omega_D}{\epsilon_F} \left\{ 2 \frac{\partial|\Delta|}{\partial t} \frac{\tau_\varepsilon}{T} N^0_{\omega_{\mathbf{q}}} \eta_1 \right\} \rho(\omega_{\mathbf{q}}) d\omega_{\mathbf{q}}. \qquad (7.65)$$

Thus, any variation in the order parameter modulus is accompanied by the exchange of phonons between the superconductor and the heat-bath (i.e., the emission or absorption of phonons is taking place).

7.2   Interference Current

An expression for a nonstationary current enters the set of TDGL-equations. As we will see in this section, the current in nonequilibrium superconductors in the vicinity of T_c consists of superfluid, normal, and interference components.

7.2.1 Usadel Approximation

The expression for the current may be derived from (4.85) by the method of analytical continuation (see the discussion at the end of Sect. 5.3). In our notation it has the form

$$
\begin{aligned}
\mathbf{j}(\mathbf{r}, t) &= -\frac{1}{2} N(0) \int_{-\infty}^{\infty} \frac{d\varepsilon}{4\pi i} \int \frac{d\Omega_{\mathbf{p}}}{4\pi} \frac{\mathbf{p}}{m} \mathrm{Tr} \left[\hat{\sigma}_z \frac{\mathbf{p}}{m} \widehat{\mathbf{g}}_p(\mathbf{r}, t) \right] \\
&= i \frac{N(0) p_F^2}{12\pi m^2} \int_{-\infty}^{\infty} d\varepsilon \, \mathrm{Tr} \left[\hat{\sigma}_z \widehat{\mathbf{g}}_p(\mathbf{r}, t) \right],
\end{aligned}
\tag{7.66}
$$

where $\widehat{\mathbf{g}}_p(\mathbf{r}, t)$ is the Keldysh vector-part of the energy-integrated matrix Green-Gor'kov $\breve{g}$ function (7.2). In the Usadel approximation [22], the $\breve{g}$ function may be assumed to be in the form ($\breve{g}_S$ is the isotropic part of $\breve{g}$)

$$
\breve{g} = \breve{g}_S + \frac{\mathbf{p}}{m} \cdot \breve{\mathbf{g}}_p.
\tag{7.67}
$$

Because the self-energy parts of the interaction of electrons with impurities may be written as (see Sect. 4.1):

$$
\widehat{\Sigma}_{\mathrm{imp}}^{R(A)} = \frac{1}{2\pi\tau} \int \frac{d\Omega_{\mathbf{p}}}{4\pi} \widehat{g}^{R(A)},
\tag{7.68}
$$

where τ is the transport mean free path time, one can show that in the adopted normalization (7.8) the solution of kinetic equation (7.1) for the vector harmonic $\breve{\mathbf{g}}_p$ is expressed as

$$
\breve{\mathbf{g}}_p = -\frac{i}{\pi} \tau \left(\breve{g}_S * \breve{\partial} * \breve{g}_S + \pi^2 \breve{\partial} \right),
\tag{7.69}
$$

where

$$
\breve{\partial} = \breve{1} \frac{\partial}{\partial \mathbf{r}} - i \breve{\sigma}_z \mathbf{A},
\tag{7.70}
$$

and the isotropic part $\breve{g}_S$ in (7.69) obeys the relations

$$
\breve{g}_S * \breve{g}_S = -\pi^2 \cdot \breve{1},
\tag{7.71}
$$

$$
\left[\breve{g}_S * \breve{\mathbf{g}}_p \right]_- = \breve{0}.
\tag{7.72}
$$

On the base of (7.66) to (7.72) we have

$$
\begin{aligned}
\mathbf{j} = -\frac{N(0) D}{4\pi^2} \int_{-\infty}^{\infty} d\varepsilon \, \mathrm{Tr} \, \hat{\sigma}_z \{ \widehat{g}^R * \widehat{\partial} * \widehat{g}^R * \widehat{a} - \widehat{g}^R * \widehat{\partial} * \widehat{a} * \widehat{g}^A \\
+ \widehat{g}^R * \widehat{a} * \widehat{\partial} * \widehat{g}^A - \widehat{a} * \widehat{g}^A * \widehat{\partial} * \widehat{g}^A \},
\end{aligned}
\tag{7.73}
$$

where

$$\hat{\partial} = \hat{1} \cdot \frac{\partial}{\partial \mathbf{r}} - i\mathbf{A}\hat{\sigma}_z, \tag{7.74}$$

and the spectral functions $\hat{g}^{R(A)}$ (5.66), according to (7.16), and (7.19) to (7.21), are[5]

$$
\begin{aligned}
\hat{g}^{R(A)} &= \overset{+}{_{(-)}} \frac{i\pi}{\sqrt{\left(\varepsilon \overset{+}{_{(-)}} i\gamma\right)^2 - |\Delta|^2}} \begin{pmatrix} \varepsilon \overset{+}{_{(-)}} i\gamma & \Delta \\ -\Delta^* & -(\varepsilon \overset{+}{_{(-)}} i\gamma) \end{pmatrix} \\
&\equiv \overset{+}{_{(-)}} \begin{pmatrix} N_1 \overset{+}{_{(-)}} i R_1 & e^{i\theta}(R_2 \overset{-}{_{(+)}} i N_2) \\ e^{-i\theta}(R_2 \overset{-}{_{(+)}} i N_2) & -(N_1 \overset{+}{_{(-)}} i R_1) \end{pmatrix}.
\end{aligned} \tag{7.75}
$$

In further transformations it is assumed that T is close to T_c, so the following inequalities are held:

$$(\gamma, |\Delta|) \ll T. \tag{7.76}$$

This means in particular that the function γ does not depend on ε. We also assume that the functions f_1 and f_2 do not explicitly depend on time, the terms with the derivatives $R_{i,\varepsilon}$, $N_{i,\varepsilon}$, $\nabla f_{1,\varepsilon}$ and terms with higher order derivatives and their products (whose contributions to the current are small) are omitted. The symmetry properties of the integrand are taken into account (R_i is an odd function of ε, and N_i is an even function of ε). Note also that in calculating the trace in (7.73) several of the terms can be reduced to total differentials, which vanish upon integration. Furthermore, as follows directly from (7.76), the following identities hold:

$$N_1^2 + N_2^2 - R_1^2 - R_2^2 = 1, \qquad R_1 N_1 + R_2 N_2 = 0. \tag{7.77}$$

On the basis of the above arguments, one finds the resulting expression for the significant (even in ε) part of the trace in (7.73):

$$
\begin{aligned}
\mathrm{Tr}(\ldots) = -4\pi^2 \Big\{ &\left(\mathbf{A} - \frac{1}{2}\nabla\theta\right)\left[4 R_1 N_1 f_1 - f_{1,\varepsilon}(\dot{R}_2^2 - \dot{N}_2^2)\right] \\
&- \left(\dot{\mathbf{A}} - \frac{1}{2}\nabla\dot\theta\right) f_{1,\varepsilon}(N_1^2 + R_2^2) + \left(\nabla f_2 - \frac{1}{2} f_{1,\varepsilon}\nabla\dot\theta\right)(N_1^2 + N_2^2) \Big\}, \tag{7.78}
\end{aligned}
$$

where the upper dot denotes partial time derivative, and $\dot{a}^2 \equiv \partial(a^2)/\partial t$. Defining the superfluid momentum by the usual relation

$$\mathbf{Q} = 2m\mathbf{v}_s = \nabla\theta - 2\mathbf{A}, \tag{7.79}$$

---

[5] In writing the spectral functions (7.75) we have completely ignored the influence of external fields $\mathbf{A}$ and φ, thus the expression (7.75) corresponds actually to the gauge $\varphi = 0$. For an arbitrary gauge with $\varphi \neq 0$ the function $\hat{g}^{(R,A)}$ (and, in particular, N_1) alter (see Sect. 5.3). This, however, produces no substantial changes in the expression for the current in quasiclassical approximation.

the expression for the current can be presented as

$$
\mathbf{j} = \sigma_n \int_{-\infty}^{\infty} d\varepsilon \left\{ \mathbf{Q} R_2 N_2 f_1 + \frac{1}{4} \mathbf{Q} (N_1^2 + N_2^2) f_{1,\varepsilon} \right.
$$
$$
\left. + \frac{1}{2} (N_1^2 + N_2^2)(\nabla f_2 - \frac{1}{2} f_{1,\varepsilon} \nabla \dot{\theta}) + \frac{1}{4} f_{1,\varepsilon} \frac{\partial}{\partial t} \left[\mathbf{Q}(R_2^2 - N_2^2) \right] \right\}, \quad (7.80)
$$

where the normal conductivity σ_n is

$$
\sigma_n = 2N(0)D = \frac{2}{3} N(0) v_F^2 \tau. \tag{7.81}
$$

At this stage we see that in the gauge $\dot{\theta} = 0$ expression (7.80) coincides with Schmid's result [9]. The last term in (7.80) with the time derivative (which was omitted in [2–6]) vanishes, if the dispersion dependence of $f_{1,\varepsilon}$ is ignored. Substituting the equilibrium value $f_1 = f_1^0(\varepsilon)$ into this term produces a nonzero result, which contains an additional small factor $|\Delta|/T$. Since this term is also proportional to another small parameter ω/T, we omit it below. Expression (7.80) is fundamental for further analysis. Because it has been derived here in an arbitrary gauge, one can be assured that the calculation scheme is self-consistent. The functions $f_1(\varepsilon) = (1 - n_\varepsilon - n_{-\varepsilon}) \text{sign}\varepsilon$ and $f_2(\varepsilon) = -\text{sign}\varepsilon \, (n_\varepsilon - n_{-\varepsilon})/N_1$ in (7.80) should generally be determined from the kinetic equation for the distribution of the nonequilibrium electron-hole excitations n_ε. In many cases, however, it is sufficient to substitute the equilibrium function n_ε^0 into (7.80). As was noticed in [23] this procedure was not carried out in [2–6, 9] sufficiently correct. Thus, certain terms whose contribution is sometimes not small were omitted from the final equation for the current. We will analyze the situation in more details below.

To transform the terms containing $\dot{\theta}$ and ∇f_2 in (7.80), we use the definitions of the gauge-invariant potential

$$
\mu = \frac{1}{2}\dot{\theta} + \varphi \tag{7.82}
$$

and the associated electric field

$$
\mathbf{E} = -\dot{\mathbf{A}} - \nabla\varphi = \frac{1}{2}\dot{\mathbf{Q}} - \nabla\mu. \tag{7.83}
$$

As it follows from (5.77), in the presence of a potential φ, the function f_2 is nonzero and for $\varepsilon \gg |\Delta|$ is equal to

$$
f_2 = -\varphi f_{1,\varepsilon}. \tag{7.84}
$$

Substitution of (7.84) into (7.14) leads to

$$
\mathbf{j} = \sigma_n \int_{-\infty}^{\infty} d\varepsilon \left\{ \mathbf{Q} R_2 N_2 f_1 + \frac{1}{2} f_{1,\varepsilon} (N_1^2 + N_2^2) \mathbf{E} \right\}. \tag{7.85}
$$

In equilibrium theory, the current in dirty superconductors is given by the first term in (7.85), where one should make $f_1 = f_1^0(\varepsilon) = \tanh(\varepsilon/2T)$ (i.e., in an equilibrium situation, the term, which is proportional to T, vanishes). In the nonequilibrium case, two additional groups of terms arise if one inserts the equilibrium function $f_1^0(\varepsilon)$ into (7.85). The reason for this is the relation:

$$N_1^2 + N_2^2 = \frac{1}{2}\left\{1 + \left[1 - \left(\frac{2\varepsilon|\Delta|}{\varepsilon^2 + \gamma^2 + |\Delta|^2}\right)^2\right]^{-1/2}\right\}, \qquad (7.86)$$

which follows directly from (7.16) and (7.10) to (7.21). The integral (7.85), taking into account (7.86) and inequalities in (7.76), can be evaluated in analytic form. In the time-dependent theory, it is necessary to evaluate this integral for an arbitrary ratio of $|\Delta|$ and γ. The equilibrium value of $|\Delta|$ in a "finite-gap" superconductor is large in comparison with γ, but in the dynamic case $|\Delta(\mathbf{r}, t)|$ may sometime vanish at some points!

We will inspect the integrals in (7.85) in more details. If $\gamma \ll |\Delta|$, the factor $R_2 N_2$ acts in fact as the δ-function of the argument ($\varepsilon \pm |\Delta|$):

$$N_2 R_2 \approx \frac{\pi}{2}|\Delta|\varepsilon\delta(\varepsilon^2 - |\Delta|^2), \qquad (7.87)$$

and thus the first term in (7.85) gives

$$\sigma_n \mathbf{Q} \int_{-\infty}^{\infty} d\varepsilon \, R_2 N_2 f_1^0 \approx \frac{\sigma\pi}{4T}\mathbf{Q}|\Delta|^2 \qquad (7.88)$$

(this result does not depend on $\gamma/|\Delta|$ and holds for arbitrary $|\Delta|$ and γ, even if δ-function becomes "smeared"). The second term in (7.85), taking into account (7.86), takes the form

$$\frac{\sigma_n \mathbf{E}}{2}\int_{-\infty}^{\infty} d\varepsilon \, f_{1,\varepsilon}^0(N_1^2 + N_2^2) = \frac{\sigma \mathbf{E}}{2T}\int \frac{d\varepsilon \, [\varepsilon^2 + (\gamma^2 + |\Delta|^2)]}{\cosh^2(\varepsilon/2T)[\varepsilon^4 + 2\varepsilon^2(\gamma^2 - |\Delta|^2) + (\gamma^2 + |\Delta|^2)^2]^{1/2}}. \qquad (7.89)$$

Expression (7.89) can be treated as the sum of two integrals

$$\frac{\sigma_n \mathbf{E}}{2T}\left\{\int_0^{\varepsilon^*} d\varepsilon \, M(\varepsilon) + \int_{\varepsilon^*}^{\infty} d\varepsilon \, M(\varepsilon)\right\}, \qquad (7.90)$$

$$M(\varepsilon) = \frac{\varepsilon^2 + \gamma^2 + |\Delta|^2}{\cosh^2(\varepsilon/2T)[\varepsilon^4 + 2\varepsilon^2(\gamma^2 - |\Delta|^2) + (\gamma^2 + |\Delta|^2)^2]^{1/2}}, \qquad (7.91)$$

where ε^* satisfies the relation (recall, that $T \gg |\Delta|, \gamma$):

$$(\gamma^2 + |\Delta|^2)^{1/2} \ll \varepsilon^* \ll T. \qquad (7.92)$$

Using the relation (7.92), one can expand $\cosh^2(\varepsilon/2T)$ in the first integral in (7.90), keeping only the lowest order term in a small parameter $\varepsilon/2T$, and use for another integral the approximation

$$\frac{\varepsilon^2 + \gamma^2 + |\Delta|^2}{[\varepsilon^4 + 2\varepsilon^2(\gamma^2 - |\Delta|^2) + (\gamma^2 + |\Delta|^2)^2]^{1/2}} \approx 1, \quad \varepsilon^* \le \varepsilon < \infty. \tag{7.93}$$

Thus the second term in (7.85) takes the form

$$\frac{\sigma_n \mathbf{E}}{2} \int_{-\infty}^{\infty} d\varepsilon \, f_{1,\varepsilon}^0 (N_1^2 + N_2^2)$$

$$= \frac{\sigma_n \mathbf{E}}{2T} \left\{ \int_0^{\varepsilon^*} \frac{d\varepsilon[\varepsilon^2 + \gamma^2 + |\Delta|^2]}{[\varepsilon^4 + 2\varepsilon^2(\gamma^2 - |\Delta|^2) + (\gamma^2 + |\Delta|^2)^2]^{1/2}} + \int_{\varepsilon^*}^{\infty} \frac{d\varepsilon}{\cosh^2(\varepsilon/2T)} \right\}. \tag{7.94}$$

One can now integrate (7.94) directly and find for (7.95):

$$\mathbf{j} = \frac{\pi \sigma_n}{4T} \mathbf{Q}|\Delta|^2 + \sigma_n \mathbf{E} \left\{ 1 + \frac{\sqrt{|\Delta|^2 + \gamma^2}}{2T} \left[K\left(\frac{|\Delta|}{\sqrt{|\Delta|^2 + \gamma^2}} \right) - E\left(\frac{|\Delta|}{\sqrt{|\Delta|^2 + \gamma^2}} \right) \right] \right\}. \tag{7.95}$$

where $K(x)$ and $E(x)$ are the complete elliptic integrals of the first and second type, respectively. In the limiting case they have the following asymptotic forms:

$$x \ll 1 : \quad \begin{aligned} K(x) &\simeq \frac{\pi}{2} \left(1 + \frac{x^2}{4} + \cdots \right), \\ E(x) &\simeq \frac{\pi}{2} \left(1 - \frac{x^2}{4} + \cdots \right), \end{aligned} \tag{7.96}$$

and

$$x \simeq 1 : \quad \begin{aligned} K(x) &\simeq \ln \frac{4}{\sqrt{1-x^2}} + \cdots, \\ E(x) &\simeq 1 - \frac{1}{2}(1-x)\ln(1-x) + \cdots. \end{aligned} \tag{7.97}$$

A good approximation for the difference function is:

$$K(x) - E(x) \cong \frac{\ln(1+x) - \ln(1-x)}{2} + (1-x)\ln(1-x) \tag{7.98}$$

$$\equiv \frac{\ln(1-x^2)}{2} - x\ln(1-x).$$

This is illustrated in Fig. 7.1.

Fig. 7.1 Logarythmic approximation (7.98) to exact difference of elliptic functions

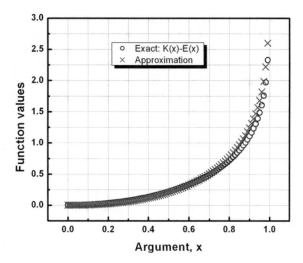

7.2.2 Normal Flow Contribution to Interference Current

Expression (7.95) may now be written in the form

$$\mathbf{j} = \mathbf{j}_s + \mathbf{j}_n + \mathbf{j}_{\text{int}}, \tag{7.99}$$

where the superfluid and normal components of the current are given by the standard relations

$$\mathbf{j}_s = \frac{\pi \sigma_n}{4T} \mathbf{Q} |\Delta|^2, \qquad \mathbf{j}_n = \sigma_n \mathbf{E}, \tag{7.100}$$

and the "interference" component is

$$\mathbf{j}_{\text{int}} = \sigma_n \mathbf{E} \frac{\sqrt{|\Delta|^2 + \gamma^2}}{2T} \left[K\left(\frac{|\Delta|}{\sqrt{|\Delta|^2 + \gamma^2}} \right) - E\left(\frac{|\Delta|}{\sqrt{|\Delta|^2 + \gamma^2}} \right) \right]. \tag{7.101}$$

The quantity $\mathbf{j}_{\text{int}}$ (7.101) has properties of both the superconducting condensate and the normal excitations. In fact, it describes some interference of two types of motion occurring in the electron subsystem of the superconductor.

A comparison of (7.101) with (7.100) shows that the interference component of the current is not always negligible. Using the asymptotic forms (7.96) and (7.97) of the elliptic integrals, one can easily show that (7.95) takes the following forms in the specified limiting cases

$$\mathbf{j} = \frac{\pi \sigma_n}{4T} |\Delta|^2 \mathbf{Q} + \sigma_n \mathbf{E} \left\{ 1 + \frac{|\Delta|}{2T} \left(\ln \frac{4|\Delta|}{\gamma} - 1 \right) \right\}, \quad \gamma \ll |\Delta|, \tag{7.102}$$

$$\mathbf{j} = \frac{\pi \sigma_n}{4T} |\Delta|^2 \mathbf{Q} + \sigma_n \mathbf{E}, \qquad \gamma \gg |\Delta|. \tag{7.103}$$

Equation (7.103) is similar to the expression for a gapless superconductor [see (4.121)]; in this case the current consists of the normal and superconducting components only. Thus, the interference term in the "finite-gap" superconductor stems from the strong correlation between the system of single-particle excitations and the pair condensate. This correlation vanishes in a gapless regime.

Using (7.102) and (7.103), we can represent (7.101) in the following approximation:

$$\mathbf{j}_{\text{int}} = \sigma_n \mathbf{E} \frac{|\Delta|}{2T} \left[\frac{\ln(1 - x^2)}{2x} - \ln(1 - x) \right], \quad x = \frac{|\Delta|}{\sqrt{|\Delta|^2 + \gamma^2}}. \tag{7.104}$$

This approximation is convenient for practical calculations.

Note, that a logarithmic renormalization of conductivity, analogous to (7.102), appears in the theory of both linear [24, 25] and nonlinear [26, 27] responses of a superconductor in a time-varying external electromagnetic field of the frequency ω; for example,

$$\sigma_*(\omega_0) = \sigma_n \left(1 + \frac{|\Delta|}{2T} \ln \Lambda \right), \quad \Lambda = \max(\omega_0 \tau_{\text{imp}}, \frac{\omega_0}{|\Delta|}). \tag{7.105}$$

Such a logarithmic renormalization of conductivity also reflects the interference between normal and superfluid motions. Although the parameter $|\Delta|/T$ near T_c is small, the corrections might be not negligible, because the logarithmic factor can, in principle, be large. We should also mention that the interference described above is closely related to the "dragging" process, investigated by Shelankov [28].

7.2.3   Condensate Contribution to Interference Current

We used above the equilibrium approximation for the functions f_1 and f_2. To find these functions [(7.42) and (7.43)] in the time-dependent theory, the nonequilibrium contributions must be taken into account. They may be expressed in the form

$$f_1 = f_1^0(\varepsilon) + \delta f_1(\varepsilon), \quad \delta f_1(\varepsilon) = -f_{1,\varepsilon}^0 \frac{R_2}{N_1} \frac{2}{\gamma} \frac{\partial |\Delta|}{\partial t}, \tag{7.106}$$

$$f_2 = -\varphi f_1^0(\varepsilon) + \delta f_2(\varepsilon), \quad \delta f_2(\varepsilon) = -2\mu \frac{N_2 \tau_\varepsilon |\Delta|}{N_1 + 2N_2 \tau_\varepsilon |\Delta|} f_{1,\varepsilon}^0. \tag{7.107}$$

The current component due to the function $\delta f_2(\varepsilon)$ in (7.85) is vanishingly small and can be ignored. However, the function $\delta f_1(\varepsilon)$, whose contribution though small in comparison with $\mathbf{j}_s$, is dissipative. In general, this component need not be small in comparison with $\mathbf{j}_n$. The resulting current is given by the following expression in the "local equilibrium approximation":

$$\mathbf{j} = \frac{\pi\sigma_n}{4T}\mathbf{Q}\left(|\Delta|^2 - \frac{\partial|\Delta|^2}{\gamma\partial t}\right)$$
$$+ \sigma_n\mathbf{E}\left\{1 + \frac{\sqrt{|\Delta|^2 + \gamma^2}}{2T}\left[K\left(\frac{|\Delta|}{\sqrt{|\Delta|^2 + \gamma^2}}\right) - E\left(\frac{|\Delta|}{\sqrt{|\Delta|^2 + \gamma^2}}\right)\right]\right\}. \quad (7.108)$$

This expression should be used in the Ginzburg–Landau equations instead of those presented in [2–6].

Thus, in the expression for current in the Ginzburg–Landau regime

$$\mathbf{j} = \mathbf{j}_s + \mathbf{j}_n + \mathbf{j}_{\text{int}}, \quad (7.109)$$

in addition to nondissipative quantum electronic motion

$$\mathbf{j}_s = \frac{\pi\sigma_n}{4T}\mathbf{Q}|\Delta|^2, \quad (7.110)$$

and dissipative normal motion of electrons

$$\mathbf{j}_n = \sigma_n\mathbf{E}, \quad (7.111)$$

we have motion of charges which has interference of quantum features and dissipative features:

$$\mathbf{j}_{\text{int}} = -\frac{\pi\sigma_n}{4T}\frac{\mathbf{Q}}{\gamma}\frac{\partial|\Delta|^2}{\partial t} + \sigma_n\mathbf{E}\frac{\sqrt{|\Delta|^2 + \gamma^2}}{2T}\left[K\left(\frac{|\Delta|}{\sqrt{|\Delta|^2 + \gamma^2}}\right) - E\left(\frac{|\Delta|}{\sqrt{|\Delta|^2 + \gamma^2}}\right)\right].$$
$$(7.112)$$

This current (7.109) enters the Maxwell set of equations, which should supplement the equation for the order parameter Δ, (7.44). After separating the real and imaginary parts of (7.44) one finds[6]:

$$-\frac{\pi}{8T_c}\sqrt{1 + (2\tau_\varepsilon|\Delta|)^2}\frac{\partial|\Delta|}{\partial t} + \frac{\pi}{8T_c}D(\nabla^2 - \mathbf{Q}^2)|\Delta|$$
$$+ \left[\frac{T_c - T}{T_c} - 7\zeta(3)\frac{|\Delta|^2 + 2\mu^2}{8(\pi T_c)^2}\right]|\Delta| = 0, \quad (7.113)$$

$$-\frac{2|\Delta|^2}{\sqrt{1 + (2\tau_\varepsilon|\Delta|)^2}}\mu + D\,\text{div}(\mathbf{Q}|\Delta|^2) = 0. \quad (7.114)$$

Note that in the equilibrium Ginzburg–Landau scheme (7.114) coincides with the continuity equation because $\mu \equiv 0$, $\mathbf{j} \equiv \mathbf{j}_s$, and $\dot{\rho} \equiv 0$ in that case. In nonequilibrium conditions, (7.114) and the continuity equation $\text{div}\,\mathbf{j} + \dot{\rho} = 0$ are independent. In the waste majority of cases we still can consider $\dot{\rho} \equiv 0$ (since any charge deviation equilibrates to $\rho = 0$ at time scales defined by the plasma frequency, while the super-

---

[6]We omit further the term $P(|\Delta|)$.

conducting time scales are much longer) and approximate the continuity equation
by

$$\operatorname{div}\mathbf{j} = 0. \tag{7.115}$$

By the same reason, one can drop the displacement current and present the Maxwell
equation as

$$\operatorname{curl}\mathbf{B} = 4\pi\mathbf{j}. \tag{7.116}$$

In both of these equations $\mathbf{j}$ enters in the form of (7.109).

7.2.4 Boundary Conditions

This set of equations must be supplemented by the boundary conditions, which may
differ in various problems. For instance, at the boundary between a superconductor
and a normal metal, one can write

$$\left.\frac{\partial\Delta}{\partial\mathbf{n}}\right|_S = \beta\Delta|_S, \tag{7.117}$$

where β is some constant, usually taken as $\beta = (\alpha\xi_0)^{-1}$ ($\alpha \approx 0.81$ in equilibrium
approximation, ξ_0 is the coherence length). In nonequilibrium conditions, α may differ from this value (see [29]), but remains of an order of unity. At the superconductor-
vacuum boundary, the following conditions are reasonable:

$$\frac{\partial\Delta}{\partial\mathbf{n}} = 0, \quad Q_n = 0, \quad E_n = 0, \tag{7.118}$$

where $\mathbf{n}$ is the vector normal to the superconductor's surface. One should also require
the continuity of the magnetic field $\mathbf{B}$ and of the tangential component of electric
field $\mathbf{E}$ [30]. Other boundary conditions are also plausible.

We conclude this section by mentioning that in the gapless ($\tau_\varepsilon \to 0$) case the
interference current disappears. Then the structure of TDGL equations coincides
with that of the equations derived in Chap. 4 (with somewhat different coefficients).
This limit, $\tau_\varepsilon \to 0$, was used for solutions in Part I of this book.

7.3 Fluctuations

We will consider here some characteristic features of fluctuational correction to self-
consistent treatments of superconductivity, such as GL or BCS theory. This reveals
the applicability limits of the self-consistent approach.

7.3.1 Ginzburg's Number

To elucidate the role of fluctuations, we will go back to the free energy functional considered in Sect. 3.2. For simplicity we will perform calculations at $T \gtrsim T_c$ for the normal phase, where the equilibrium value is $\Psi_0 \equiv 0$. Then it is convenient to denote the fluctuating value of the order parameter as Ψ. The fluctuation probability is governed by the expression

$$W \propto \exp(-\delta\mathcal{F}/T), \tag{7.119}$$

where $\delta\mathcal{F}$ is defined by (3.46), (3.45), and (3.31), with $F_n^0 = 0$. Since we expect the fluctuations to be small, it is sufficient to keep the second order expansion terms in the free energy functional:

$$\delta\mathcal{F} = \int_{V_0} \left\{ \alpha |\Psi|^2 + \frac{\hbar^2}{2m_*} \left| \frac{\partial \Psi}{\partial \mathbf{r}} \right|^2 \right\} d^3\mathbf{r}. \tag{7.120}$$

Both terms are positive in (7.120), since $T > T_c$.

Let us now make a Fourier expansion of the fluctuating quantities in the volume V_0 (for simplicity we will take $V_0 \equiv 1$ below):

$$|\Psi(\mathbf{r})| = \sum_{\mathbf{k}} \Psi_{\mathbf{k}} e^{i\mathbf{k}\mathbf{r}}, \quad \left| \frac{\partial \Psi}{\partial \mathbf{r}} \right| = \sum_{\mathbf{k}} i\mathbf{k}\Psi_{\mathbf{k}} e^{i\mathbf{k}\mathbf{r}}. \tag{7.121}$$

Since $|\Psi(\mathbf{r})|$ is real, $\Psi_{-\mathbf{k}} = \Psi_{\mathbf{k}}^*$. Substituting (7.121) into (7.120) and integrating over the volume, we find ($\epsilon_{\mathbf{k}} \equiv \hbar^2 \mathbf{k}^2/2m_*$):

$$\delta\mathcal{F} = \sum_{\mathbf{k}} (\alpha + \epsilon_{\mathbf{k}}) |\Psi_{\mathbf{k}}|^2 \equiv \sum_{\mathbf{k}} \delta\mathcal{F}_{\mathbf{k}}. \tag{7.122}$$

As follows from (7.122), (7.119), and (7.122), fluctuations with different values of $\mathbf{k}$ are statistically independent.

Let us consider now the sum over states (a "partition function"), caused by the fluctuations:

$$Z^{fl} = \sum_{\Psi_{\mathbf{k}}} \exp(-\delta\mathcal{F}/T). \tag{7.123}$$

This yields the fluctuational contribution to the free energy of the system:

$$\mathcal{F}^{fl} = -T \ln Z^{fl} = -T \ln \sum_{\Psi_{\mathbf{k}}} \exp \left(\frac{-\sum_{\mathbf{k}} (\alpha + \epsilon_{\mathbf{k}}) |\Psi_{\mathbf{k}}|^2}{T} \right). \tag{7.124}$$

Performing straightforward transformations, we obtain:

$$\mathcal{F}^{fl} = -T \ln \prod_{\mathbf{k}} \int_0^{\infty} \exp\left(-\frac{(\alpha + \epsilon_{\mathbf{k}})\,|\Psi_{\mathbf{k}}|^2}{T}\right) d\,\mathrm{Im}\,\Psi_{\mathbf{k}}\,d\,\mathrm{Re}\,\Psi_{\mathbf{k}}$$

$$= -T \sum_{\mathbf{k}} \ln\left[\pi \int_0^{\infty} \exp\left(-\frac{(\alpha + \epsilon_{\mathbf{k}})\,|\Psi_{\mathbf{k}}|^2}{T}\right) d\,|\Psi_{\mathbf{k}}|^2\right]$$

$$= -T \sum_{\mathbf{k}} \ln \frac{\pi T}{(\alpha + \epsilon_{\mathbf{k}})} \tag{7.125}$$

[in writing (7.125) we took into account the relation $d\,\mathrm{Im}\,\Psi_{\mathbf{k}}\,d\,\mathrm{Re}\,\Psi_{\mathbf{k}} = 2\pi\,|\Psi_{\mathbf{k}}|\,d\,|\Psi_{\mathbf{k}}|$]. To evaluate the role of the order parameter fluctuations, one can calculate the fluctuational contribution to the heat capacity C^{fl}, which is defined via the general relation

$$C = -T(\partial^2 \mathcal{F}/\partial T^2). \tag{7.126}$$

Since in (7.125) in a variation of T the most important contribution comes from the temperature dependence of α, one can write:

$$C^{fl} \approx -T_c \left(\frac{\partial \alpha}{\partial T}\right)_{T_c}^2 \frac{\partial^2 \mathcal{F}^{fl}}{\partial \alpha^2}, \tag{7.127}$$

or, taking into account (7.125):

$$C^{fl} \approx \left[T_c \left(\frac{\partial \alpha}{\partial T}\right)_{T_c}\right]^2 \sum_{\mathbf{k}} \frac{1}{(\alpha + \epsilon_{\mathbf{k}})^2} = \frac{\left[T_c\,(\partial \alpha/\partial T)_{T_c}\right]^2}{(2\pi)^3} \int \frac{d^3 \mathbf{k}}{(\alpha + \epsilon_{\mathbf{k}})^2}$$

$$= const \, \frac{\left[T_c\,(\partial \alpha/\partial T)_{T_c}\right]^{3/2}}{(\hbar^2/2m_*)^{3/2}} \, |\varepsilon|^{-1/2}, \quad \varepsilon \equiv \left(\frac{T - T_c}{T_c}\right), \tag{7.128}$$

where the constant is a number ~ 1. [One should note that the long-wavelength fluctuations play the most important role in (7.128). Also, hereafter we will use the absolute values of $\delta T \to |T - T_c|$. The symmetry of the behavior of fluctuating quantities within this Ornstein-Zernicke description can be confirmed by direct calculations at $T \lesssim T_c$].

We can now compare C^{fl} with some characteristic equilibrium value, such as the jump in heat capacity $\delta C = C_S - C_N$ at the transition point from a superconducting to a normal state. Using (7.126), (3.40), and (3.46), one can calculate

$$\delta C = \frac{2\left[T_c\,(\partial \alpha/\partial T)_{T_c}\right]^2}{\beta T_c}. \tag{7.129}$$

Thus the fluctuations are small if

$$Gi \equiv \frac{C^{fl}}{\delta C} \simeq \frac{m^3 \beta^2 T_c}{\hbar^6 (\partial \alpha / \partial T)_{T_c}} \ll |\varepsilon| \ll 1. \tag{7.130}$$

As follows from (7.130), the mean-field theory has an applicability range only at small values of the parameter Gi (usually called the "Ginzburg number"). Fortunately, the Gi-number is very small for conventional, low temperature superconductors, and thus the mean-field theory is well applicable even very close vicinity of T_c. Indeed, using for "clean" superconductors the values of $(\partial \alpha / \partial T)_{T_c}$ and β, which follow from (3.161) and (3.166), we find

$$Gi_{\text{clean}} \simeq \left(\frac{T_c}{\epsilon_F}\right)^4 \tag{7.131}$$

[in writing (7.131) we used (3.166), in which the density N of electrons may be expressed as $N = 2[(4/3)\pi p_F^3]/(2\pi\hbar)^3 = p_F^3/3\pi^2\hbar^3]$. Usually $(T_c/\epsilon_F) \sim 10^{-3}$, so that Ginzburg's number is incredibly small for superconductors. To estimate Gi in the case of "dirty" superconductors, we again need the microscopic values of phenomenological parameters (3.36) and (3.38). For these values we will compare (3.48) and (3.51) with (7.44) and (7.100), respectively. It follows then [for completeness we also provide here the relationship between Ψ and Δ, which is analogous to (3.165) for the "dirty" case] that:

$$\left(\frac{\partial \alpha}{\partial T}\right)_{T_c} = \frac{6}{\pi} \frac{\hbar}{\tau_{\text{imp}}} \frac{m}{p_F^2} = \frac{2\hbar}{\pi m D}, \tag{7.132}$$

$$\frac{(\partial \alpha / \partial T)_{T_c}}{\beta} = \frac{2\pi^2}{7\varsigma(3)} \frac{\tau_{\text{imp}}}{\hbar} N, \tag{7.133}$$

$$\Psi(\mathbf{r}) = (\pi/4)^{1/2} (N\tau_{\text{imp}}/\hbar T_c)^{1/2} \Delta(\mathbf{r}), \tag{7.134}$$

and for Gi we obtain

$$Gi_{\text{dirty}} \simeq \left(\frac{\hbar}{\tau_{\text{imp}}}\right)^3 \frac{T_c}{\epsilon_F^4}, \tag{7.135}$$

which is also very small, so that usually the range of temperature fluctuations is not of practical importance. It is worth mentioning once again that the smallness of the Gi-parameter permits us to apply the Ginzburg–Landau type approach to the description of superconductors. At the same time, it is wrong to conclude that the smallness of Gi rules out the possibility of experimental observation of fluctuational phenomena in superconductors: fluctuations may reveal themselves in one- or two-dimensional samples [31–33]. We will treat different mechanisms of resistivity fluctuations in the next Section.

7.3.2 Paraconductivity

Let us suppose that $T \gtrsim T_c$ and there is a constant electric field $\mathbf{E}$ applied to the metal. One can expect then that spontaneous fluctuations of the order parameter create droplets of finite superfluid density, which will be accelerated by the electric field, raising the normal conductivity σ_n. Actually, the change of the conductivity is small: $\delta\sigma \ll \sigma_n$, but the temperature dependence $\delta\sigma(T)$ is peculiar and thus could be detected. Fluctuations of the order parameter may lead also to the specific temperature dependence of the heat capacity in small superconducting particles [34].

Following Schmid [35], we first treat the average (in thermodynamic sense) current, coupled to the applied field, via the relation:

$$\langle \mathbf{j} \rangle = \sigma_S \mathbf{E}, \tag{7.136}$$

where (still hypothetical) conductivity equals to

$$\sigma_S = \frac{e_*^2 N_S \tau^0}{m_*}. \tag{7.137}$$

In (7.137) N_S is the density of electrons fluctuating between normal and superconducting states: $N_S = \langle |\Psi|^2 \rangle$, and τ^0 is the lifetime of electrons in the superconducting state. As noted, fluctuations at different wavelengths that contribute to the free energy (7.122) are statistically independent. In view of that[7]:

$$\langle |\Psi_{\mathbf{k}}|^2 \rangle = \frac{\int_0^\infty |\Psi_{\mathbf{k}}|^2 \exp\left[-(\alpha + \epsilon_{\mathbf{k}}) |\Psi_{\mathbf{k}}|^2 / T\right] \, d\,|\Psi_{\mathbf{k}}|^2}{\int_0^\infty \exp\left[-(\alpha + \epsilon_{\mathbf{k}}) |\Psi_{\mathbf{k}}|^2 / T\right] d\,|\Psi_{\mathbf{k}}|^2} = \frac{T}{(\alpha + \epsilon_{\mathbf{k}})}. \tag{7.138}$$

To obtain the value of τ^0 one should consider TDGL equation (7.42). In the fluctuational regime, $\tau_\varepsilon |\Delta| \ll 1$. Thus all the nonlinear terms, including contributions from the vector potential, as well as the phonon term $P(|\Delta|)$, could be neglected, yielding

$$\eta_0 \left(\frac{\partial}{\partial t} + i \frac{e_*}{\hbar} \varphi \right) \Psi - (\alpha - \frac{\hbar^2}{2m_*} \nabla^2) \Psi = 0, \tag{7.139}$$

where $\eta_0 = [T_c(\partial\alpha/\partial T)_{T_c}](\pi/8T_c) = (4mD)^{-1}$ for "dirty" superconductors. Since in the linear approximation the relaxation time should not depend on the electric field applied, one can discard the scalar potential φ in (7.139), and obtain for the Fourier

---

[7] Actually, there is a degeneracy in the system described by (7.122): the states with $\mathbf{k}$ and $-\mathbf{k}$ are physically identical. Thus in the expression (7.119) for fluctuational probability of $|\Psi_{\mathbf{k}}|^2$ the value of $\delta\mathcal{F}$ doubles. This causes the value of amplitude $\langle |\Psi_{\mathbf{k}}|^2 \rangle$ in (7.138) to be two times smaller: $\langle |\Psi_{\mathbf{k}}|^2 \rangle = T/2(\alpha + \epsilon_{\mathbf{k}})$ (cf., e.g., [36]). To use the explicit form (7.138), one should perform in expressions like (7.153), (7.154) the subsequent integration over $\mathbf{k}$ over a single hemisphere of its values (cf. [37]).

component $\Psi_{\mathbf{k}}$ the equation:

$$\eta_0 \frac{\partial}{\partial t} \Psi_{\mathbf{k}} = -(\epsilon_{\mathbf{k}} + \alpha) \Psi_{\mathbf{k}}. \tag{7.140}$$

For the relaxation time of fluctuating components it follows then that:

$$\tau_{\mathbf{k}}^0 = \frac{\eta_0}{\epsilon_{\mathbf{k}} + \alpha}. \tag{7.141}$$

To find the value of σ_S (7.137), one should compute

$$N_S \tau^0 = \sum_{\mathbf{k}} \langle |\Psi_{\mathbf{k}}|^2 \rangle \tau_{\mathbf{k}}^0 \tag{7.142}$$

As was demonstrated by Aslamazov and Larkin [38], the result of this summation depends on the samples dimensionality. Indeed, since the minimal distance (the "unit length") in this scheme of calculation is restricted by the coherence length $\xi(T)$, the values of $\{k_x, k_y, k_z\}$ in (7.142) are restricted by the condition

$$k_i \left(\equiv \frac{2\pi}{L_i} n \right) \lesssim \xi(T)^{-1}, \tag{7.143}$$

(n is an integer) and when the characteristic length L_i along the i-axis is smaller than $\xi(T)$, only the term $n = 0$ contributes substantially ($|\Psi_{\mathbf{k}}|^2$ is homogeneous along that direction). Thus for bulk samples

$$\sigma_S \propto N_S \tau^0 = \int_0^\infty \frac{d^3 \mathbf{k}}{(2\pi)^3} \frac{T \gamma_0}{(\alpha + \epsilon_{\mathbf{k}})^2} \propto |\varepsilon|^{-1/2} \equiv \left| \frac{T_c - T}{T_c} \right|^{-1/2}, \tag{7.144}$$

while for thin films

$$\sigma_S \propto \frac{1}{L} \int_0^\infty \frac{d^2 \mathbf{k}}{(2\pi)^2} \frac{T \gamma_0}{(\alpha + \epsilon_{\mathbf{k}})^2} \propto |\varepsilon|^{-1} \equiv \left| \frac{T_c - T}{T_c} \right|^{-1}, \tag{7.145}$$

so that in samples with smaller dimensionality, the fluctuations near T_c are more pronounced. For the one-dimensional case:

$$\sigma_S \propto |\varepsilon|^{-3/2} \equiv \left| \frac{T_c - T}{T_c} \right|^{-3/2}. \tag{7.146}$$

This phenomenon is called "paraconductivity" and was first described theoretically by Aslamazov and Larkin [38]. The Green's function technique was used and the diagram for the current-current correlation function (shown in Fig. 7.2a) was considered. Later Maki [39, 40] and Thompson [41] took into account another diagram

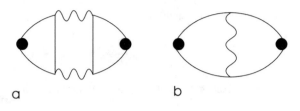

(shown in Fig. 7.2b), which yields a different contribution that is dominant in some conditions. We will consider both mechanisms without referring to these slightly mysterious diagrams, but to the much more transparent TDGL-scheme.

7.3.3 Aslamazov-Larkin Mechanism

The physics of the fluctuations outlined by (7.136) is rather transparent. At the same time, an important question still remains open, namely: how to justify (7.136) itself? The fact is that in thermodynamic equilibrium the superconducting current has a form (3.54), which is associated with the vector $\mathbf{Q} \propto \mathbf{v}_s$ rather than with the vector $\mathbf{E}$. To proceed with this problem one should bear in mind the relation (3.51). In view of the gauge $\mathbf{A} = 0$, adopted earlier for (7.139), it becomes clear that both the modulus $|\Psi|$ and the phase θ of the wave function are fluctuating, so that $\langle \nabla \theta \rangle$ is proportional to $\mathbf{E}$.[8]

One can rewrite (3.51) in the form:

$$\mathbf{j}_S = -\frac{ie_*\hbar}{2m_*} \langle \Psi^* \nabla \Psi - \Psi \nabla \Psi^* \rangle = \sum_{\mathbf{k}} \frac{e_*\mathbf{k}}{m_*} \langle |\Psi_{\mathbf{k}}|^2 \rangle \tag{7.147}$$

Following Abrikosov [36], one can represent $\Psi_{\mathbf{k}}$ in the form

$$\Psi_{\mathbf{k}} = \Psi_{\mathbf{k}}^{(0)} + \Psi_{\mathbf{k}}^{(1)} \tag{7.148}$$

and then use the TDGL equation in the form (7.139) to derive the value of $\Psi_{\mathbf{k}}^{(1)}$ based on the known value of $\Psi_{\mathbf{k}}^{(0)}$ (7.138). For homogeneous electric field, $\varphi = -\mathbf{E} \cdot \mathbf{r}$. Since the field $\mathbf{E}$ is a static one, only low frequency fluctuations contribute to the response, so one can omit the time derivative in (7.139). Taking into account that at the Fourier transformation $\mathbf{r}\Psi(\mathbf{r}) \doteq i(\partial \Psi_{\mathbf{k}}/\partial \mathbf{k})$, one obtains from (7.139):

$$\Psi_{\mathbf{k}}^{(1)} = -\frac{\gamma_0 e_*}{\alpha + \epsilon_{\mathbf{k}}} \mathbf{E} \cdot \frac{\partial \Psi_{\mathbf{k}}^{(0)}}{\partial \mathbf{k}}, \tag{7.149}$$

so that

---

[8] A good insight into this problem was made by Abrahams and Woo [42].

$$\mathbf{j} = \frac{T\gamma_0 e_*^2}{m_*^2} \sum_\mathbf{k} \mathbf{k} \frac{\mathbf{k} \cdot \mathbf{E}}{(\alpha + \epsilon_\mathbf{k})^3} \equiv \mathbf{E} \sum_\mathbf{k} \sigma_{AL}(\mathbf{k}). \tag{7.150}$$

We will focus our attention to the most interesting case of "dirty" superconductors. Considering first the case of bulk samples, one can write

$$\sigma_{AL}(3D) = \int \frac{d^3 k}{(2\pi)^3} \sigma_{AL}(\mathbf{k}). \tag{7.151}$$

Using the values for α and η_0 [see (7.139), (7.132), and (3.37)], we arrive at

$$\sigma_{AL}(\mathbf{k}) = 2\pi e^2 \xi^4 k^2 \cos^2\theta \, /[|\varepsilon| + (\xi k)^2]^3, \tag{7.152}$$

where $\xi = (\pi D/8T_c)^{1/2}$, $\hbar \equiv 1$, and θ is the angle between the vectors $\mathbf{k}$ and $\mathbf{E}$. For the bulk sample we obtain

$$\sigma_{AL}(3D) = \int_0^1 d\cos\theta \int \frac{k^2 dk}{4\pi^2} \sigma_{AL}(\mathbf{k}) = (e^2/32\xi) \, |\varepsilon|^{-1/2} \tag{7.153}$$

in accordance with [41, 43]. For thin films the result is completely independent of the material parameters

$$\sigma_{AL}(2D) = \frac{1}{L} \int \frac{d^2 k}{(2\pi)^2} \sigma_{AL}(\mathbf{k}) = \frac{1}{L} \int_0^\pi \frac{d\theta}{2\pi} \int \frac{dk^2}{4\pi} \sigma_{AL}(\mathbf{k})$$
$$= (e^2/16L) \, |\varepsilon|^{-1} \tag{7.154}$$

and is determined only by the value of the film's thickness and the closeness to the critical temperature.[9] We should note here the accord between (7.144) and (7.145), and (7.153) and (7.154), respectively. It is interesting to mention that in the 2-dimensional case, a consideration [35] based on (7.136) and (7.137) provides the same numerical coefficient as the proper diagrammatic treatment! [38].

7.3.4 Maki-Thompson Mechanism

In the preceding consideration of the mechanism of paraconductivity we referred to the superfluid component of the current (7.147), which resulted in a term, proportional to $\mathbf{E}$.

---

[9]We refer to the following values of integrals

$$\int_0^\infty dx x^4 (1 + x^2)^{-3} = (3\pi/16) \text{ and } \int_0^\infty dx x(1 + x)^{-3} = 1/2,$$

arising at the calculation of (7.153) and (7.154).

Meanwhile, [44], an expression of the same type follows directly from the interference term in the nonequilibrium current (7.112). Indeed, one can rewrite (7.112) in the fluctuational limit $|\Delta| \ll \gamma$ as

$$
\begin{aligned}
\mathbf{j}_{\text{int}} &= \mathbf{E} \left\{ \frac{\gamma}{2T_c} \sigma_n \left[\mathbf{K}\left(\frac{|\Delta|}{\sqrt{|\Delta|^2 + \gamma^2}} \right) - \mathbf{E}\left(\frac{|\Delta|}{\sqrt{|\Delta|^2 + \gamma^2}} \right) \right] \right\} \\
&\cong \mathbf{E}\sigma_n \frac{\pi}{8T_c\gamma} |\Delta|^2.
\end{aligned}
\tag{7.155}
$$

The physical meaning of this term is, as was discussed in Sect. 7.2, in the interference between normal and superfluid motions of the electrons. As a result of the interference, the normal motion described by the relation $\mathbf{j}_n = \sigma_n \mathbf{E}$, acquires an addition [cf. (7.142)]:

$$
\langle \mathbf{j}_{\text{int}} \rangle = \mathbf{E}\sigma_n \frac{\pi}{8T_c\gamma} \left\langle |\Delta|^2 \right\rangle = \mathbf{E}\left(\sigma_n \frac{\pi}{8T_c} \right) \sum_{\mathbf{k}} \frac{\langle |\Delta(\mathbf{k})|^2 \rangle}{\gamma_{\mathbf{k}}}.
\tag{7.156}
$$

In accordance with Sect. 7.2, the parameter γ, which smears out the BCS-singularity in the single-particle density of states, should be taken as the maximum of possible depairing factors related to $1/2\tau_\varepsilon$, $Dk^2/2$, τ_s^{-1}, etc. Taking $\gamma \simeq Dk^2/2$ (cf. [4]), and using (7.134) for the case of "dirty" superconductors, we arrive at

$$
\langle \mathbf{j}_{\text{int}} \rangle = \mathbf{E}\left(\frac{\sigma_n}{DN\tau_{\text{imp}}} \right) \sum_{\mathbf{k}} \frac{\langle |\Psi_{\mathbf{k}}|^2 \rangle}{k^2}.
\tag{7.157}
$$

Substituting (7.137) into (7.157), one can confirm that the resulting expression has exactly the same form[10]

$$
\langle \mathbf{j}_{\text{int}} \rangle = \mathbf{E}\frac{e^2\pi}{2} \sum_{\mathbf{k}} \frac{1}{[\varepsilon + (\xi k)^2]k^2},
\tag{7.158}
$$

as was used by Thompson [41]. It was pointed out in [44] that this leads to the Maki-Thompson conductivity σ_{MT}. Moving from (7.158) to integration over all values of $\mathbf{k}$ (as was done in [41]), we will get for the bulk sample:

$$
\sigma_{\text{MT}}^{\text{dirty}}(3D) = (e^2/8\xi) |\varepsilon|^{-1/2},
\tag{7.159}
$$

which means that in this case[11]

---

[10]To make the comparison easier one should replace in the expression (19) of Thompson [41] the derivative of digamma function by its numerical value: $\psi'(1/2) = \pi^2/2$.

[11]In view of the footnote (Sect. 7.3.2) the values of σ_{MT} should be twice smaller than given below for both $3D$ and $2D$ cases.

$$\sigma_{MT}^{dirty}(3D) = 4\sigma_{AL}^{dirty}(3D). \tag{7.160}$$

For the samples of lower dimensionality, the value of σ_{MT}^{dirty}, which follows from (7.158), is divergent: for thin films one should deal with expressions of the type

$$\sigma_{MT}(2D) = \frac{e^2}{4L} \int_{k(\to 0)}^{\infty} \frac{dk}{k[(T - T_c)/T_c + (\xi k)^2]}, \tag{7.161}$$

which demands a low-momentum cutoff $k_{min} = k_0$. In view of this factor, one can obtain [41]:

$$\sigma_{MT}^{dirty}(2D) = 2\sigma_{AL}^{dirty}(2D) \ln\{[\xi^{-2}|\varepsilon| + k_0]/k_0\}. \tag{7.162}$$

Generally, the cutoff may be caused by internal or external pair-breaking, owing for example, to inelastic energy relaxation ($Dk^2/2 \to \tau_\varepsilon^{-1}$), or the influence of the magnetic field [$Dk^2/2 \to (4eDH/c)(n + 1/2)$, $n = 0, \pm 1, \pm 2, \ldots$]. In the case of small cutoff ($\xi^2 k_0^2 \ll |T - T_c|/T_c$), the value of (7.161) may exceed (7.153) by an order of magnitude. In the opposite limit of strong pair breaking (or very close to T_c), σ_{MT} tends to zero, as follows from (7.160). One might expect such behavior since in the gapless regime the interference current components disappear in the general TDGL description. It is important to note that the regularization procedure for the case of restricted dimensionality is not trivial, even in absence of external pair breaking: Keller and Korenman [45] and Patton [46] came to the conclusion that the dominant contribution to this cutoff mechanism comes from the nonlinear self-influence of the fluctuations of the pair-field. The related scattering of electrons is more effective here, than the inelastic single-particle scattering. Later the corresponding process got an analog in localization theory [47], from where the electron phase-relaxation time τ_ϕ (so that $k_0^2 \equiv \pi/8\xi^{-2}T_c\tau_\phi$) migrated into this area. We will not consider this problem in more detail, nor different limiting cases for more complicated physical situations (see, in particular [48–82]). Instead, we refer the reader to the very interesting discussion presented by Reizer [83].

7.4 Longitudinal Electric Field in Superconductors

The appearance of new physical quantity μ introduces a new characteristic length into the theory of nonequilibrium superconductivity. Recall that in Chap. 3, in discussing the static description of superconductors, two characteristic lengths were mentioned—the penetration length of the magnetic field λ_L, and the coherence length $\xi(T)$, which characterizes the spatial variation of the order parameter modulus. A new characteristic value related to the presence of μ determines the penetration length of the longitudinal electric field $\mathbf{E}$ in a nonequilibrium superconductor. We emphasize that the field $\mathbf{E}$ is not incorporated in the equilibrium theory, so such a feature does not arise there.

7.4.1   Tinkham-Clark Gauge-Invariant Potential

As was shown in Sect. 5.3, the Fourier-transform of the charge density in superconductors has the form[12] (here $e = 1$):

$$\rho_\omega(\mathbf{k}) = -N(0)\left[2\varphi_\omega(\mathbf{k}) + \frac{1}{2}\text{Tr}\int_{-\infty}^{\infty}\frac{d\varepsilon}{4\pi}\frac{d\Omega_\mathbf{p}}{4\pi}\widehat{g}_{\varepsilon\varepsilon-\omega}(\mathbf{p},\mathbf{k})\right]. \tag{7.163}$$

Using (5.99), (5.77), (5.79), and (5.80), one can establish that in superconductors the charge density (7.163) must have the form

$$\rho = -2N(0)\left\{\varphi + \frac{1}{4}\int_{-\infty}^{\infty}d\varepsilon\left[N_1(f_1 + f_2) + \overline{N}_1(f_1 - f_2)\right]\right\} \tag{7.164}$$

which is explicitly gauge-invariant. [Note that in [2–6] less general expression for ρ is given; it can be obtained from (7.164) if $N_1 = -\overline{N}_1$, which is not fulfilled at $\varphi \neq 0$]. We recall now the gauge-invariant potential μ:

$$\mu = \varphi + \frac{\dot\theta}{2}, \tag{7.165}$$

where θ is the phase of the complex order parameter

$$\Delta = |\Delta|\exp(i\theta). \tag{7.166}$$

In case of the gauge transformation (3.118), θ transforms as $\theta \to \theta + \chi$. The condition of the superconductor's charge neutrality, $\rho = 0$, taking into account (7.164), provides the relation (in the first order in $|\Delta|/\epsilon_F$):

$$\varphi = -\frac{1}{4}\int_{-\infty}^{\infty}d\varepsilon\left[N_1(f_1 + f_2) + \overline{N}_1(f_1 - f_2)\right]. \tag{7.167}$$

Writing (7.167) in the $\theta = 0$, using the relation (7.165) and iterating over φ, one finds in the first approximation

$$\mu = \int_{|\Delta|}^{\infty}(n_\varepsilon - n_{-\varepsilon})\,d\varepsilon. \tag{7.168}$$

This is a familiar expression for the experimentally observed potential μ, introduced by Tinkham [84]. This formula, as is clear from the derivation above, is only

---

[12]To avoid misunderstanding we emphasize the difference between notations for the Fourier component of the charge density $\rho_\omega(\mathbf{k})$ and the function $\rho(\omega_\mathbf{k})$ related to the photon density of states.

the first approximation to more general equations of the theory.[13] At the same time, treating $(-\dot{\theta}/2)$ in the expression (7.165) as the chemical potential of the paired electrons and referring to φ as the chemical potential of normal electrons, one can interpret μ, in general, as the difference between the potentials of the normal and superfluid components of the electron liquid.

7.4.2 Normal Metal–Superconductor Interface

As in the cases of lengths λ_L and $\xi(T)$, we turn to the Ginzburg–Landau equations (generalized for nonstationary problems), and study the process of current flow across the boundary between the superconductor and a normal metal (this problem was considered by Rieger et al. [86]).

We will consider first the case of a gapless superconductor. The equation for an order parameter in this case has the form (see Sect. 4.3):

$$- 12\tau_0 \left(\frac{\partial}{\partial t} + 2i\varphi \right) \Delta + \xi^2(T)\nabla^2 \Delta + \left(1 - \frac{|\Delta|^2}{\Delta_0^2} \right) \Delta = 0, \qquad (7.169)$$

where

$$\tau_0 = (2\tau_s \Delta_0^2)^{-1}, \quad \Delta_0^2 = 2\pi^2(T_c^2 - T^2). \qquad (7.170)$$

Let the superconductor occupy the region $x > 0$, and the normal metal $x < 0$ (Fig. 7.3). The equation for $|\Delta|$ in the stationary case of interest coincides with the static Ginzburg–Landau equation:

$$\xi^2(T)\frac{\partial^2 |\Delta(x)|}{\partial x^2} + |\Delta(x)| \left(1 - \frac{|\Delta(x)|^2}{\Delta_0^2} \right) = 0. \qquad (7.171)$$

Its solution, obeying the boundary condition $|\Delta(x = 0)| = 0$, is the function

$$|\Delta(x)| = \Delta_\infty \tanh \frac{x}{\sqrt{2}\xi(T)}, \qquad (7.172)$$

where Δ_∞ is the value of the order parameter modulus $|\Delta(x)|$ at $x = \infty$. The imaginary part of (7.169) coincides with the continuity equation, which has the form

$$12\sigma\mu\frac{|\Delta(x)|^2}{\Delta_0^2} = \xi^2(T)\frac{\partial j_s}{\partial x} = -\xi^2(T)\frac{\partial j_n}{\partial x} \qquad (7.173)$$

---

[13]Using the substitution $\xi = \sqrt{\varepsilon^2 - |\Delta|^2}\mathrm{sign}\varepsilon$, the expression (7.168) may be presented in a form $\mu = \int_0^\infty d\xi(n_\xi - n_{-\xi})\xi/\varepsilon$. Because the value of ξ/ε is commonly identified with the excitation's charge in superconductors (see Sect. 3.1), usually the gauge-invariant potential is related to the charge-imbalance [85].

Fig. 7.3 Appearance of an
electric field in the
superconductor bounding a
normal metal: the
NS-junction. a—current
flowing across the
NS-boundary; b—the
electric field $\mathbf{E}$ and the order
parameter modulus $|\Delta|$ as
functions of the x-coordinate
for the current flow in the
gapless superconductors;
c—the same for finite-gap
superconductors

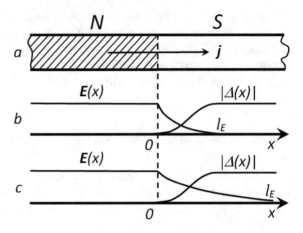

in the gauge $\dot{\theta} = 0$ [in this gauge, according to (7.165), $\mu = \varphi$]. In deriving (7.173),
the expression for total current density in a superconductor was used:

$$j = j_n + j_s, \tag{7.174}$$

subject to (4.121):

$$j_s = \frac{\sigma_n}{4ie\tau_0\Delta_0^2}\left(\Delta^*\frac{\partial\Delta}{\partial x} - \Delta\frac{\partial\Delta^*}{\partial x}\right). \tag{7.175}$$

Using the relation (4.121) for j_n:

$$j_n = \sigma_n E = -\sigma_n\frac{\partial\varphi}{\partial x} = -\sigma_n\frac{\partial\mu}{\partial x} \tag{7.176}$$

and the dependence (7.172) for $|\Delta(x)|$, one finds at $x > 0$ the equation for the
potential μ:

$$12\tanh^2\left(\frac{x}{\xi(T)\sqrt{2}}\right)\mu = \xi^2(T)\frac{\partial^2\mu}{\partial x^2}. \tag{7.177}$$

We will not seek the explicit solutions of the equation (7.177).[14] The form of this
equation itself shows that the potential μ and the related electric field $\mathbf{E}$ descend at
a distance an order of $\xi(E)$. Hence the characteristic length of electric field descend
(usually denoted as l_E) for gapless superconductors is of the order

$$l_E \sim \xi(T). \tag{7.178}$$

---

[14]The exact solution is given via the hypergeometric function [87].

7.4.3 New Characteristic Length in Superconductors

We will estimate now l_E for finite gap superconductors. To tackle this problem, we have to use the set of dynamic Ginzburg–Landau type equations for pure superconductors, which are derived in Sect. 7.1. Separating in (7.44) the real and imaginary parts in analogy with (7.171) and (7.173), one obtains:

$$-\frac{\pi}{8T_c}\sqrt{1+(2\tau_\varepsilon|\Delta|)^2}\frac{\partial|\Delta|}{\partial t} + \frac{\pi}{8T_c}D(\nabla^2 - Q^2)|\Delta|$$
$$\times\left[\frac{T_c - T}{T_c} - 7\zeta(3)\frac{(|\Delta|^2 + 2\mu^2)}{8(\pi T_c)^2}\right]|\Delta| = 0, \tag{7.179}$$

$$\frac{2|\Delta|^2}{\sqrt{1+(2\tau_\varepsilon|\Delta|)^2}}\mu - D\,\text{div}(|\Delta|^2\mathbf{Q}) = 0. \tag{7.180}$$

We emphasize that in the present case the continuity equation

$$\text{div}\,\mathbf{j} + \dot{\rho} = 0 \tag{7.181}$$

and (7.180) are independent. Also, using the expressions (7.109)–(7.112) for the current $\mathbf{j}$, and considering the stationary case, when $\partial|\Delta|/\partial t \equiv 0$, one finds an equation for the potential μ in the form[15]

$$\frac{2|\Delta|^2}{\sqrt{1+(2\tau_\varepsilon|\Delta|)^2}}\mu - D\frac{4T}{\pi}\frac{\partial^2\mu}{\partial x^2} = 0. \tag{7.182}$$

Because we are interested in the case of finite-gap superconductors, we can put in (7.182)

$$\tau_\varepsilon|\Delta| \gg 1. \tag{7.183}$$

[Recall, that τ_ε^{-1} in (7.182) is the energy damping of single-electron excitations, which in the case of interest must be significantly less than the gap $|\Delta|$ in the electron energy spectrum.] Thus (7.182) may be presented in the form

$$l_E^2\frac{\partial^2\mu}{\partial x^2} = \mu, \tag{7.184}$$

where

$$l_E = \sqrt{D\tau_\varepsilon\frac{4T}{\pi|\Delta|}}. \tag{7.185}$$

---

[15] We neglect here the contribution provided by the interference current, since the parameter $|\Delta|/T$ is considered very small.

Hence, the penetration depth of an electric field into the superconductor l_E in vicinity of T_c exceeds the length of the energy relaxation of electron excitations,

$$l_E \gg l_\varepsilon \left(\equiv \sqrt{D\tau_\varepsilon} \right), \qquad (7.186)$$

which in its turn may be substantially larger than $\xi(T)$ and λ_L (Fig. 7.3).[16]

Because (7.185) contains $|\Delta(T)|$ in the denominator, $l_E \to \infty$ at $T \to T_c$, and hence the electric field penetrates further and further into the bulk superconductor. Thus a natural transition from a superconducting to a normal state occurs at $T = T_c$.

Concluding this section, we would like to emphasize that in accordance with definition (7.165), which is the basic one for the value μ, both the single-particle electron excitations and the Cooper condensate contribute to the creation of the electric field in superconductors. It would be wrong to state that the potential μ arises as a consequence of the branch population imbalance only, as would follow from (7.168). This expression was derived in a fixed gauge and is a consequence of the assumptions made in Sect. 7.4. Note also, that in thermodynamic equilibrium $\mu = 0$: this value corresponds to the absolute minimum of the free energy.

7.4.4 Carlson-Goldman Modes

The existence of weakly decaying collective excitations in superconductors came under discussion immediately after the appearance of the BCS microscopic theory. In particular, the weakly damping oscillations of the order parameter, which have a sound-like spectrum in a neutral Fermi-liquid, were discussed by Bogoljubov (see, e.g., [88]) and Anderson [89, 90]. Later it was realized that these oscillations are connected with the vibrations of electron density. So it became necessary to account for the Coulomb interaction. The Coulomb interaction shifts these oscillations into the range of plasma frequency. Consequently, the specific superconducting characteristics can not be important to these oscillations, because the scale of superconducting energies is much less than the plasma one.

In the two-fluid hydrodynamics of superfluid helium, certain kinds of weakly damping collective excitations are known [91]. Among them the first, second and fourth sounds represent three-dimensional oscillations with sound-like spectra.

In superfluid helium, the first sound is connected with the density oscillations of normal and superfluid components. In the charged superfluid system, the frequency of these oscillations, alongside with the Bogoljubov-Anderson modes, would be displaced toward the plasma frequency region. The same occurs with the fourth sound, which is connected with the oscillations of the superfluid component. The second sound represents the oscillations of temperature (entropy), not the density oscillations of the electron liquid and, in principle, might be detected in superconductors.

---

[16]For example, in aluminum $\tau_\varepsilon \sim 10^{-8}$ s and thus in a pure metal $l_E \sim 1$ mm.

However, as the investigations of Ginzburg [92] and Bardeen [93] have shown, the damping of the second sound is very large in practically any real experimental conditions.

The appearance of the potential μ (and of the related electric field) in superconductors brings into existence a new type of a sound mode that has no analogy in superfluid helium. Such collective oscillations, reported first by Carlson and Goldman [94], reveal themselves in the high-frequency range

$$\omega \gg \tau_\varepsilon^{-1}. \tag{7.187}$$

During such oscillations the total current density equals zero, i.e., the normal and superconducting currents are oppositely directed. The zero value of the total current and hence of the magnetic field, makes it possible for these oscillations to exist in the depth of a superconductor, because there are now no restrictions related to the Meissner effect. With the Carlson-Goldman oscillations, the longitudinal electric field

$$\mathbf{E} = -\nabla\mu + \frac{\partial \mathbf{Q}}{\partial t} \tag{7.188}$$

appears in the superconductor, although the value of $\mathbf{E}$ is small: $E \ll |\nabla\mu|$—this ensures the weak damping of oscillations.

7.4.5 Dispersion of Charge-Imbalance Mode

A simple description of the Carlson-Goldman oscillating mode, which is based on the generalized dynamic equation (7.180), was introduced by Schmid [9] and Schön [95]. We will reproduce this approach in some detail. In the limit of finite-gap superconductors, (7.180) takes the form

$$-\frac{1}{\tau_\varepsilon}\mu + \frac{1}{|\Delta|} D \operatorname{div}(|\Delta|^2 \mathbf{Q}) = 0. \tag{7.189}$$

The dynamic equation (7.189) was obtained on the assumption of small characteristic frequencies

$$\omega\tau_\varepsilon \ll 1. \tag{7.190}$$

In the opposite to (7.190) limit, it may be rewritten as

$$-\frac{\partial \mu}{\partial t} + \frac{1}{|\Delta|} D \operatorname{div}(|\Delta|^2 \mathbf{Q}) = 0. \tag{7.191}$$

As before, we assume the condition $\omega \ll |\Delta|$, which was used in deriving of (7.189). Taking into account expressions (7.111) for $\mathbf{j}_s$ and (7.112) for $\mathbf{j}_n$, and also the above-mentioned relation

$$\mathbf{j}_s + \mathbf{j}_n = 0, \tag{7.192}$$

we find from (7.191) a dispersion equation [9, 95] for the Carlson-Goldman mode

$$\omega_q = -\frac{1}{2}i\Gamma \pm \sqrt{a^2 q^2 - \Gamma^2/4}, \tag{7.193}$$

which corresponds to propagation of the wave $\exp\left[-i(\omega_q t - qx)\right]$ with the velocity a and the damping Γ:

$$a^2 = \frac{N_s}{N} \frac{4T}{\pi|\Delta|} \frac{v_F^2}{3}, \qquad \Gamma = \frac{N_s}{N} \frac{1}{\tau_{\text{imp}}}. \tag{7.194}$$

We note that the velocity of propagation of the Carlson-Goldman mode is greater than the velocity of the second sound by the factor $(4T/\pi|\Delta|)^{1/2}$.

Concluding Remark Collective modes in superconductors described above are relatively high-frequency phenomena. To describe phenomena in this range of frequencies, one should go beyond the limits of TDGL equations. However, in their range of applicability, TDGL equations are fully adequate for describing a great variety of phenomena taking place in superconductors. Two examples, Maki-Thompson and Aslamazov-Larkin mechanisms considered in Section "Paraconductivity" demonstrate the equivalence of TDGL approach with first-principle calculations based on Green's functions technique. In general, combination of TDGL with COMSOL promises numerous results of modeling of real practical tasks, and elucidates the essence of ongoing physical processes.

References

1. M.P. Kemoklidze, L.P. Pitaevskii, Dynamics of a superfluid Fermi gas at finite temperatures. Sov. Phys. JETP **25**(6), 1036–1049 (1967) [Zh. Eksp. i Teor. Fiz. **52**(6), 1556–1569 (1967)]
2. A.A. Golub, Dynamic properties of short superconducting filaments. Sov. Phys. JETP **44**(1), 178–181 (1976) [Zh. Eksp. i Teor. Fiz. **71**[**1**(**7**)], 314–347 (1976)]
3. L. Kramer, R.J. Watts-Tobin, Theory of dissipative current-carrying states in superconducting filaments. Phys. Rev. Lett. **40**(15), 1041–1043 (1978)
4. G. Schön, V. Ambegaokar, Collective modes and nonequilibrium effect in current-carrying superconductors. Phys. Rev. B **19**(7), 3515–3528 (1979)
5. C.-R. Hu, New set of time-dependent Ginzburg-Landau equations for dirty superconductors. Phys. Rev. B **21**(7), 2775–2798 (1980)

6. R.G. Watts-Tobin, Y. Krähenbühl, L. Kramer, Nonequilibrium theory of dirty, current-carrying superconductors: phase-slip oscillations in narrow filaments near T_c. J. Low Temp. Phys. **42**(5/6), 459–501 (1981)

7. S.N. Artemenko, A.F. Volkov, Electric field and collective oscillations in superconductors. Sov. Phys. Uspekhi **22**(5), 295–310 (1979) [Usp. Fiz. Nauk **128**(1), 3–30 (1979)]

8. V.F. Elesin, Yu.V. Kopaev, Superconductors with excess quasiparticles. Sov. Phys. Uspekhi **24**(2), 116–141 (1981) [Usp. Fiz. Nauk **133**(2), 259–307 (1981)]

9. A. Schmid, Kinetic equations for dirty superconductors, in *Nonequilibrium Superconductivity, Phonons and Kapitza Boundaries,* ed. by K.E. Gray (Plenum Press, New York, 1981), pp. 423–480

10. J.A. Pals, K. Weiss, P.M.T. van Attekum, R.E. Horstman, J. Wolter, Nonequilibrium superconductivity in homogeneous thin films. Phys. Rep. **80**(4), 323–390 (1982)

11. A.M. Gulian, G.F. Zharkov, G.M. Sergoyan, Dynamic generalization of Ginzburg-Landau equations, in *Problems of Theoretical Physisc and Astrophysics* (The volume devoted to the 70-th anniversary of V.L. Ginzburg), ed. by L.V. Keldysh, V.Ya. Fainberg (Nauka, Moscow, 1989), pp. 145–163 (in Russian)

12. G.M. Eliashberg, Inelastic electron collisions and nonequilibrium stationary states in superconductors. Sov. Phys. JETP **34**(3), 668–676 (1972) [Zh. Eksp. i Teor. Fiz. **61** [3(9)], 1254–1272 (1971)]

13. A.I. Larkin, Yu.N. Ovchinnikov, Nonlinear effects during the motion of vortices in superconductors. Sov. Phys. JETP **46**(1), 155–162 (1977) [Zh. Eksp. i Teor. Fiz. **73**[1(7)], 299–312 (1977)]

14. I.E. Bulyzhenkov, B.I. Ivlev, Nonequilibrium phenomena in superconductor junctions. Sov. Phys. JETP **47**(1), 115–120 (1978) [Zh. Eksp. i Teor. Fiz. **74**(1), 224–235 (1978)]

15. V.G. Valeev, G.F. Zharkov, YuA Kukharenko, Kinetic theory of nonequilibrium processes in superconductors, in *Nonequilibrium Superconductivity*, ed. by V.L. Ginzburg (Nova Science, New York, 1988), pp. 203–282

16. V.P. Galaiko, Microscopic theory of resistive current states in superconducting channels. Sov. Phys. JETP **41**(1), 108–114 (1975) [Zh. Eksp. i Teor. Fiz. **68**(1), 223–237 (1975)]

17. V.P. Galayko, Features of the volt-ampere characteristics and oscillations of the electric potential in superconducting channels. Sov. Phys. JETP **44**(1), 141–148 (1976) [Zh. Eksp. i Teor. Fiz. **71**[1(7)], 273–285 (1976)]

18. U. Eckern, A. Schmid, M. Schmutz, G. Schön, Stability of superconducting states out of thermal equilibrium. J. Low Temp. Phys. **36**(5/6), 643–687 (1979)

19. S.R. De Groot, *Thermodynamics of Irreversible Processes* (North-Holland, Amsterdam, 1951), pp. 195–207

20. J.-J. Chang, Kinetic equations in superconducting thin films, in *Nonequilibrium Superconductivity*, ed. by D.N. Langenberg and A.I Larkin (North-Holland, Amsterdam, 1985), pp. 453–492

21. J.-J. Chang, D.J. Scalapino, Kinetic equation approach to nonequilibrium superconductivity. Phys. Rev. B **15**(5), 2651–2670 (1977)

22. K.D. Usadel, Generalized diffusion equation for superconducting alloys. Phys. Rev. Lett. **25**(8), 507–508 (1970)

23. A.M. Gulian, G.F. Zharkov, G.M. Sergoyan, Interference current in nonequilibrium superconductors. Sov. Phys. JETP **65**(1), 107–111 (1987) [Zh. Eksp. i Teor. Fiz. **92**(1), 190–199 (1987)]

24. Yu.N. Ovchinnikov, Properties of thin superconducting films in high-frequency fields. Sov. Phys. JETP **32**(1), 72–78 (1971) [Zh. Eksp. i Teor. Fiz. **59**(7), 128–141 (1970)]

25. Yu.N. Ovchinnikov, A.R. Isahakyan, Electromagnetic field absorption in superconducting films. Sov. Phys. JETP **47**(1), 91–94 (1978) [Zh. Eksp. i Teor. Fiz. **74**(1), 178–183 (1978)]

26. L.P. Gor'kov, G.M. Eliashberg, The behavior of a superconductor in a variable field. Sov. Phys. JETP **28**(4), 1291–1297 (1969) [Zh. Eksp. i Teor. Fiz. **56**(4), 1297–1308 (1969)]

27. A.G. Aronov, YuM Gal'perin, V.L. Gurevich, V.I. Kozub, Nonequilibrium properties of superconductors (transport equation approach), in *Nonequilibrium Superconductivity*, ed. by D.N. Langenberg and A.I Larkin (North-Holland, Amsterdam, 1985), pp. 325–376

28. A.L. Shelankov, Dragging of normal components by the condensate in nonequilibrium super-conductors. Sov. Phys. JETP **51**(6), 1186–1193 (1980) [Zh. Eksp. i Teor. Fiz. **78**(6), 2359–2379 (1980)]

29. V.F. Elesin, V.A. Kashurnikov, A.V. Kharlamov, Boundary condition for nonequilibrium super-conductors. Sov. J. low Temp. Phys. **12**(7), 392–395 (1986) [Fiz. Nizk. Temp. **12**(7), 694–700 (1978)]

30. L.P. Gor'kov, G.M. Eliashberg, Generalization of the Ginzburg-Landau equations for non-stationary problems in the case of alloys with paramagnetic impurities. Sov. Phys. JETP **27**(3), 328–334 (1968) [Zh. Eksp. i Teor. Fiz. **54**(2), 612–625 (1968)]

31. J.S. Shier, D.M. Ginzberg, Superconducting transitions of amorphous bismuth alloys. Phys. Rev. **147**(1), 384–391 (1966)

32. R.E. Glower, Ideal resistive transition of a superconductor. Phys. Lett. A **25**(7), 542–544 (1967)

33. M. Strongin, O.F. Kramer, J. Crow, R.S. Thompson, H.L. Fine, "Curie-Weiss" behavior and fluctuation phenomena in the resistive transitions of dirty superconductors. Phys. Rev. Lett. **20**(17), 922–925 (1968)

34. V.V. Shmidt, Phase transition in superconductors of small size. JETP Lett. **3**(3), 91–93 (1966) [Pis'ma v Zh. Eksp. i Teor. Fiz. **3**(3), 141–145 (1966)]

35. A. Schmid, The resistivity of a superconductor in its normal state. Z. Phys. **215**, 210–212 (1968)

36. A.A. Abrikosov, *Fundamentals of the Theory of Metals* (North-Holland, Amsterdam, 1988), p. 594

37. E.M. Lifshitz, L.P. Pitaevskii, *Statistical Physics, Part 1* (Pergamon Press, Oxford, 1980), pp. 471–478

38. L.G. Aslamazov, A.I. Larkin, Effect of fluctuations on the properties of a superconductor above the critical temperature. Sov. Phys. Solid State **10**(4), 875–880 (1968) [Fiz. Tverd. Tela **10**(4), 1104–1111 (1968)]

39. K. Maki, The critical fluctuation of the order parameter in type-II superconductors. Progr. Theor. Phys. **39**(4), 897–911 (1968)

40. K. Maki, Critical fluctuation of the order parameter in a superconductor. Progr. Theor. Phys. **40**(2), 193–200 (1968)

41. R.S. Thompson, Microwave, flux flow, and fluctuation resistance of dirty type-II superconduc-tors. Phys. Rev. B **1**(1), 327–333 (1970)

42. E. Abrahams, J.W. Woo, Phenomenological theory of the rounding of the resistive transition of superconductors. Phys. Lett. A **27**(2), 117–118 (1968)

43. H. Schmidt, The onset of superconductivity in the time dependent Ginzburg-Landau theory. Z. Phys. **216**, 336–345 (1968)

44. A.M. Gulian, Time-dependent Ginzburg-Landau equations for finite-gap superconductors and the problem of paraconductivity. Phys. Lett. A **200**, 201–204 (1995)

45. J. Keller, V. Korenman, Fluctuation-induced conductivity of superconductors above the transi-tion temperature: regularization of the Maki diagram. Phys. Rev. B **5**(11), 4367–4375 (1972)

46. B.E. Patton, Fluctuation theory of the superconducting transition in restricted dimensionality. Phys. Rev. Lett. **27**(19), 1273–1276 (1971)

47. P.A. Lee, T.V. Ramakrishnan, Disordered electric systems. Rev. Mod. Phys. **57**(2), 287–337 (1985)

48. C. Caroli, K. Maki, Fluctuations of the order parameter in type-II superconductors I. Dirty limit. Phys. Rev. **159**, 306–315 (1967)

49. C. Caroli, K. Maki, Fluctuations of the order parameter in type-II superconductors II. Pure limit. Phys. Rev. **159**, 316–326 (1967)

50. K.D. Usadel, The influence of a static magnetic field on the fluctuation superconductivity. Z. Phys. **227**, 260–270 (1969)

51. K.D. Usadel, Fluctuations in superconductors above T_c in the high field region. Phys. Lett. A **29**(9), 501–502 (1969)

52. G. Bergman, Superconducting fluctuations in a magnetic field. Z. Phys. **225**, 430–443 (1969)

53. E. Abrahams, M. Redi, J.W. Woo, Effect of fluctuations on electric properties above the super-conducting transition. Phys. Rev. B **1**(1), 208–213 (1970)

54. K. Kajimura, N. Mikoshiba, Fluctuations in the resistive transition in aluminum films. J. Low Temp. Phys. **4**(3), 331–348 (1971)

55. E. Abrahams, R.E. Prange, M.J. Stephen, Effect of a magnetic field on fluctuations above T_c. Physica **55**, 230–233 (1971)

56. R.S. Thompson, The influence of magnetic fields on the paraconductivity due to fluctuations in thin films. Physica **55**, 296–302 (1971)

57. R.A. Craven, G.A. Thomas, R.D. Parks, Fluctuation induced conductivity of a superconductor above the transition temperature. Phys. Rev. B **7**(7), 157–165 (1973)

58. K. Maki, Thermoelectric power above superconducting transition. J. Low Temp. Phys. **14**(5/6), 419–432 (1974)

59. L.G. Aslamazov, A.A. Varlamov, Fluctuation conductivity in intercalated superconductors. J. Low Temp. Phys. **38**(2), 223–241 (1980)

60. A.I. Larkin, Reluctance of two-dimensional systems. JETP Lett. **31**(4), 219–223 (1980) [Pis'ma Zh. Eksp. i Teor. Fiz. **31**(4), 239–243 (1980)]

61. S. Hikami, A.I. Larkin, Magnetoresistance of high-temperature superconductors. Mod. Phys. Lett. B **2**(5), 693–698 (1988)

62. A.G. Aronov, S. Hikami, A.I. Larkin, Zeeman effect on magnetoresistance in high-temperature superconductors. Phys. Rev. Lett. **62**(8), 965–968 (1989)

63. K. Maki, S. Thompson, Fluctuation conductivity of high-temperature superconductors. Phys. Rev. B **39**(4), 2767–2774 (1989)

64. V.V. Gridin, T.W. Krause, W.R. Datars, Two-dimensional paraconductivity in superconducting $Bi_{1.6}Pb_{0.4}Sr_2Ca_2Cu_3O_y$. J. Appl. Phys. **68**, 675–678 (1990)

65. S.-K. Yip, Fluctuations in an impure unconventional superconductor. Phys. Rev. B **41**(13), 2612–2615 (1990)

66. A.G. Aronov, H.S. Hikami, Skew-scattering effect on the Hall-conductance fluctuation in high-temperature superconductors. Phys. Rev. B **41**, 9548–9550 (1990)

67. Q.Y. Ying, H.S. Kwok, Kosterlitz-Thouless transition and conductivity fluctuations in Y-Ba-Cu-O thin films. Phys. Rev. B **42**, 2242–2245 (1990)

68. S. Ullah, A.T. Dorsey, Critical fluctuations in high-temperature superconductors and the Ettingshausen effect. Phys. Rev. Lett. **65**(16), 2066–2069 (1990)

69. J.B. Bieri, K. Maki, R.S. Thompson, Non-local effect in magnetoconductivity of high-T_c superconductors. Phys. Rev. B **44**(9), 4709–4711 (1991)

70. V.A. Gasparov, Berezinskii-Kosterlitz-Thouless transition and fluctuation paraconductivity in $Y_1 Ba_2 Cu_3 O_7$ single crystal films. Physica C **178**, 449–455 (1991)

71. J.B. Bieri, K. Maki, Magnetoresistance of high-T_c superconductors in the fluctuation regime. Phys. Rev. B **42**(7), 4854–4856 (1990)

72. R. Hopfengärtner B. Hensel, G. Saeman-Ischenko, Analysis of the fluctuation-induced excess conductivity of epitaxial $Y Ba_2 Cu_3 O_7$ films: influence of a short-wavelength cutoff in the fluctuation spectrum. Phys. Rev. B **44**(2), 741–749 (1991)

73. S. Ullah, A.T. Dorsey, Effect of fluctuations on the transport. Phys. Rev. B **44**(1), 262–273 (1991)

74. M. Ausloos, F. Gillet, Ch. Laurent, P. Clippe, High-temperature crossover in paraconductivity of granular $Y_1 Ba_2 Cu_3 O_{7-x}$. Z. Phys. B **84**, 13–16 (1991)

75. A.B. Kaiser, G. Mountjoy, Consistency with anomalous electron-phonon interactions of the thermopower of high-T_c superconductors. Phys. Rev. B **43**(7), 6266–6269 (1991)

76. M. Anderson, Ö. Rapp, Magnetoresistance measurements on polycrystalline $Y Ba_2 Cu_3 O_{7-\delta}$. Phys. Rev. B **44**(14), 7722–7725 (1991)

77. A.A. Varlamov, D.V. Livanov, The effect of fluctuations on the Hall-effect in high T_c superconductors. Phys. Lett. A **157**(8/9), 519–522 (1991)

78. A.A. Varlamov, D.V. Livanov, The effect of fluctuations on thermomagnetic phenomena in high T_c superconductors. Phys. Lett. A **157**(8/9), 523–526 (1991)

79. K. Semba, T. Ishii, A. Matsuda, Absence of the Zeeman effect on the Maki-Thompson fluctuation in magnetoresistance of $Y Ba_2 Cu_3 O_7$ single crystals. Phys. Rev. Lett. **67**(6), 769–772 (1991)

80. B.N. Narozhny, Fluctuation conductivity in strong-coupling superconductors. Sov. Phys. JETP **77**(2), 301–306 (1993) [Zh. Eksp. i Teor. Fiz. **104**(2), 2825–2837 (1993)]

81. W. Holm, Yu. Eltsev, Ö. Rapp, Paraconductivity along the a and b axes in $YBa_2Cu_3O_{7-\delta}$ single crystals. Phys. Rev. B **51**(17), 11992–11995 (1995)

82. J. Axnäs, W. Holm, Yu. Eltsev, Ö. Rapp, Sign change of c-axis magnetoconductivity in $YBa_2Cu_3O_{7-\delta}$ single crystals. Phys. Rev. Lett. **77**(11), 2280–2283 (1996)

83. MYu. Reizer, Fluctuation conductivity above the superconducting transition: regularization of the Maki-Thompson term. Phys. Rev. B **45**(22), 12949–12958 (1992)

84. M. Tinkham, Tunneling generation, relaxation and tunneling detection of hole-electron imbalance in superconductors. Phys. Rev. B **6**(5), 1747–1756 (1972)

85. G.J. Pethick, H. Smith, Charge imbalance in nonequilibrium superconductors. J. Phys. C: Solid State Phys. **13**, 6313–6347 (1980)

86. T.J. Rieger, D.J. Scalapino, J.E. Mercereau, Charge conservation and chemical potentials in time-dependent Ginzburg-Landau theory. Phys. Rev. Lett. **27**(26), 1787–1790 (1971)

87. A.F. Volkov, Theory of the current-voltage characteristics of one-dimensional S-N-S and S-N-junctions. Sov. Phys. JETP **39**(2), 366–369 (1974) [Zh. Eksp. i Teor. Fiz. **66**(2), 758–765 (1974)]

88. N.N. Bogolyubov, V.V. Tolmachov, D.V. Shirkov, *in A New Method in the Theory of Super-conductivity* (Consultants Bureau, New York, 1959), pp. 31–99

89. P.W. Anderson, Coherent excited states in the theory of superconductivity: gauge invariance and the Meissner effect. Phys. Rev. **110**(4), 827–835 (1958)

90. P.W. Anderson, Random-phase approximation in the theory of superconductivity. Phys. Rev. **112**(6), 1900–1916 (1958)

91. I.M. Khalatnikov, *An Introduction to the Theory of Superfluidity* (W.A. Benjamin, New York, 1965), pp. 72–77

92. V.L. Ginzburg, Second sound, the convective heat transfer mechanism, and exciton excitations in superconductors. Sov. Phys. JETP **14**(3), 594–598 (1962) [Zh. Eksp. i Teor. Fiz. **41**[3(9)], 828–834 (1961)]

93. J. Bardeen, Two-fluid model of superconductivity. Phys. Rev. Lett. **1**(2), 399–402 (1958)

94. A.V. Carlson, A.M. Goldman, Superconducting order parameter fluctuations below T_c. Phys. Rev. Lett. **31**(5), 880–882 (1973)

95. G. Schön, Collective modes in superconductors, in *Nonequilibrium Superconductivity*, ed. by D.N. Langenberg and A.I Larkin (North-Holland, Amsterdam, 1985), pp. 589–641

Index

A

Abrikosov vortices, 51, 57, 86
  vortex and anti-vortex, 51
Analytical continuation, 166, 168, 175, 181–
    184, 202, 212, 216, 221, 244
Andreev reflection, 118, 119

B

BCS
  approach, 145
  attraction, 42
  electronic density of states, 231, 260
  equations, 145
  Gor'kov model, 129, 137, 179
  Gor'kov theory, 28, 129, 179
  interaction potential, 137
  mechanism, 47
  model, 42, 44, 132, 134, 137
  picture, 137
  potential, 38, 130, 142
  theory, 28, 38, 113, 129, 135, 137, 143,
    252, 266
  work, 47
Bogolyubov–De Gennes equations, 116, 118
Born approximation, 152, 201
Branch imbalance, 197

C

Canonical collision integral
  phonon-electron, 226
Carlson-Goldman modes, 266, 268
Causality principle, nonlinear, 164

Channel
  nondiagonal, 197, 199, 200, 235
  of imbalance relaxation, 207
  of inelastic relaxation, 238
Coherence
  factors, 120, 121, 222
  length, 52, 143, 252, 257, 261
  phase, 131
Coherent behavior, 52
Collision integral, 189, 194
  canonical form, 197, 201
    electron-electron, 205
    electron-phonon, 219
    electron-phonon no branch imbal-
      ance, 222
    electron-photon, 222
    phonon-electron, 221, 227
    phonon-electron branch symmetric,
      226
  effective, 200
  generalized, 197
  in terms of electron-hole distribution
    function, 200
  meaning, 207
  self-energies, 203
Conductivity
  ideal, 114
  of normal electrons, 55
  small addition, 68
  very poor, 62
Cooper
  condensate, 25, 34, 52, 145, 197, 231,
    266
  pairing, 142
  pairs, 25, 47, 142, 143, 146, 149, 231
    breaking, 224, 229
    density, 57, 64, 72, 75

© Springer Nature Switzerland AG 2020
A. Gulian, *Shortcut to Superconductivity*,
https://doi.org/10.1007/978-3-030-23486-7

out of equilibrium, 175
quantum statistics, 25
recuperation, 229
size, 26

D

Density of states
of electrons in normal metals, 44
of phonons, 227
of photons, 223
role in kinetic processes, 121
singularity in superconductors, 260
singularity smearing, 164
Detailed equilibrium violation, 229

E

Effective mass, 8, 22
Electron-phonon interaction, 49
Elementary acts, essence, 207
Eliashberg
analytical continuation, 164
equations, 50
field term, 224
gap enhancement mechanism, 226
model of superconductivity, 48
TDGL derivation with Gor'kov, 231
Energy gap, 44, 226
Equivalent mass approach, 116
Excitation spectrum, 120, 133, 135, 163

F

Fluctuations, 26, 36, 131, 253–258, 261
Flux
fluxon, 28
generation, 96
in hollow cylinder, 143
in the SQUID, 34
non-quantized, 28
quantization, 26–28
quantized, 28
quantum, 33, 34
regeneration, 97
SFQ, 51
SFQ propagation, 51
vortices, 57
Free energy, 46, 121–123, 126, 129, 241,
253, 256, 266
Fröhlich's Hamiltonian, 175

G

Gap
and GL order parameter, 138
and magnetic impurities, 162
and nonmagnetic impurities, 156
at arbitrary temperatures, 137
determination of value, 136
energy levels redistribution, 120
finite, 107
from self-consistency equation, 133
in energy spectrum, 115
-like singularity, 116
nonequivalence with order parameter,
163
order parameter, 135
single excitations birth, 116
to Tc ratio, 137
Gapless
superconductors, 107
Gapless superconductivity, 163
Gauge
advantage, 9
choice, 28
coupling by, 30
fixed, 266
in 1D, 28, 258
invariance of TDGL current, 36
invariance of TDGL equations, 36
invariance of TDGL wave function, 37
-invariant, 262
current, 25
London gauge, 9
non-invariant vector potential, 23
transformation, 24, 36, 262
zero phase, 262
Ginzburg–Landau
approach, 55
equations, 52
derivation by Gor'kov, 143
equations for alloys with paramagnetic
impurities, 164
equations stationary, 113
GL theory, 121
nonstationary equations, 164, 166
normalization, 124
parameter, 53, 55, 59
system of equations, 124
Ginzburg's number, 255
Gor'kov
derivation of GL theory, 138
equations, 134

formulation of BCS theory, 113, 129
proof of charge doubling, 143
theory at finite temperatures, 135

I

Instability of lattice, 50
Interference current, 231, 252, 261
Isotope effect, 134

J

Josephson
effect non-stationary, 31
effects, 3, 28
effect stationary, 30
junction, 31, 51, 95

K

Keldysh technique, 164, 175, 187, 193, 197
Kinetic equations
generalized, 197, 231, 232
without integration over energies, 193
Klein paradox, 119

L

Local equilibrium, 231, 237
approximation, 236, 242, 250
generalized, 242
violated, 243
London
current, 9
equations, 5, 6, 18
gauge, 9
moment, 3
penetration depth, 6, 8, 17
Longitudinal
acoustic phonon field, 222
electric field, 267
electric field in superconductors, 261
vibrations of electron density, 266
wave, 268

M

Matsubara technique, 135, 164
Maxwell equations, 6, 8, 17, 18, 23, 39, 52,
55, 124, 125, 173, 251, 252
Meissner effect, 3, 5, 8, 17, 22, 23, 25, 125,
144, 267

Migdal
diagram approach, 177
-Eliashberg model, 137, 179
Galitzkiy, 162
vertex correction, 177
Migdal's theorem, 175

N

Nondiagonal channel, *see* Channel nondiagonal
Nonequilibrium distribution
of electrons, 227
of phonons, 220
Normalization condition, 170, 190, 233, 235
Numerical
approach, 10
calculations, 50
coefficient, 48, 259
computation, 229
criterion, 128
evaluation, 229
methods, 243
solutions, 89, 252
solving, 52

O

Order parameter, 121, 124, 127, 128, 138,
162–164, 171, 173, 175, 199, 232,
234, 236, 237, 239, 242, 243, 251,
253, 254, 256, 261–263, 266
definition, 198

P

Pair field, 140, 163
Paraconductivity mechanism
Aslmazov-Larkin, 257
Maki-Tompson, 257
Pauli matrices, 116
Penetration depth
electric field, 266
London, 126, 143, 144
Phase slip, 82
Phonon deficit effect, 226, 229
Phonon heat-bath, 180
Polarization operators, 220
Propagator
Eilenberger, energy integrated, 188
Keldysh, 193
Matsubara, 135
nondiagonal, 131

R
Relaxation time, 36, 256, 257, 261

S
Scattering
  nonmagnetic impurities, 150
  spin-flip, 158
Self-energy parts
  analytical continuation, 218
  electron-electron, 201
  electron-impurity, 200
  electron-phonon, 181
  phonon electron, 213
Source
  inequilibrium, phonon kinetic equation,
    217
  of current, 31, 88
  of phonons, 227
Spectral functions, 189, 191, 245
Superconductivity

gapless, *see* gapless superconductivity
Superconductors
  finite gap, 173, 232
    electric field penetration depth, 265
  gapless, 107, 149
  high-temperature, 8, 50
  Type I, 62
  Type II, 62
Surface energy, 126

T
TDGL equations
  finite gap, 231
  gapless, 172
Tinkham-Clark potential, 262

U
Usadel approximation, 244